2nd International Workshop on Non-Equilibrium Processes in Plasmas and Environmental Science 2008

Journal of Physics: Conference Series Volume 162

Belgrade and Novi Sad, Serbia
23-26 August 2008

Editors:

Zoran Petrovic
Gordana Malovic
Dragana Maric

ISBN: 978-1-61738-372-4

2nd International Workshop on Non-Equilibrium Processes in Plasmas and Environmental Science 2008

Journal of Physics: Conference Series Volume 162

Belgrade and Novi Sad, Serbia
23-26 August 2008

ISBN: 978-1-61738-372-4

IOP Institute of Physics

Printed from e-media with permission by:

Curran Associates, Inc.
57 Morehouse Lane
Red Hook, NY 12571
www.proceedings.com

Some format issues inherent in the e-media version may also appear in this print version.

Printed by Curran Associates, Inc. (2010)

Institute of Physics
Dirac House, Temple Back
Bristol BS1 6BE UK

Phone: 44 1 17 929 7481
Fax: 44 1 17 920 0979

techtracking@iop.org

TABLE OF CONTENTS

Study of Higher Excited States of Some Polyatomic Molecules Relevant for Plasma Physics and Environment 1
B P Marinković

Low Energy Positron Interactions - Trapping, Transport and Scattering 6
J P Sullivan, S J Buckman, A Jones, P Caradonna, C Makochekanwa, D Slaughter, Z Lj Petrović, A Banković, S Dujko, J P Marler, R D White

Positron Transport in Gases 23
M Charlton

Cross Sections and Transport Properties of Negative Ions in Rare Gases 42
J V Jovanović, Z Lj Petrović

Nonequilibrium Calculations of the Role of Electron Impact in the Production of NO and its Emissions 52
L Campbell, M J Brunger

Recent Developments on PLASMAKIN – A Software Package to Model the Kinetics in Gas Discharges 62
N R Pinhão

Hollow Cathode Discharges: Volt-Ampere Characteristics and Space-Time Resolved Structure of the Discharge 73
D Marić, N Škoro, G Malović, Z Lj Petrović, V Mihailov, R Djulgerova

An Overview of Modelling of Low-Pressure Post-Discharge Systems used for Plasma Sterilization 83
K Kutasi, C D Pintassilgo, J Loureiro

Electric Propulsion using Ion-Ion Plasmas 96
A Aanesland, A Meige, P Chabert

The Electrical Asymmetry Effect - A Novel and Simple Method for Separate Control of Ion Energy and Flux in Capacitively Coupled RF Discharges 109
U Czarnetzki, B G Heil, J Schulze, Z Donkó, T Mussenbrock, R P Brinkmann

Space and Phase Resolved Optical Emission in Mode Transitions of Radio-Frequency Inductively Coupled Plasmas 125
D O'Connell, K Niemi, M Zaka-Ul-Islam, T Gans

Influence of Plasma and Radio Wave Treatment on Paper Structure 133
I I Filatova, V V Azharonok, V A Dlugunovich, O Tsaruk, I V Voschula, N A Fomin, N B Bazylev

The Dynamics of Radio-Frequency Driven Atmospheric Pressure Plasma Jets 140
L Schaper, S Reuter, J Waskoenig, K Niemi, V Schulz-Von Der Gathen, T Gans

Progress Report: Direct Injection of Liquids into Low-Pressure Plasmas 148
M Goeckner, D Ogawa, I Saraf, L Overzet

Surface Functionalization of Organic Materials by Weakly Ionized Highly Dissociated Oxygen Plasma 157
A Vesel, M Mozetic

Two-Dimensional Dusty Plasma Crystals and Liquids 177
Z Donkó, P Hartmann, G J Kalman

Effects of Friction on Modes in Collisional Multicomponent Plasmas 190
J Vranjes, S Poedts, M Kono, M Y Tanaka

Source Apportionment of Atmospheric Bulk Deposition in the Belgrade Urban Area using Positive Matrix Factorization 200
M Tasić, Z Mijić, S Rajšić, A Stojić, M Radenković, J Joksić

Determination of Chemical Composition of Individual Airborne Particles by SEM/EDX and Micro-Raman Spectrometry: A Review 212
E A Stefaniak, A Buczynska, V Novakovic, R Kuduk, R Van Grieken

Author Index

INTRODUCTION

This book aims to give a cross section from a wide range of phenomena that, to different degrees, fall under the heading of non-equilibrium phenomenology. The selection is, of course, biased by the interests of the members of the scientific committee and of the FP6 Project 026328 IPB-CNP Reinforcing Experimental Centre for Non-equilibrium Studies with Application in Nano-technologies, Etching of Integrated Circuits and Environmental Research. Some of the papers included here are texts based on selected lectures presented at the Second International Workshop on Non-equilibrium Processes in Plasmas and Environmental Science. However, this volume is not just the proceedings of that conference as it contains a number of papers from authors that did not attend the conference. The goal was to put together a volume that would cover the interests of the project and support further work. It is published in the Institute of Physics journal *Journal of Physics: Conference Series* to ensure a wide accessibility of the articles.

The texts presented here range from in-depth reviews of the current status and past achievements to progress reports of currently developed experimental devices and recently obtained still unpublished results. All papers have been refereed twice, first when speakers were selected based on their reputation and recently published results, and second after the paper was submitted both by the editorial board and individual assigned referees according to the standards of the conference and of the journal. Nevertheless, we still leave the responsibility (and honours) for the contents of the papers to the authors. The papers in this book are review articles that give a summary of the already published work or present the work in progress that will be published in full at a later date (or both).

In the introduction to the first volume, in order to show how far reaching, ubiquitous and important non-equilibrium phenomena are, we claimed that ever since the early cosmos collapsed from the uniform plasma stage into stars and empty space, practically nothing is in real equilibrium only in local equilibrium. How wrong we were. As our focus turned to anti particles, positrons and positronium, we realized that even in those early stages there was major non-equilibrium between matter and anti matter originating from the earliest stages of the Big Bang. Thus it is safe to correct the famous quote by the renowned natural philosopher Sheldon Cooper into: 'If you know the laws of [non-equilibrium] physics anything is possible'. From the matter–anti-matter ratio in the universe to life itself. But do we really need such farfetched introductory remarks to justify our scientific choices? It suffices to focus on non-equilibrium plasmas and transport of pollutants in the air and see how many new methods for diagnostics and treatment have been proposed for medicine in the past 10 years. So in addition to the past major achievements such as plasma etching for integrated circuit production, the field is full of possibilities and truly, almost anything is possible. We hope that some of the papers presented here summarize well how we learn about the laws of non-equilibrium physics in the given context of plasmas and air pollution and how we open new possibilities for further understanding and further applications.

A wide range of topics is covered in this volume. This time we start with elementary collisional processes and a review of the data for excitation of polyatomic molecules obtained by the binary collision experiments carried out at the Institute of Physics in Belgrade by the group of Bratislav Marinković. A wide range of activities on the foundation of gaseous positronics ranging from new measurements in the binary regime to the simulation of collective transport in dense gases is presented by James Sullivan and coworkers. This work encompasses three continents, half a dozen groups and several lectures at the workshops while also covering a lot of material that was not presented as a lecture at the workshop. Michael Charlton has written a major review of the past work on the transport of positrons in gases. This review is a thorough summary of the field which more importantly looks at the future and invites a continuation of activities while providing an excellent foundation for the new experiments and modeling.

The next paper submitted by Jasmina Jovanović covers the ongoing activities in the Gaseous Electronics Laboratory in Belgrade to prepare sets of data for ions that are required for modeling of gas discharges based on cross sections rather than interaction potentials. In many situations direct application of swarm physics modeling is possible, one such example is in the upper layers of the atmosphere – how this is done in the case of NO production and emissions from NO is shown in a paper by Laurence Campbell from Flinders University. Self-consistent coupling of electron kinetics as described by the solution to the Boltzmann equation and chemical and excited state kinetics in gases is described by Nuno Pinhao.

From swarms to gas discharges the transition is made through gas breakdown. Studies of the development of the anatomy of a hollow cathode discharge obtained in collaboration between groups from Bulgaria and Serbia are presented by Dragana Marić. Remote treatment by plasmas is an option in biomedical applications and one such example is given by Kinga Kutasi, presenting results of a modeling of a well established plasma sterilizer. Another interesting application of plasmas is for the propulsion of satellites in vacuum where intelligent design (of plasma geometry and operating conditions) proves to be the most efficient method of controlling the orbits. Some new results combining experiments and modeling of plasma propulsion devices from Ecole Polytechnique in Paris are presented here by Ane Aanesland.

Just how much can the studies inspired by the practical needs of plasma technologies lead to new fundamental understanding is illustrated well in the paper by Uwe Czarnetzki which describes a new method for separate control of flux and energy of ions reaching the surface of electrodes. Deborah O'Connell from Belfast has shown space and phase resolved mode transitions in rf inductively coupled plasmas obtained by optical emission measurements. At the same time an application of a similar rf discharge for the treatment of paper was presented by Irina Filatova from Belarus.

Many applications of non-equilibrium plasmas depend on the development of plasma sources operating at atmospheric pressure and one such source that promises to be prominent in medicine is described by Timo Gans. In a similar way, practical considerations require studies of the injection of liquids into plasmas and progress on the development of one such source is described by Mathew Goeckner and his colleagues from Dallas. From the Institute Jožef Štefan in Slovenia and the group of Miran Mozetič we have a detailed review of their work on functionalization of organic materials by oxygen plasmas.

Even higher density plasmas, where the collective phenomena dominate, show different degrees of non-equilibrium and one example presented here by Zoltan Donko deals with two dimensional plasma dust crystals and liquids, while the lecture by Jovo Vranješ from Belgium deals with the treatment of collisions in multicomponent plasmas.

Finally we have papers on the transport of pollutants. The association of the two fields started initially through joint interest in some of the methods for removal of NO_x and SO_x, from electrostatic precipitation of industrial dust to dielectric barrier discharges. The joint work continued on the application of flowing afterglow plasma combined with a hollow cathode discharge in order to achieve a proton transfer mass analysis of organic volatile compounds and also on the possibilities of applying similar methods for solving transport equations. In this volume we have the presentation of monitoring of the deposition of airborne particles by the group from Belgrade led by Mirjana Tasić, and a study of such particles by elemental analysis by van Grieken and his colleagues from Belgium.

We hope that the continuation of our workshops and the publication of our books will contribute to finding a common thread that connects different topics, even different fields, that share some aspects of the phenomena associated with non-equilibrium. As Anton Chekhov once stated 'Only entropy comes easy' so any work aimed at bringing order into the field is difficult. Organization of the workshop and publication of the book are of course not as hard as the pursuit of knowledge itself but we hope that it is, to some degree, a minor contribution to the everlasting human struggle against the entropy. And while we, of course, agree with scientists that are much better than we are that thermodynamics will never be overthrown, it is only human to try to cheat it. Doing the related science is allowing us to achieve exactly that and it is a source of numerous practical applications.

The editors are grateful to all the members of the Gaseous Electronics Laboratory for organization of the workshop, in particular the members of the organizing committee and the staff of the Academy of Science and Institute of Physics. Finally and above all we acknowledge great efforts of all the participants who have invested a lot of funds, their time and effort to join us, sometimes travelling from distant continents. This book exists, however, mainly thanks to the efforts of all the authors who have invested their time and experience to write the papers. We also acknowledge the contribution by Professor Rastko Ćirić whose rendering of Maxwell's demon remains as symbol of our meeting and our publications.

Perhaps the most chaotic aspect of human society, as our current experience teaches us, is the flow of funds and several agencies helped us get the needed funds to continue. The conference and this book were primarily supported by the COE Centre for Non-equilibrium processes and the Ministry of Science of the Republic of Serbia. Additional funding and facilities were provided by the Academy of Sciences and Arts of Serbia, Institute of Physics Belgrade (project No 141025) and Hiden Analytical.

The editors
Zoran Petrovic, Dragana Maric and Gordana Malovic

Journal of Physics: Conference Series **162** (2009) 012001 doi:10.1088/1742-6596/162/1/012001

Study of higher excited states of some polyatomic molecules relevant for plasma physics and environment

B P Marinković

The Institute of Physics, Belgrade 11080, Pregrevica 118, Serbia, and
College for Electrical Engineering and Computing, Belgrade11010, Vojvode Stepe 283, Serbia

bratislav.marinkovic@phy.bg.ac.yu

Abstract. Studies of higher excited states of some polyatomic molecules relevant for plasma physics and environment have been presented. Spectra of chlorofluorocarbons are discussed together with their influence on ozone layer depletion and global warming. Tetrahydrofuran molecule was studied by photoabsorption and electron energy loss spectroscopy while the states are assigned following extensive *ab initio* calculations. Nitrous oxide and hydrogen sulphide spectra are discussed in terms of identifying valence and Rydberg character of excited states.

1. Introduction

Since the extensive studies of polyatomic molecular spectra by Herzberg [1] and Robin [2] there have not been attempts for systematic research of higher excited states and elucidation of electronic structure that would lead to comprehensive understanding of the behaviour of bound electrons in molecules. However, a tremendous advancement in both experimental methods and theoretical analysis of particular cases became evident trough a number of published papers covering all aspects of complex molecular spectra and their manifolds. On the other hand, the role of electronically excited states of polyatomic molecules had been profoundly investigated when the specific questions had been raised, like the influence of chlorofluorocarbons (CFC) on ozone layer depletion and global warming [3-6], the replacement for plasma etching molecules [5,7] or radiation damage of DNA deoxyribose analogue molecules [8,9].

Certain difficulty in studying the higher excited states of polyatomic molecules represents the need for identification of the nature of the state; is it of valence type or Rydberg type. Very often the character of the state is mixed and it is hard to characterize it even by looking the spatial distribution of electron density function. The common knowledge that governs the experimentalist to assign the corresponding state as a Rydberg one is that a state frequency ν fits equation $h\nu=E_i-R/(n-\delta)^2$ where E_i is the ionization potential toward which the series converges with higher n's, R is the Rydberg constant, and δ is quantum defect of the series. It is a general rule that the oscillator strengths fall off as n^{-3}. Since Rydberg orbitals are largely nonbonding, the transition from bonding molecular orbital to Rydberg orbital will in general be narrower than that from the same bonding molecular orbital to an antibonding valence shell molecular orbital. For the interpretation of spectra, the determination of the Rydberg o valence shell character of the upper electronic state is of primary importance.

Journal of Physics: Conference Series **162** (2009) 012001 doi:10.1088/1742-6596/162/1/012001

In this paper the studies of higher excited states for chlorofluorocarbon (CFC) molecules, tetrahydrofuran (THF) molecule and small, triatomic molecules hydrogen sulphide (H_2S) and nitrous oxide (N_2O) are highlighted.

2. Higher excited stets of chlorofluorocarbon molecules

Chlorofluorocarbon molecules under photolysis are a source of atmospheric radicals and therefore are widely recognized to contribute significantly to stratospheric ozone depletion. CFC molecules are also strong greenhouse gases and they usually have a residence time in the atmosphere of about 100 years and an estimated global warming potential of 10^4. Two main mechanisms of destruction in terrestrial atmosphere have been identified, that are: UV photo-dissociation in the stratosphere and mesosphere, and electron attachment and ion-molecule reactions in mesosphere. Comprehensive studies of these molecules and their influence on global warming have been performed by N. J. Mason and coworkers. Dichlorodifluoromethane (CCl_2F_2) molecule has been investigated by Limão Vieira *et al* [3] who determined the photo-absorption cross sections using synchrotron radiation in the range 5.5 – 11 eV. The electronic state assignment had been proposed for several observed bands incorporating both valence and Rydberg transitions. The strongest peak was found at the energy of 9.801 eV and was assigned as 4p Rydberg transition from $4b_2$ molecular orbital and quantum defect of 1.65. The measured cross-sections were used to derive the photolysis rate and lifetime of this molecule as a function of altitude in the terrestrial atmosphere.

Eden *et al* [4] studied photoabsorption in CF_3Cl and other fluro-alkanes at high-resolution in the energy range 3.9 – 11.8 eV. The source of this molecule is purely anthropogenic and it is banned under terms of Montreal Protocol. The authors identified a valence state with the maximum cross section of 0.019 Mb at energy of 7.90 eV and two Rydberg type states assigned to the n=4 members of an ns and np series with the peaks at 9.679 eV and 10.63 eV with cross sections of 34.06 Mb and 13.07 Mb, respectively. Kitajima *et al* [5] studied electron energy spectra of several halogen-containing molecules, the CF_3Cl being among them. They identified the low-lying transition as ${}^1A_1 \rightarrow {}^1E$ transition governed by quadruple allowed interactions. They also observed higher members of two Rydberg series converging to ionization limits $6e^{-1}$ and $7a_1^{-1}$.

3. Higher excited stets of tetrahydrofuran molecule

The excited states of tetrahydrofuran (THF), C_4H_8O, molecule have been extensively studied by Giuliani *et al* [9] very recently. The THF is a five membered heterocyclic ring molecule. The electronic configuration of THF in its electronic ground state and in the C_{2v} formalism is: $(1a_1)^2$ $(1b_1)^2$ $(2a_1)^2$ $(3a_1)^2$ $(2b_1)^2$ $(4a_1)^2$ $(5a_1)^2$ $(3b_1)^2$ $(6a_1)^2$ $(4b_1)^2$ $(1b_2)^2$ $(7a_1)^2$ $(1a_2)^2$ $(2b_2)^2$ $(5b_1)^2$ $(8a_1)^2$ $(6b_1)^2$ $(9a_1)^2$ $(2a_2)^2$ $(3b_2)^2$: $\tilde{X}\,{}^1A_1$. The C_{2v} correlates unambiguously to the other lower geometries. In the C_2 point group, the C_{2v} a_1 and a_2 become a and b_1 and b_2 correlate to b. For the C_s geometry, a_1 and b_1 become a′ and a_2 and b_2 become a″. Ab initio calculations were used to determine vertical excitation energies of the electronic states. The equilibrium geometry of the C_2 and C_s conformers were considered.

The lowest excited states of the molecule were identified as the series converging to the ionic ground electronic state. The 3s terms are spanning from 6.04 eV to 6.88 eV while the vertical energy for the 3s(C_s) transition is estimated to be at 6.353 eV. The 3p terms are spanning from 6.8 eV to 8.4 eV while the vertical energy for the 3p(C_s) transition is estimated to be at 7.154 eV. The 3d terms are predicted to contain transitions from both conformers (C_2 and C_s) while the spectral region extends from 7.40 eV to 8.15 eV. A broad feature was observed at 7.483 eV, not well resolved from the underlying background but which agrees with the theoretical value of 7.474 eV for the excitation of the 3d(C_2) state. Additionally two intense features are observed at 7.730 eV and 7.813 eV, the values match the predicted transition energies to two 3d(C2) states at 7.715 eV and 7.754 eV.

A Rydberg series analysis by Doucet *et al* [10], who measured the electronic spectrum of tetrahydrofuran between 2000 and 1200 Å in the gas phase, reported a single ns series (δ = 0.94), two np series (δ = 0.64 and 0.52) and a single nd series (δ = 0.08).

The spectrum of higher excited states studied by Giuliani *et al* [9] is spanning from 8.2 eV to 10.6 eV. The transitions were attributed to the Rydberg series converging to three ionization limits, E_i, of both conformers; the first ionic limit being 9.433 eV, the second ionic limit at 11.512 eV and the third at 11.964 eV. There were identified 13 transitions coming from C_2 conformer and 17 from C_s conformer that converge to the first ionization limit. Four transitions were observed only in C_2 conformer that converge to the second ionization limit, while one transition for both conformers was identified to belong to the third limit.

4. Higher states of some triatomic molecules

Herzberg [1] had divided the consideration of triatomic molecules into three groups: di-hydrides, mono-hydrides and non-hydrides. The mostly studied di-hydride molecule is water molecule due its importance. A comprehensive review of electron cross sections for water molecule has been given recently by Itikawa and Mason [11] while the threshold electron impact spectroscopy has been presented by Jureta [12].

If the molecule has a closed-shell ground state, then the upper Rydberg states will in general appear as singlet-triplet splitting pairs. At the Hartree Fock level, the singlet-triplet splitting is 2K, where K is the exchange integral between valence shell and Rydberg orbitals. Because of the compact nature of core wavefunction, ψ_c, and the extended nature of Rydberg wavefunction, ψ_R, K will be small.

4.1.1. *Higher states of nitrous oxide molecule.* The electron spectroscopy of nitrous oxide molecule has been extensively performed in Laboratory for Atomic Collision Processes at The Institute of Physics (LACPIP), Belgrade [13]. Both electron energy loss and electron threshold spectra were recorded and many valence and Rydberg states as well as higher excited states were identified. Rydberg states arise not only from the excitation of electron from the most loosely bound orbital in the ground state but also exciting it from any other orbital occupied in the ground state. The stability of these Rydberg states depends on the (non-, anti-) bonding character of the orbital from which electron was excited from.

Photoabsorption spectra of N_2O have been extensively studied by UV sources and electron dipole spectroscopy [14]. Also, preliminary high resolution data has been recently reported between 4.13 eV (300 nm) and 11.3 eV (110 nm) [15]. Nitrous oxide does absorb in the near ultraviolet. Absorption spectrum shows two broad features followed by the series of higher excited states.

4.1.2. *Higher states of hydrogen sulphide molecule.* Sulphur gases are known to be precursors of sulphate aerosol particles and cloud condensation nuclei over remote parts of the oceans and could also act as a feedback mechanism in climate regulation which affects the Earth's radiative balance by direct scattering of solar radiation. The ground state electronic configuration of H_2S is $(1a_1)^2(2a_1)^2(1b_2)^2(3a_1)^2(1b_1)^2(4a_1)^2(2b_2)^2(5a_1)^2(2b_1)^2$ X^1A_1. Most of the excited states are formed from the excitation of an electron from $2b_1$ orbital. The $2b_1$ and $5a_1$ orbitals are nonbonding, while the $6a_1$ orbital is anti-bonding. Since the $2b_1$ orbital is nonbonding, the ground state of H_2S^+ with one electron removed from this orbital is expected to have nearly the same stability as neutral molecule and the same applies for Rydberg states which converge to the ground state of ion. The vibrational frequencies of neutral and ion molecule differ only slightly. That is why 0-0 bands are the strongest bands in each series corresponding to transitions from the ground stat to Rydberg states. The first excited states $^{3,1}A_1$ and $^{3,1}B_1$ are repulsive states and responsible for the broad maximum in the spectra.

The electron spectroscopy studies of H_2S have been performed in LACPIP Belgrade, both electron energy loss and electron threshold spectra were recorded. Energy loss spectra were recorded at higher impact energies and smaller scattering angles that favour the optically allowed transitions and at smaller impact energies and higher scattering angles where the optically forbidden states show up.

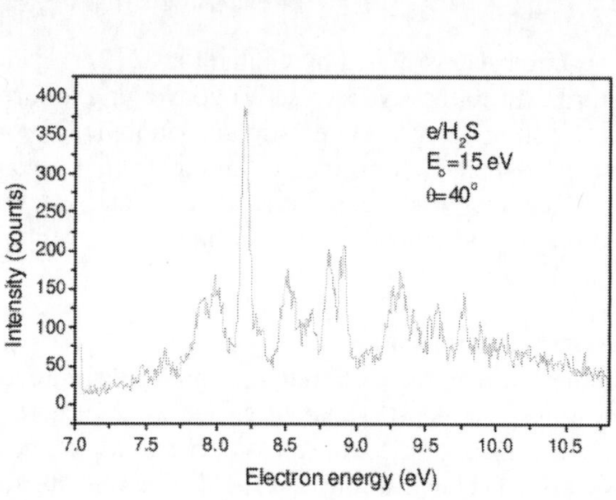

Figure 1. Electron energy loss spectrum of higher excited sates of H_2S recorded at 15 eV impact energy and 40° scattering angle

5. Conclusions

The higher excited states of some chlorofluorocarbon molecules, tetrahydrofuran molecule and triatomic nitrous oxide and hydrogen sulphate molecules have been presented and discussed. The electron energy loss spectrum of H_2S has been recorded at low impact energy of 15eV and relatively large scattering angle of 40°.

Acknowledgements

This work has been performed within the MSTD project 141011 and ESF/COST Actions CM0601 (ECCL – Electron Controlled Chemical Lithography) and .P9 (RADAM – Radiation Damage in Bimolecular Systems).

References

[1] Herzberg G 1966 *Molecular Spectra and Molecular Structure Volume III, "Molecular Spectra and Molecular Structure of Polyatomic Molecules"* (Van Nostrand Reinhold, Princeton, New Jersey)

[2] Robin A 1974 *Higher Excited States of Polyatomic Molecules Volume I* (Academic Press, New York and London)

[3] Limão Vieira P, Eden S, Kendall P A, Mason N J and Hoffmann S V 2002 VUV photo-absorption cross-section for CCl_2F_2, *Chem. Phys. Lett.* **364** 535–41

[4] Eden S, Limão-Vieira P, Hoffmann S V and Mason N J 2006 VUV photoabsorption in CF_3X (X = Cl, Br, I) fluoro-alkanes, *Chem. Phys.* **323** 313–33

[5] Kitajima M, Suzuki R, Tanaka H, Pichl L and Cho H 2003 Electron impact excitation and dissociation of halogen-containing molecules, *Nukleonika* **48** 89–93

[6] Limão Vieira P, Eden S, Kendall P A, Mason N J, Giuliani A, Heinesch J, Hubin-Franskin M-J, Delwiche J and HoffmannS V 2004 An experimental study of SF_5CF_3 by electron energy loss spectroscopy, VUV photo-absorption and photoelectron spectroscopy, *Int. J. Mass Spetrom.* **233** 335–41

Mason N J, Dawes A, Mukerji R, Drage E A, Vasekova E, Webb S M and Limão-Vieira P 2005 Atmospheric chemistry with synchrotron radiation, *J. Phys.B: At. Mol. Opt. Phys.* **38** S893–911

Limão-Vieira P, Vasekova E, Raja Sekhar N B, Mason N J and Hoffmann S V 2006 VUV electronic state spectroscopy of 1,1-difluoroethene and difluorochloromethane by high resolution synchrotron radiation, *Phys. Chem. Chem. Phys.* **8** 4766–72

[7] Ohtake H and Samukawa S 2002 Charging-damage-free and precise dielectric etching in pulsed

C_2F_4/CF_3I plasma, *J. Vac. Sci. Technol.*B **20** 1026–30

Mason N J, Limão-Vieira P, Eden S, Kendall P, Pathak S, Dawes A, Tennyson J, Tegeder P, Kitajima M, Okamoto M, Sunohara K, Tanaka H, Cho H, Samukawa S, Hoffmann S V, Newnham D and Spyrou S M 2003 VUV and low energy electron impact study of electronic state spectroscopy of CF_3I, *Int. J. Mass Spetrom.* **223-224** 647–60

[8] Giuliani A, Delwiche J, Hoffmann S V, Limão-Vieira P, Mason N J and Hubin-Franskin M.-J 2003 2-methyl furan: An experimental study of the excited electronic levels by electron energy loss spectroscopy, vacuum ultraviolet photoabsorption, and photoelectron spectroscopy, *J. Chem. Phys.* **119** 3670-80

Walker I C, Palmer M H, Delwiche J, Hoffmann S V, Limão Vieira P, Mason N J, Guest M F, Hubin-Franskin M-J, Heinesch J and Giuliani A 2004 The electronic states of isoxazole studied by VUV absorption, electron energy-loss spectroscopies and ab initio multi-reference configuration interaction calculations", *Chem. Phys.* **297** 289–306

Milosavljević A R 2008 Electron interaction with DNA deoxyribose analogue molecules, *AIP Conf. Proc.* **876**, *The physics of ionized gases: 23rd Summer school and international symposium* Eds. Hadžievski Lj, Marinković B P and Simonović N S, pp 80-7

Milosavljević A R, Šević D and Marinković B P 2008 Electron interaction with deoxyribose analogue molecules in gaseous phase, *J. Phys: Conf. Ser.* **101**, *Radiation damage in biomolecular systems (RADAM07)* Ed. McGuigan K J, 012014

[9] Giuliani A, Limão-Vieira P, Duflot D, Milosavljević A R, Marinković B P, Hoffmann S V, Mason N J, Delwiche J and Hubin Franskin M.-J 2009 Electronic states of neutral and ionized tetrahydrofuran studied by VUV spectroscopy and ab initio calculations" *E. Phys. J.* D **51** 97-108

[10] Doucet J, Sauvageau P and Sandorfy C 1972 The vacuum ultraviolet spectrum of tetrahydrofuran, *Chem. Phys. Lett.* **17** 316–9

[11] Itikawa Y and Mason N J 2005 Cross sections for electron collisions with water molecules, *J. Phys. Chem. Ref. Data* **34** 1–22

[12] Jureta J J 2005 The threshold electron impact spctrum of H_2O, *Eur. Phys. J. D* **32** 319–28

[13] Marinković B, Szmykowski Cz, Pejčev V, FilipovićD and Vušković L 1986 Differential cross sections for elastic and inelastic scattering of electrons by N_2O in the range from 10 to 80 eV, *J. Phys. B: At. Mol. Phys.* **19** 2365–75

Čubrić D, Cvejanović D, Jureta J, Cvejanović S, Hammond P, King G C and Read F H 1986 Threshold electron impact spectrum of N_2O, *J. Phys. B: At. Mol. Phys.* **19** 4225–39

Marinković B, Panajotović R, Pešić Z D,. Filipović D M, Felfli Z and Msezane A Z 1999 Normalization of the measured relative electron differential cross sections for $2^1\Sigma^+$ and $^1\Pi$ sates of N_2O, *J. Phys. B: At. Mol. Phys.* **32** 1949–57

[14] Rabalais J W, Mcdonald J M, Scherr V and Mcglynn S P 1971 Electronic spectroscopy of isoelectronic molecules. II. Linear triatomic groupings containing sixteen valence electrons, *Chem. Rev.* **71** 73–108

Shaw D A, Holland D M P, MacDonald M A Hopkirk A, Hayes M A and McSweeney S M 1992 A study of the absolute photoabsorption cross section and the photoionization quantum efficiency of nitrous oxide from the ionization threshold to 480 Å, *Chem. Phys.* **163** 387–404

Chan W F, Cooper G and Brion C E 1994 Discrete and continuum photoabsorption oscillator strengths for the electronic spectrum of nitrous oxide (5.5-203 eV), *Chem. Phys.* **180** 77–88

[15] Marinković B P, Jones N C, Walker I C and Mason N J 2008 An experimental study of the excited electronic levels of N_2O molecule by electron energy loss spectroscopy and vacuum ultraviolet absorption, *Proc.* 15th *Central European Workshop on Quantum Optics* (CEWQO 2008), Eds. Bozić M and Arsenović D, (Institute of Physics, Belgrade, 2008). ISBN 978-86-82441-23-6, Poster, p.56.

Journal of Physics: Conference Series **162** (2009) 012002 doi:10.1088/1742-6596/162/1/012002

Low energy positron interactions – trapping, transport and scattering

J. P. Sullivan[1], S. J. Buckman[1], A. Jones[1], P. Caradonna[1], C. Makochekanwa[1], D. Slaughter[1], Z. Lj. Petrović[2], A. Banković[2], S. Dujko[2,3], J. P. Marler[2,4] and R. D. White[3]

[1] ARC Centre for Antimatter-Matter Studies, Research School of Physical Sciences and Engineering, The Australian National University, Canberra, ACT, Australia

[2] Institute of Physics Belgrade, University of Belgrade, Pregrevica 118, POB 68, 11080 Zemun Serbia

[3] ARC Centre for Antimatter-Matter Studies, School of Mathematics, Physics and IT, James Cook University, Townsville 4810, QLD, Australia

[4] University of Aarhus, Aarhus Denmark

jps107@rsphysse.anu.edu.au

Abstract. Recent experiments, theory and modelling of positron interactions with atoms and molecules are discussed. The first half of the paper is devoted to binary collisions between positrons and crossed beams of atoms or molecules (in this case neon) and the second half deals with ensembles of non-interacting positrons, otherwise known as swarms which are transported through the background gas. We review the recent results on measurements of the cross sections based on obtained from collisional positron traps and subsequent calculations of transport properties of positron swarms that may be used to model thermalization experiments, collisional traps and possible applications of positrons in materials science and biomedicine. It was found that kinetic phenomena occur in positron transport that are mainly the result of the positronium (Ps) formation which has a larger cross section than elastic scattering in most gases and at the same time is a non-conservative process. Most importantly negative differential conductivity (NDC) occurs only for the bulk drift velocity while it does not exist for the flux property, a phenomenon that has not been observed for electrons.

1. Introduction

Over the past 40 years or so electron scattering data have been obtained experimentally by the application of two techniques. Quasi-monoenergetic beams were employed to obtain the energy dependence of scattering processes, angular differential cross section data and the fine structure and contribution to the cross sections of negative ion resonances [e.g. 1, 2]. However, many of these approaches lacked the ability to produce good quality absolute data and comprehensive data sets that included all processes. On the other hand, measurements of transport properties of 'swarms' of particles in gases (a 'swarm' is an ensemble of particles that do not interact with each other, their behaviour being determined by an external electric field and their collisions with the unperturbed background gas) provided data which can be directly applicable in the modelling of low temperature (non-equilibrium) plasmas. These approaches also gave the necessary data to obtain a full set of cross sections for a particular gas through an iterative approach, consisting of solutions to the Boltzmann equation and subsequent modifications of the cross sections in order to reduce the disagreement between the calculated and the measured transport data. The swarm derived cross sections [3, 4, 5] gave good absolute values and total summed balances of number, energy and momentum albeit with a poorer resolution, non-uniqueness and with limits in applicability to lower mean energies (typically below ~10 eV). Most importantly the result of the swarm analysis was obtained by averaging over a very broad distribution of particles that is exactly the same as in most non-equilibrium plasmas. Advantages and limitations of the two techniques have been discussed in the past and, if one weighs the pros and cons of both approaches, a joint, complementary approach seems to be the best strategy and this is still valid today. In some cases contemporary beam measurements [6, 1] can match the accuracy of swarm derived cross sections with the flexibility in energy resolution and range, while swarm calculations have became the backbone of the models of collisional plasmas [7]. However there remain, unfortunately, very few joint efforts to produce new cross section data.

In the case of positron interactions with atoms and molecules neither of the two approaches outlined above were easily implemented in the past. This is as a result of significant difficulties in performing positron swarm measurements (see [8]). Positron beam experiments have provided the bulk of the experimental cross section information but these too have been limited in applicability due to low beam fluxes and insufficient energy resolution (see for example [9] and references therein). In this paper we attempt to illustrate how recent advances in beam techniques have enabled a reopening of the positron transport studies, which may eventually benefit those devices that use positrons (e.g. in materials and biomedical research). Here we combine the efforts of both the beam and transport groups from the Australian Centre for Antimatter and Matter Studies with the input of their associated colleagues from Belgrade in an attempt to show the present status of positron scattering and transport studies, and point out the need for future developments and combined efforts.

2. Low energy positron scattering from neon

Since the development of the Surko buffer-gas trap for storing positrons and forming a high resolution, low energy positron beam [10, 11], a number of new and improved experiments have been possible in the field of positron scattering. High resolution beams and new scattering techniques have begun to provide new insight into low energy positron interactions with atoms and molecules [12, 13]. In particular, for the noble gases, it is now possible to perform measurements of scattering with a degree of precision that is far in excess of that previously possible. Improved energy resolution also makes possible the examination of threshold effects and the search for positron scattering resonances, the elusive analogue to negative ion resonances in electron scattering.

In addition, there is now sufficient experimental and theoretical data available to start modelling the transport of positrons on gases, a process that can be expected to proceed in a manner quite different to electron transport – which has been studied for many years. In particular, by modelling the trapping process, it may be possible to gain new insight into the mechanisms that give rise to the effectiveness of buffer gas trapping.

Surko traps are based on a modified Penning-Malmberg trap, with positrons confined axially in a magnetic field of up to 1.5 kG. Confinement in the axial direction is achieved by introducing the positrons to an electrostatic potential well, filled with a mixture of N_2 and CF_4 gases. Positrons lose energy initially by exciting an electronic state of the N_2 (principally the a1Π state) and become trapped in the potential well structure. After becoming trapped, they continue to lose energy through inelastic collisions on the buffer gas, through electronic, vibrational and rotational excitation. In this way, the positrons fall to the bottom of the potential well, and thermalise to the gas temperature, typically room temperature. This provides a reservoir of positrons with a temperature of approximately 40 meV, which can then be used to form a magnetised positron beam.

The apparatus used for the experiments presented in this paper is based on this technique, and its operation has been described in detail elsewhere [14], only a brief overview will be given here. Positrons from a neon-moderated ^{22}Na source are directed into an electrode assembly contained in a 500 gauss magnetic field. The electrodes contain a buffer gas mixture of N_2 and CF_4, and positrons exciting electronic transitions in the gas are trapped and subsequently cool to room temperature, as described above. The cooled positrons are emitted from the trap as a pulsed beam, with an energy spread of approximately 60 meV, and are directed to a scattering cell (also in a 500 gauss magnetic field) which contains the target gas. The relative potential of the final stage of the trap and the scattering cell determines the collision energy, which can be tuned from ~0.5 to 100 eV. Scattering events redistribute energy in the components of the positron motion parallel and perpendicular to the confining magnetic field. The beam is then directed to a retarding potential analyser to analyse the energy distribution within the beam, before being directed to an microchannel plate assembly for detection. Analysis of the energy distribution gives the cross section values, with normalisation dependent only on the length of the gas cell and the pressure of the target gas inside the cell.

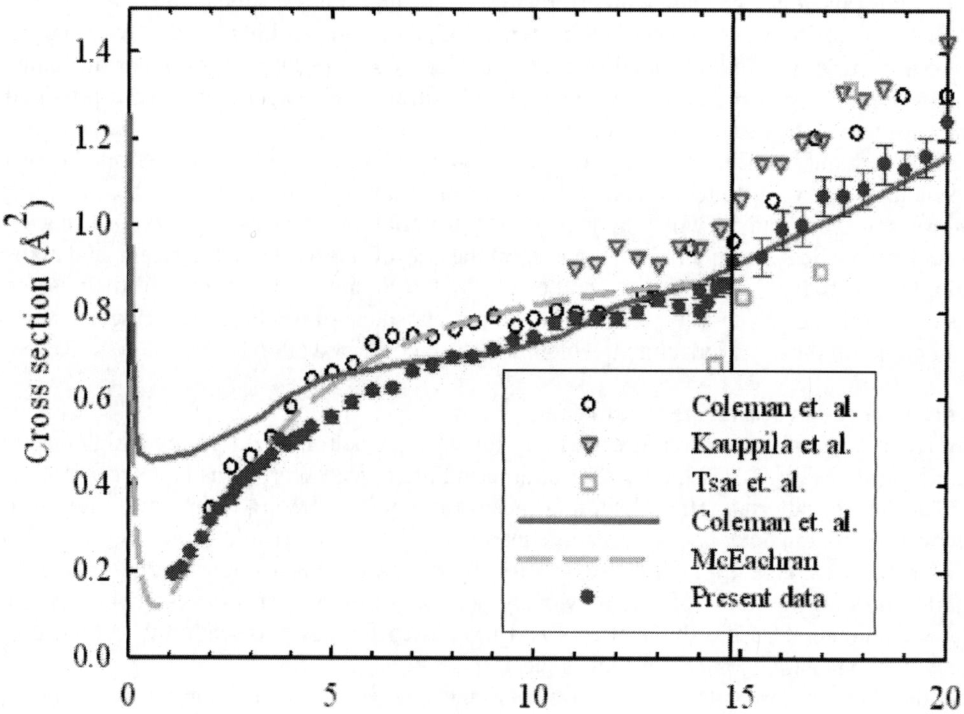

Figure 1: Low energy grand total cross section for positron scattering from neon. Present experiments are the solid circles. Data is compared to previous measurements of Coleman et al. [15], Kauppila et al. [16, 17] and Tsai et al. [18] as well as calculations by Coleman et al. [19] and McEachran [20], as detailed in the legend.

Presented here are preliminary results for positron scattering from neon. The grand total cross section, from 1 to 60 eV has been measured, as has the positronium formation cross section, from threshold to 60 eV. The grand total cross section for the low energy region is shown in Figure 1.

Below the positronium formation threshold, this is equivalent to the elastic scattering cross section. The data is compared with a number of recent theoretical calculations and previous experiments. For the lowest energies, below about 5 eV, the agreement between the recent theory of McEachran [20] and the experimental data is very good, showing the characteristic fall of the cross section associated with the Ramsauer-Townsend minimum. Between 5 eV and the positronium formation threshold, the experiment and theory disagree by up to 10%, well outside the experimental (absolute) error. There is also disagreement in this energy range between the other two experimental measurements, which falls outside the combined errors of the measurements. Conversely, agreement with the theory of Coleman et al. [19] is better at higher energies, above about 8 eV.

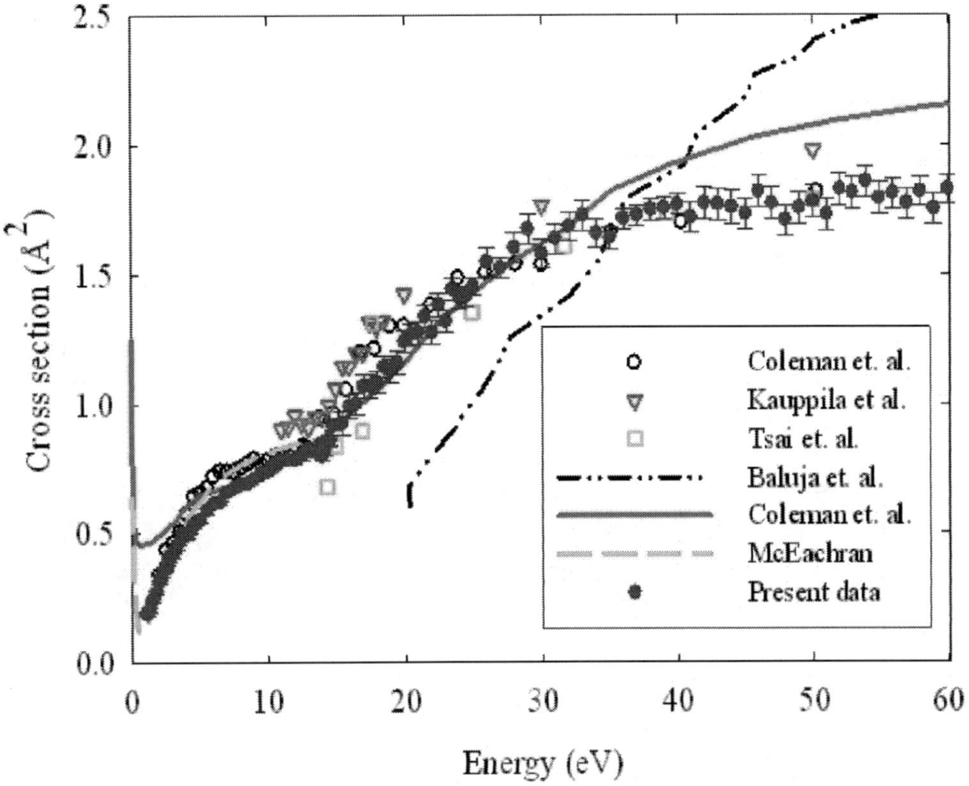

Figure 2: Grand total cross section for positron scattering from neon up to 60 eV incident energy. The data on the plot is as for Figure 1, with the addition of the theory of Baluja et al. [21].

Figure 2 shows the extended energy range measurements. Over this energy range, agreement with previous measurements is generally good, with the exception of the data from the Detroit group [16, 17] which is systematically higher than the present measurements. Agreement with the theory of Coleman falls away from about 30 eV, and the agreement with the calculations of Baluja et al. [21] is quite poor. Generally, while agreement between experiments for the grand total cross section is reasonable, there are still substantial discrepancies that remain and which obviate against claiming these results to represent a 'benchmark' standard of cross section. The recent theory of McEachran appears to be a much better description of the scattering system below the Ps threshold, and further measurements at lower energies are required to examine the Ramsauer-Townsend region in more detail.

The positronium formation cross section is presented in Figure 3, along with two previous experimental measurements and two theoretical calculations. While all experiments agree at energies close to the threshold, there is a divergence between the two previous measurements above about 30 eV scattering energy. The present measurements are in better agreement with the data of Marler et al. [22], although given the present error, it is difficult to make a definitive statement in this regard. The two calculations disagree with all experiments in the near threshold region and, at higher energies the calculation of McAlinden and Walters [23] is favoured by the present experimental results over the calculations of Gilmore [24]. The agreement between the measurements of Marler et al. and the present results suggest that speculation of systematic errors in the data of Laricchia et al. [25] may be correct. There has been disagreement between the two experimental groups for Ps formation cross sections for positron scattering from all the noble gases, and systematic studies to compare with all previous measurements may shed further light on this matter. It is clear that theoretical descriptions of this problem need to be improved to match the observations, and it may be that this will also provide an avenue by which to resolve the present disagreements.

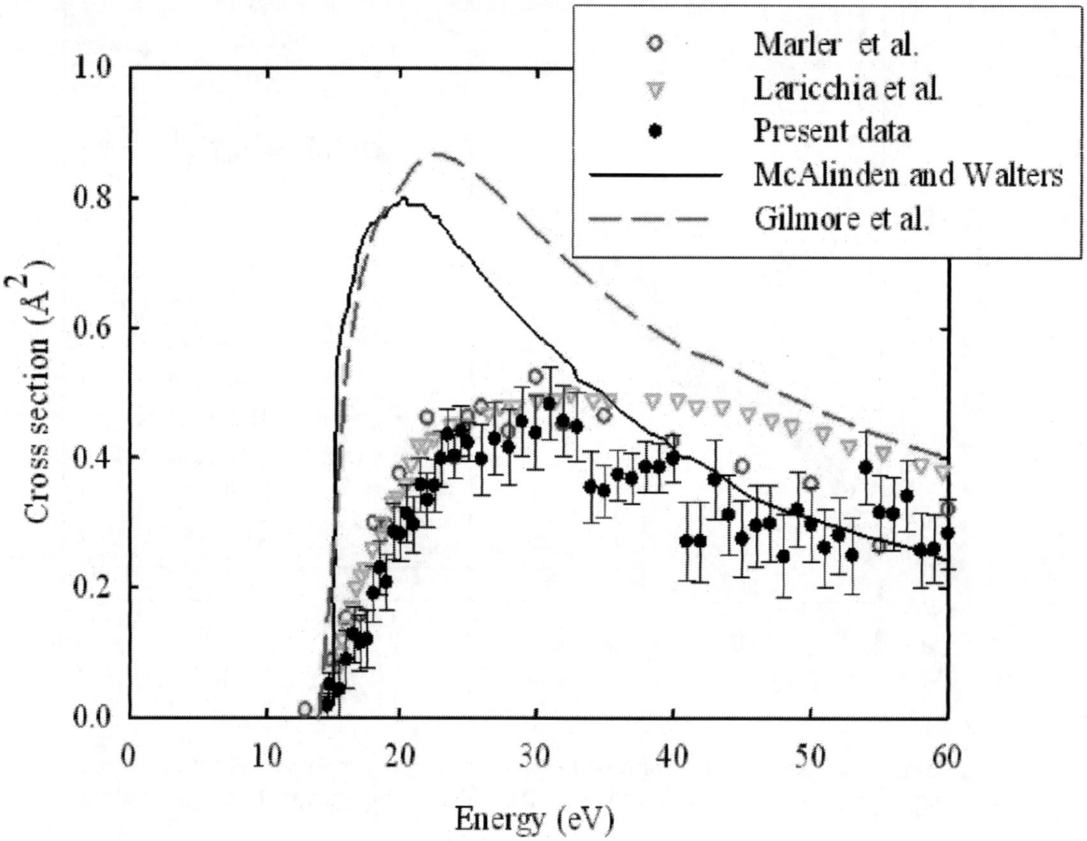

Figure 3: Positronium formation cross section for neon. The present data is compared to the previous measurements of Marler et al [22] and Laricchia et al [25] and the calculations of McAlinden and Walters [23] and Gilmore et al. [24]

3. Positron transport and thermalization

3.1. Input data

The intrinsic property of charged particle swarm calculations is that they HAVE to include complete processes responsible for number changing, momentum transfer and energy transfer collisions. Complete does not mean every little detail but that global balance should be made regardless of how the partial contributions are distributed among the processes. This may lead to non-uniqueness which may only be resolved by the input of additional data, either transport coefficients or groups of cross sections from other sources that are assumed to be accurate. But most importantly, the cross section sets used in (or obtained from) the swarm analysis MUST have a complete set of cross sections in the relevant energy range and until recently such data were no available. Recently, with the extended use of the Surko trap and other techniques (Recent advances in experimental techniques have enabled measurements of a wide range of inelastic cross sections for positron scattering on atoms and molecules [26, 27, 28, 29].) it became possible to make reasonably complete sets in the region of energies below the threshold for ionization and while there were no experimental transport data to compare to it was worth performing calculations to see what kind of phenomena may be found in positron transport.

In this paper we will mainly show the data for positrons in argon (as described in paper by Šuvakov et al [30]) and to make a complete set we have used

a) The elastic momentum-transfer cross section (MTCS) and isotropic scattering of McEachran [31] as this calculation is in reasonable agreement with the existing experimental data for total cross sections [32, 33].
b) Experimental results for the electronic excitation of the two lowest lying 3p4s J = 1 levels of argon and theoretical results were used to form effective cross sections [34, 35].
c) Excitation of higher singlet levels based on that for electron impact excitation of argon [36].
d) The positronium (Ps) formation and ionization cross sections were taken from the work of Marler et al [22]. These results are in good quantitative agreement with an alternative approach of Laricchia et al [25]. It is important to note that Ps formation cross section is typically two orders of magnitude larger than the equivalent process for losses of electrons, the electron attachment. And while Ps formation is a non-conservative (number changing) process for positrons, the ionization is not which is different from electron transport.
e) Direct annihilation has been neglected as the cross section is known to be several orders of magnitude lower than that for Ps formation [37].

With this set it is reasonable to assume that we may have a very good qualitative and a reasonable quantitative prediction of the positron transport for energies up to 20 eV.

3.2. Basic definitions of transport coefficients

In modelling of the transport of charged particles the connection between experiment and theory is usually made through the equation of continuity

$$\frac{\partial n(\mathbf{r},t)}{\partial t} + \nabla \cdot \mathbf{\Gamma}(\mathbf{r},t) = S(\mathbf{r},t), \tag{1}$$

where $n(\mathbf{r},t)$ is the number density of swarm particles (positrons), while $\mathbf{\Gamma}(\mathbf{r},t) = n\langle \mathbf{v} \rangle$ is the swarm particle flux and $S(\mathbf{r},t)$ represents the production rate per unit volume per unit time arising from non-conservative collisional processes such as Ps formation or annihilation. It is important to note that the average/drift velocity of the swarm is usually several orders of magnitude smaller than the instantaneous random velocities of the particles.

Far from boundaries, sources and sinks, the hydrodynamic regime is assumed to apply. In this regime, the space-time dependence of the quantities is expressible in terms of linear functionals of $n(\mathbf{r},t)$, generally a density gradient expansion. Assuming this functional relationship, the flux $\mathbf{\Gamma}(\mathbf{r},t)$ and source term $S(\mathbf{r},t)$ in the continuity equation are expanded as:

$$\mathbf{\Gamma}(\mathbf{r},t) = \mathbf{W}_F n(\mathbf{r},t) - \mathbf{D}_F \cdot \nabla n(\mathbf{r},t) \quad \text{(Fick's law)} \tag{2}$$

$$S(\mathbf{r},t) = S^{(0)} n(\mathbf{r},t) - \mathbf{S}^{(1)} \cdot \nabla n(\mathbf{r},t) + \mathbf{S}^{(2)} : \nabla\nabla n(\mathbf{r},t) \tag{3}$$

where $\mathbf{W}_F$ and $\mathbf{D}_F$ define respectively the *flux* drift velocity and *flux* diffusion tensor. Substitution of expansion the expressions for the source and flux into the continuity equation yields the diffusion equation,

$$\frac{\partial n}{\partial t} + \mathbf{W}_B \cdot \nabla n - \mathbf{D}_B : \nabla\nabla n = -R_a n\,, \tag{4}$$

where

$$R_a = -S^{(0)} \text{ (loss rate)}, \tag{5}$$

$$\mathbf{W}_B = \mathbf{W}_F + \mathbf{S}^{(1)} \text{ (bulk drift velocity)}, \tag{6}$$

$$\mathbf{D}_B = \mathbf{D}_F + \mathbf{S}^{(2)} \text{ (bulk diffusion tensor)}. \tag{7}$$

In swarm experiments it is generally the bulk transport coefficients which are measured and tabulated. The reader is referred to Robson et al. [38] for a detailed discussion on the types and application of these transport coefficients. Physically, the bulk transport coefficients are associated with the swarm's centre of mass transport (and spread about its centre of mass) while the flux quantities are essentially averages in velocity space. The explicit influence of non-conservative collisional processes (such as Ps formation) on the swarm's centre of mass transport is described by the terms $\mathbf{S}^{(1)}$ and $\mathbf{S}^{(2)}$. Obviously, in the absence of non-conservative processes, the two sets of transport coefficients coincide. The differences between, and implications associated with, the two sets of transport coefficients will be highlighted later and the differences associated with Ps formation can be particularly large.

3.3. Calculations of transport properties

Two independent techniques for calculation of positron transport in gases under non-equilibrium conditions are the Monte-Carlo simulation and the Boltzmann equation solution. Details of the two techniques have been reviewed elsewhere, and here we give a brief overview of each of the two techniques.

3.3.1. *Monte-Carlo simulation:*

The Monte Carlo simulation technique for purely electric field and for the orthogonal configuration of electric and magnetic fields has been discussed in our previous papers [39, 40].

The actual code used in this calculation has been described briefly in [41, 30] as is based on similar principles as our previous work. It has been tested to satisfy all the benchmarks as our previous code and to give even better results than before that only limited by the statistical uncertainty.

We employ the coordinate system where the electric field **E** defines the z-axis, while the magnetic field **B** lies in the y-z plane, making an angle ψ with respect to the electric field **E**. Hence, the equations of electron motion can be given analytically and the reader is referred to [42] for an explicit form of these equations.

The only unknown in the MC procedure is the time step for the determination of the exact moment of the next collision. This is determined by solving the equation for the collision probability, using either the null-collision [43] or integration technique [44]. In our code, the latter approach is employed. All dynamic properties of each electron such as the position, velocity and energy can be updated and the electron motion through the neutral gas can be followed between collisions.

The angular distributions of both elastic and inelastic collisions are taken to be isotropic. It must be emphasized that this approximation is generally valid for relatively low E/N but needs to be corrected when forward scattering becomes dominant. In that respect, Biagi has adopted the method proposed by Longo and Capitelli [45] in his Monte Carlo code to include the angular distribution of both elastic and inelastic collisions [42]. For CF_4, it was shown by Vasenkov [46], the errors in transport coefficients, resulting from the assumption of isotropic scattering are less than about 3% for E/N below 0.1 Td. For higher E/N, the errors increase rapidly up to 35%. In addition, it is necessary to re-consider the influence of superelastic collisions in an effort to account properly for the behavior of electron swarms at low energies. However, since we attempt to scan a wide range of transport coefficients and to provide a database for plasma modeling, we will not focus on these issues in the present paper.

3.3.2. Transport coefficients, definitions and calculations

Transport coefficients up to diffusion are determined after relaxation to the steady state from the formulae described in [47, 48, 49]. The number changing reaction rate is defined by

$$\alpha = -\frac{d}{dt}(\ln N) \quad , \tag{8}$$

the bulk drift velocity by

$$\mathbf{W}_B = \frac{d}{dt}\langle \mathbf{r} \rangle, \tag{9}$$

and the bulk diffusion tensor by

$$\mathbf{D}_B = \frac{1}{2!}\frac{d}{dt}\langle \mathbf{r}^*\mathbf{r}^* \rangle. \tag{10}$$

where $r^* = r - \langle r \rangle$. These are the so-called "bulk" transport coefficients that are established from the mean position of the electron swarm in the configuration space [50].

In the absence of the non-conservative collisions, one may avoid the differentiation required to obtain the drift velocity and components of the diffusion tensor by using the so-called "flux" quantities. Hence the flux drift velocity and the flux diagonal elements of the diffusion tensor are given by:

$$W_{F,i} = \left\langle \frac{dx_i}{dt} \right\rangle = \langle v_i \rangle, \; (i = x, y, z), \tag{11}$$

$$D_{F,xx} = \langle x v_x \rangle - \langle x \rangle \langle v_x \rangle, \tag{12}$$

$$D_{F,yy} = \langle y v_y \rangle - \langle y \rangle \langle v_y \rangle, \tag{13}$$

$$D_{F,zz} = \langle z v_z \rangle - \langle z \rangle \langle v_z \rangle. \tag{14}$$

It follows from (9) that bulk drift velocity is displacement of the mean position of the electron swarm and it describes the motion of the centre of mass of the total ensemble of electrons. On the other hand, the flux drift velocity is the mean velocity of electrons. These two sets of transport coefficients are equal by definition, as mentioned above, in the absence of non-conservative collisions (ionization/attachment), however in the presence of non-conservative processes they may differ. The distinction between these two sets of transport coefficients was discussed at length in the 1980's (see [50]), but has been ignored in the majority of previous work in plasma modeling community. This has lead to a potentially serious mismatch between input swarm data (generally the bulk transport properties) and the parameters (often the flux transport properties) required in many plasma fluid models [51,52]. Note that only theory, i.e., Boltzmann equation calculations and/or Monte Carlo simulations, can resolve any such mismatch, by providing both flux and bulk transport coefficients.

Typically we follow 10 000 to 1000 000 particles through a large number of collisions. In all cases our results were in excellent agreement with those obtained by numerical solution to the Boltzmann equation.

3.3.3. Boltzmann equation solution:

The starting point of any kinetic theory calculation is the Boltzmann equation, which can be written in generic form as

$$\left[\frac{\partial}{\partial t} + \mathbf{v}.\nabla + \frac{e}{m}(\mathbf{E} + \mathbf{v} \times \mathbf{B}).\ \frac{\partial}{\partial \mathbf{v}}\right] f \; = J(f) \tag{15}$$

This is essentially a continuity equation for f in phase-space. For swarms, only one species of charged particle is usually present in the gas, **E** and **B** are prescribed external fields, and the electron-

electron (or ion-ion) mutual interaction term is neglected in the collision operator $J(f)$. The kinetic equation is thus linear in f.

Solution of the Boltzmann equation is generally made through representations of the space and velocity dependence of $f(\mathbf{r}, \mathbf{v}, t)$. Under hydrodynamic conditions the space-time dependence of $f(\mathbf{r}, \mathbf{v}, t)$ is represented in the form of a density gradient expansion (detailed above). The velocity dependence in $f(\mathbf{r}, \mathbf{v}, t)$ is generally represented in terms of a combined spherical harmonic and modified Laguerre basis set (combined generally referred to as Burnett functions). This set is complete on velocity space. It is important to emphasize that this is a true multi-term technique (i.e. as many spherical harmonics are retained as required to achieve the accuracy criterion), thus removing the restrictions to near isotropy in velocity space generally associated with previous positron transport studies which are based on the two-term solution of Boltzmann equation. No assumptions are made on the anisotropy of the differential cross-sections. Matrix elements of the Boltzmann equation are then made, and a hierarchy of coupled matrix equations is formed for the moments of the distribution function. The transport properties and coefficients detailed above can then be represented in terms of these moments. For further details the reader is referred to the reviews of Robson and Ness [53] and White et al. [54].

4. Thermalization of positrons in N_2

In Monte Carlo simulations the particles have to start from some initial distribution. While proper swarm experiments require that the temporal relaxation be complete and spatial relaxation extends over a small portion of the gap between the two electrodes, the simulation follows the development of the swarm from the initial or boundary conditions and we may then choose to follow the space/time development of the properties or to wait until relaxation is complete to start sampling. Thus we may use the same tool to model thermalization experiments, low energy transport coefficients and also, to perhaps model some applications.

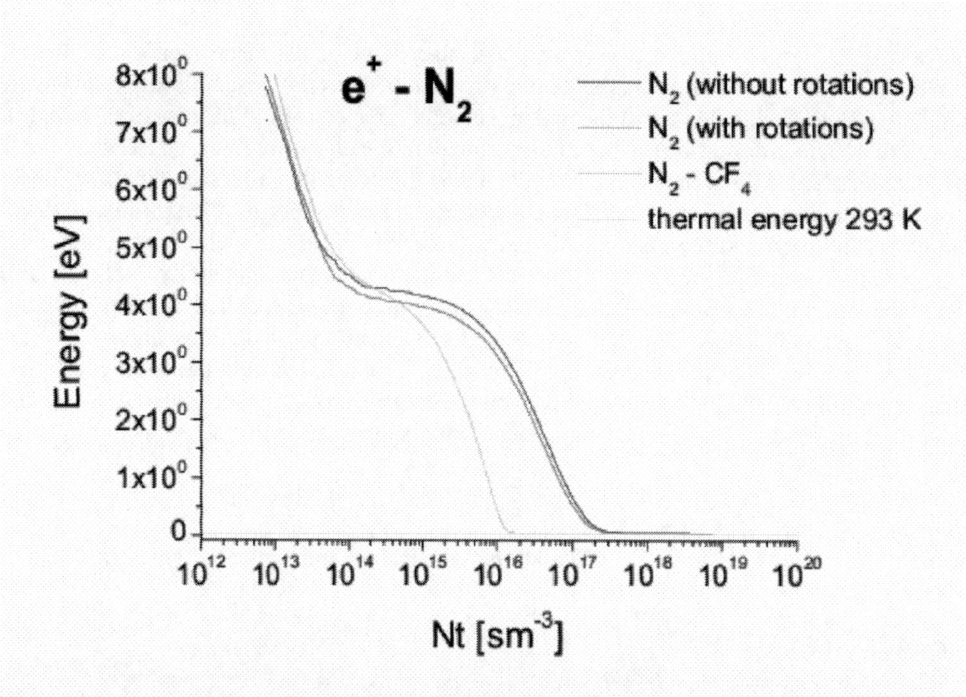

Figure 4. Thermalization of the mean energy of positrons from a Maxwelian initial distribution with a mean starting energy of 10eV in N_2 and N_2-CF_4 mixture

One example of the calculation of temporal relaxation to the thermal energy which is equivalent to modelling of a thermalization experiment is given for the case of nitrogen: In Figure 4 we show the temporal development of the mean energy of an ensemble of positrons in nitrogen (for the cross sections and other information check [55]). It is possible to isolate a time/energy region where different processes dominate the thermalization.

We have also compared the calculated values of thermalization times in different gases (see Table 1) and obtained a good agreement with the measurements of Charlton and his colleagues [56].

gas	Nt (sm^{-3}) Present results	Nt (sm^{-3}) Al-Qaradawi et al (2000)
N_2	$2.5*10^{17}$	$3.8*10^{17}$
N_2-CF_4	$1.4*10^{16}$	X
H_2	$4.5*10^{16}$	$6.5*10^{16}$

Table. Comparison between the thermalization times calculated using our Monte Carlo code and measured by Al-Qaradawi et al. [56].

Spatial relaxation was followed by simulations starting from a point with mono-energetic distribution function and allowing the swarm to spread in the electric field in hydrogen gas [57]. While one could expect that Ps formation could erase the oscillatory variation of the averaged properties we see that it is neither the case for the momentum nor for the energy relaxation.

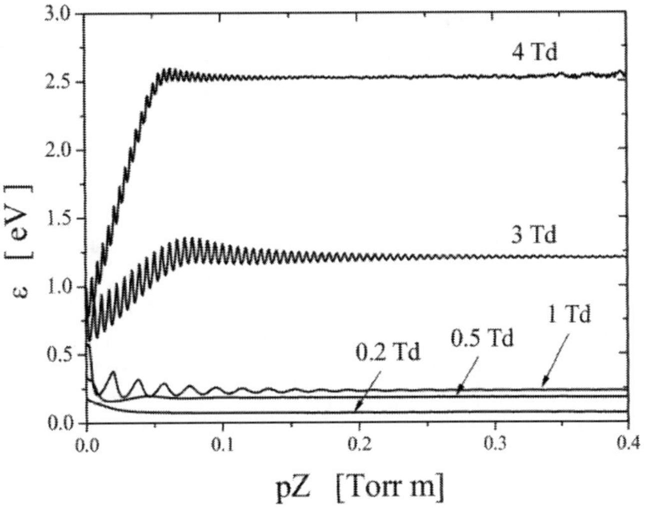

Figure 5. Spatial relaxation of the mean energy for positrons in hydrogen for different E/N as indicated in the graph [57].

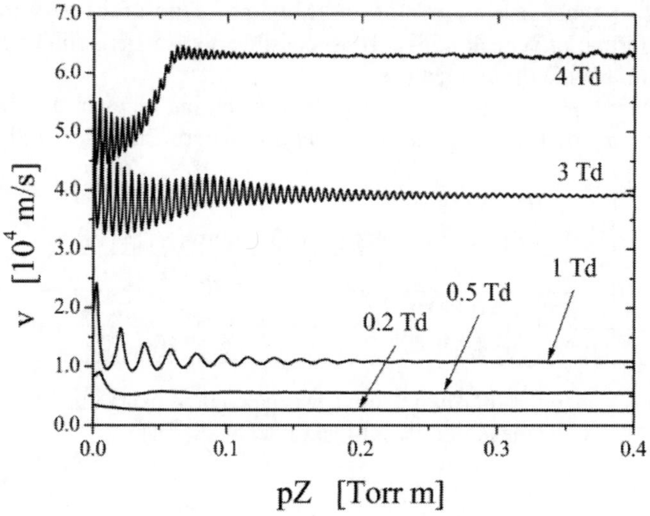

Figure 6. Spatial relaxation of the average velocity for positrons in hydrogen for different E/N as indicated in the graph [57].

Since all these calculations were made for the temporal and spatial development in a gas and the results are scaled either through Nt or Nd (or pZ), the results are relevant and directly applicable for gas phase relaxation in gas under conditions even equivalent to the collisional traps. These results however, are, as can be seen from comparisons above, directly applicable for gas thermalization measurements.

5. Non-equilibrium transport: Positron transport in Ar

Recently, there has been a renewed interest in the calculation of non-equilibrium transport properties of positron swarms in gases under the influence of electric fields [55, 58, 30]. There has been a host of striking phenomena, including the existence of negative differential conductivity (NDC) only for the bulk drift velocity in both atomic and molecular gases.

Of particular note is the magnitude of the Ps formation cross-section with a threshold of around 7eV for the case of Ar. The magnitude of the Ps formation cross-section is greater than both the elastic and excitation cross-sections over an extended energy range to approximately 30eV. As detailed in Šuvakov et al. [30], the Ps formation cross-section has a dramatic impact on the transport properties of the positron swarm due to the explicit non-conservative action of the Ps formation on the transport properties. The initial calculations for positrons in Ar were made using the Monte-Carlo simulation technique developed by the Petrović's group and the set of cross-sections detailed above. The neutral temperature is assumed to be 0K in these calculations. All scattering is assumed to be isotropic. In what follows, we present a comparison of the Boltzmann equation solution results with the MC simulation results [30] for positrons in Ar.

In Figure 7 and Figure 8, we display the mean energy and drift velocities (bulk and flux) of the swarm as a function of E/N (where N is the gas number density). Over the range of E/N we observe very good agreement between the two independent techniques for both the mean energy and the flux drift velocity. Convergence problems with the Boltzmann equation calculation of the bulk drift velocity in the vicinity of 10 Td were experienced, however the general behavior detailed in Šuvakov et al. [30] in this region are clearly reinforced.

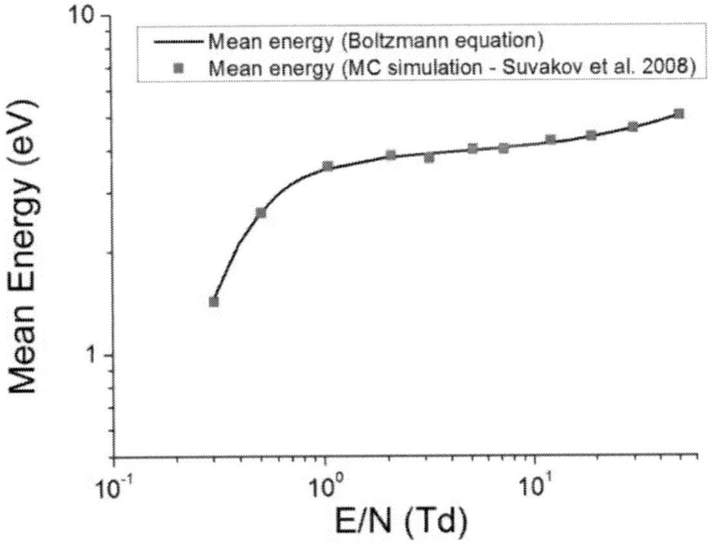

Figure 7: Comparison of the Boltzmann equation solution with the MC simulation results of Šuvakov et al. [30] for the mean energy of positron swarms in Ar as a function of E/N.

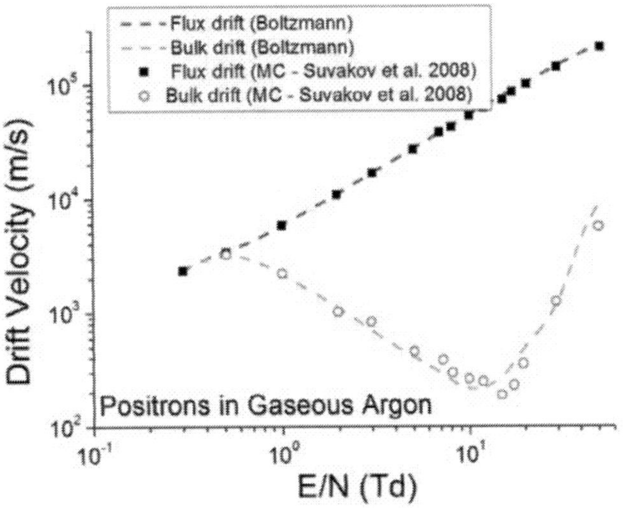

Figure 8: Comparison of the Boltzmann equation solution and the MC simulations of Šuvakov et al. [30] for the bulk and flux drift velocities of positron swarms in Ar as a function of E/N.

The striking phenomenon of NDC in the bulk drift velocity is clearly evidenced in Figure 8. The flux drift velocity does not exhibit such behavior. The NDC in the bulk drift velocity is explicitly due to the influence of Ps formation on the centre-of-mass of the positron swarm. Over the region of E/N considered, the Ps formation processes are preferentially taking positrons from the higher energy part of the energy distribution. Since positrons at the front of the swarm have generally fallen through a greater potential, they are generally higher in energy. Thus, Ps formation is preferentially taking positrons from the front of the swarm and consequently the modification to the centre of mass motion brought about by this process acts to decrease the bulk drift velocity relative to the flux drift velocity. Of course, there is an implicit effect associated with the cooling effect on the positron swarm due to the preferential loss of higher energy positrons (reducing the mean energy over what it would have been).

The effect of NDC is inherently different from that for electrons as it is solely induced by the non-conservative nature of the Ps formation. This was shown by the calculations shown in [30] where Ps formation was treated as a conservative inelastic process. While a small change in the mean energy followed from this change in the nature of the collisions both flux and bulk drift velocities were identical and very close to the flux drift velocity in the case of transport for realistic Ps formation. The possibility of NDC induced by a non-conservative process was first predicted by Vrhovac and Petrović [59] but based on the experience with the electrons it was predicted that bulk NDC was possible only if the conditions are also met for the flux, i.e. so that either NDC or at least a plateau exist for the flux drift velocity. In this case the flux drift velocity does not show any sign of NDC while the bulk property drops by two orders of magnitude.

The relationship between the flux and bulk drift velocity may be calculated if one takes into account the rate of the non-conservative process (in this case ν_{PF} the Ps formation rate). It was found [50] that the bulk drift velocity may be estimated to first order from the flux value as follows:

$$W_B = W_F - \frac{2\varepsilon}{3e}\frac{d\nu_{PF}}{dE} \tag{16}$$

where ε is the mean energy, E the electric field and in the case of particle loss the bulk property is smaller than the flux drift velocity. In Figure 9 we show how such calculation would work for positrons in argon [60].

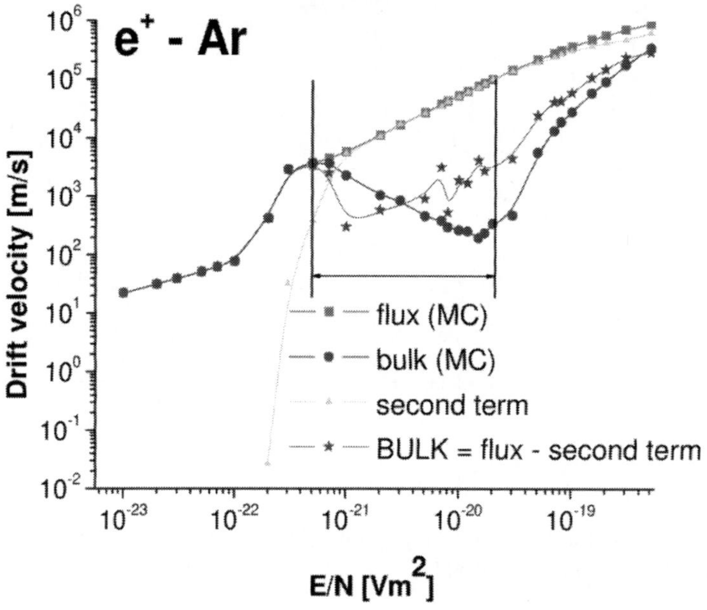

Figure 9. The calculation of the bulk drift velocity (stars) determined from the MC flux drift velocity (squares) by subtracting the second term in equation (16). The results are compared to the results of direct MC simulation (circles) [60]

The calculated bulk drift velocity agrees qualitatively with the result of direct simulation but quantitatively there are some discrepancies mainly due to very close values of the two terms that are subtracted. A much better agreement between simulations and the application of the formula was achieved for N2 [60]. It was proposed that due to a very large degree of skewing of the spatial distribution of the swarm of particles the simple, first order formula (16) may not be sufficiently accurate and more exact treatment may be required but nevertheless it provides a physical explanation of the process.

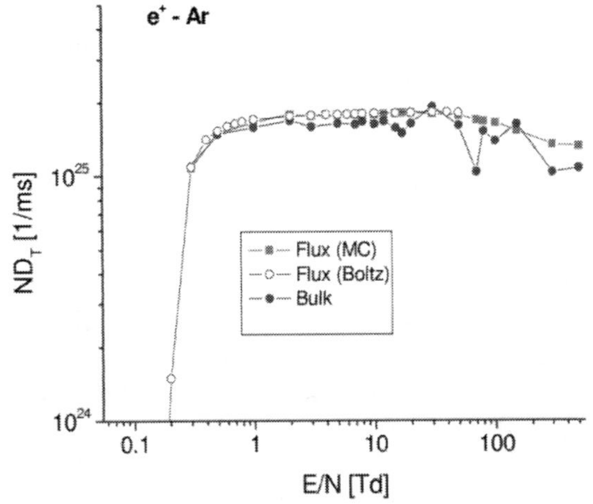

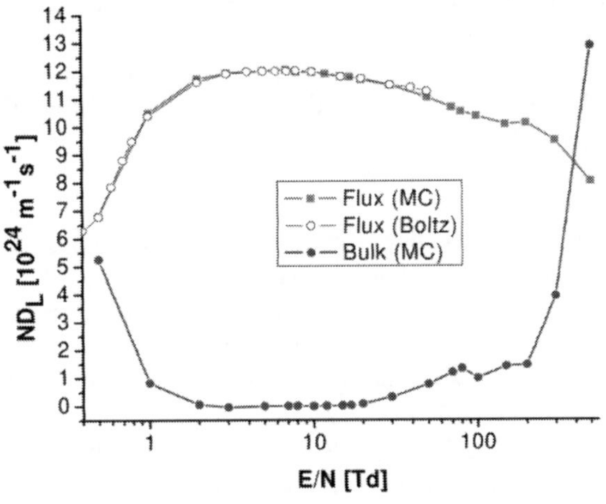

Figure 10. The results of simulations (MC) for (a) transverse and (b) longitudinal components of the diffusion tensor [30]. Results are compared with the Boltzmann equation (Bolz) results where available.

Šuvakov et al. [30] found that in an almost similar fashion the bulk diffusion coefficient in the longitudinal direction is affected by the positronium formation, while this has not been observed either for the transverse diffusion or for both flux diffusion coefficients. The results of simulation for the diffusion coefficients are shown in Figure 10. The diffusion coefficients are more difficult to obtain due to the need to sample more complex quantities and the subsequent differentiation needed to produce bulk properties may lead to somewhat more scattered results. Yet we may conclude that for all practical purposes transverse diffusion coefficients are identical, while the two order of magnitude differences observed in the longitudinal components can certainly not be explained by statistical uncertainty. The almost zero value of the diffusion coefficient along the direction of electric field occurs in a similar E/N region as the NDC yet it has a different shape and value. As could be expected this feature disappears as the Ps formation changes its magnitude.

6. Conclusion

High resolution measurements of positron scattering from neon have been compared to previous experimental and theoretical work. While the experiments have a much better resolution than previous work (with the exception of the data of Marler et al. [22]) and a high statistical accuracy, there remain unresolved discrepancies in the grand total cross section determination. In the case of the positronium formation cross section, present results favor the measurement of Marler et al., but higher statistical accuracy is required before any definitive conclusions can be made. Future work in this regard will concentrate on better quality measurements of the positronium formation cross section, as well as high resolution measurements of cross sections at both low energy (in the case of the total cross section) and around threshold regions for positronium formation, electronic excitation and ionisation. It is anticipated that a program of inelastic scattering measurements will be undertaken in the future, as well as measurements of differential scattering cross sections.

In the swarm part of the activity of our collaboration it was found that with the newly measured and calculated data reasonably complete sets of cross section data may be found. These sets provide a very good qualitative picture of the transport of swarms and predict a new set of kinetic phenomena that could not be observed for electrons because of the large difference in the magnitude of the non-conservative cross sections for the two types of particles. The most important effect is that of the NDC for the bulk drift velocity even when there is no indication of NDC for the flux drift velocity. Additional tests show that the effect is ONLY induced by the non-conservative nature of the Ps formation and it is not the result of the shape of the cross sections. This effect together with almost zero diffusion in the longitudinal direction and with a very large degree of skewness of the spatial time of flight are worthy of further investigation. For these reasons alone it would be worthwhile to reopen the positron swarm experiments which have had a large degree of difficulty in the past and presently exist only in measurements of thermalization times [8]. On the other hand with more accurate measurements of swarm parameters it would be possible to test the sets of cross sections and in particular, based on accurate measurements of some processes, renormalize the contribution of the missing processes. While Monte Carlo codes, as presented here, still have a long way to go to be applicable in modelling of applications in liquids and living tissues they are based on best available cross section data rather than semi-empirical formulae and therefore should be a better representation of the physics in a broader set of circumstances. Presently tests of codes are based on comparisons of calculated trajectories which is not sufficient. A more rigorous test, as 50 years of swarm physics teach us, would be to compare accurate values of well defined averaged transport coefficients especially if those could be collaborated by measurements from experiments.

Acknowledgements

The work presented here was supported by MNTRS grant 141025. The authors are grateful to the Australian Research Council for support under the Centers of Excellence Program and to the Australian Government Department of Innovation, Industry, Science and Research for support under the International Linkages Program. In particular we are grateful to our colleagues M. Charlton, R. Robson and C. Surko on useful comments and discussions on the topics related to this work.

Journal of Physics: Conference Series **162** (2009) 012002 doi:10.1088/1742-6596/162/1/012002

References

[1] Brunger MJ and Buckman SJ 2002 *Phys. Reports* **357** 215
[2] Buckman SJ and Clark CW 1994 *Rev. Mod. Phys.* **66** 539
[3] Huxley LG and Crompton RW 1974 *The drift and diffusion of electrons in gases* (Wiley Interscience New York)
[4] Crompton RW 1994 *Advances in Atomic Molecular and Optical Phys.* **32** 97
[5] Petrović ZLj, Šuvakov M, Nikitović Ž, Dujko S, Šašić O, Jovanović J, Malović G and Stojanović V 2007 *Plasma Sources Sci. Technol.* **16** S1–S12
[6] Christophorou LG, Olthoff JK and Rao MVVS 1996 *J. Phys. Chem. Rev. Data* **25** 1341
[7] Makabe T and Petrović ZLj 2006 *Plasma Electronics* (Taylor and Francis, New York)
[8] Charlton M *the present issue*
[9] Charlton M and Humberston JW 2001 *Positron Physics* (Cambridge University Press).
[10] Murphy T and Surko CM 1992 *Phys. Rev. A* **46** 5696
[11] Gilbert SJ, Kurz C, Greaves RG and Surko CM 1997 *Appl. Phys. Lett.* **70** 1944
[12] Sullivan JP, Gilbert SJ, Marler JP, Greaves RG, Buckman SJ and Surko CM 2002 *Phys. Rev. A* **66** 042708
[13] Sullivan JP, Gilbert SJ, and Surko CM 2001 *Phys. Rev. Lett* **86** 1494
[14] Sullivan JP, Jones A, Caradonna P, Makochekanwa C and Buckman SJ 2008 *Rev. Sci. Instr.* **79** 113105
[15] Coleman PG, McNutt JD, Diana LM, Burciaga JR 1979 *Phys. Rev. A* **20** 145
[16] Stein TS and Kauppila WE 1982 *Adv. At. Mol. Phys.* **18** 53
[17] Kauppila WE, Stein TS, Smart JH, Dababneh MS, Ho YK, Downing JP, Pol V 1981 *Phys. Rev. A* **24**, 725
[18] Tsai JS, Lebow L, Paul DAL 1976 *Can. J. Phys.* **54** 1741
[19] Coleman PG, Griffith TC, Heyland GR and Twomey TR 1976 *Appl. Phys.* **11** 321
[20] McEachran RP 2008 *private communication*
[21] Baluja KL, Jain A, Jones HW, Weatherford CA, Karim KR 1991 *J. Phys. B* **24** L93
[22] Marler JP, Sullivan JP and Surko CM 2005 *Phys. Rev. A* **71** 022701
[23] McAlinden MT and Walters HRJ 1992 *Hyperfine Interactions* **73** 65
[24] Gilmore S, Blackwood JE and Walters HRJ 2004 *Nucl. Instr. Meth. B* **221** 129
[25] Laricchia G, Van Reeth P, Szluinska M and Moxom J 2002 *J. Phys B.* **35** 2525
[26] Marler JP and Surko CM 2005 *Phys. Rev. A* **72** 062702
[27] Marler JP and Surko CM 2005 *Phys. Rev. A* **72** 062713
[28] Charlton M 1985 *Rep. Prog. Phys.* **48** 737
[29] Karwasz GP 2005 *Eur. Phys. J. D* **35** 267
[30] Šuvakov M, Petrović ZLj, Marler JP, Buckman SJ, Robson RE and Malović G 2008 *New J. Phys.* **10 No 5** 053034
[31] McEachran RP 2006 private communication
[32] Karwasz GP, Pliszka D and Brusa RS 2006 *Nucl. Instrum. Methods B* **247** 68
[33] Kauppila WE and Stein TS 1990 *Adv. At. Mol. Opt. Phys.* **26** 1
[34] Sullivan JP, Marler JP, Gilbert SJ, Buckman SJ and Surko CM 2001 *Phys. Rev. Lett.* **87** 073201
[35] Parcell LA, McEachran RP and Stauffer A 2000 *Nucl. Instrum. Methods B* **171** 113
[36] Phelps AV and Tachibana K 1985 private communication
[37] Barnes LD 2005 PhD Thesis University of California, San Diego
[38] Robson RE, White RD and Petrović ZLj 2005 *Rev. Mod. Phys.* **77** 1303
[39] Dujko S, Raspopović ZM and Petrović ZLj 2005 *J. Phys. D: Appl. Phys.* **38** 2952
[40] Raspopović ZM, Sakadžić S, Bzenić SA and Petrović ZLj 1999 *IEEE Trans. Plasma Sci.* **27** 1241
[41] Šuvakov M, Ristivojević Z, Petrović ZLj, Dujko S, Raspopović ZM, Dyatko NA and Napartovich AP 2005 *IEEE Trans. Plasma Sci.* **33** 532

[42] Biagi SF 1999 *Nucl. Instrum. Methods A* **421** 234
[43] Skullerud HR 1968 *J. Phys. D: Appl. Phys.* **1** 1567
[44] Itoh H and Musha T 1960 *J. Phys. Soc. Japan* **15** 1675
[45] Longo S and Capitelli M 1993 *Plasma Chem. Plasma Process.* **14** 1
[46] Vasenkov AV 2000 *J. Appl. Phys.* **88** 626
[47] Nolan AM, Brennan MJ, Ness KF and Wedding AB 1997 *J. Phys. D* **30** 2865
[48] White RD, Brennan MJ and Ness KF 1997 *J. Phys. D* **30** 810
[49] Raspopović ZM, Sakadžić S, Petrović ZLj and Makabe T 2000 *J. Phys. D: Appl. Phys.* **33** 1298
[50] Robson RE 1991 *Aust. J. Phys.* **50** 577
[51] White RD, Robson RE and Ness KF 1999 *Phys. Rev E* **60** 7457
[52] Robson RE 1986 *J. Chem. Phys.* **85** 4486
[53] Robson RE and Ness KF 1986 *Phys. Rev. A* **33** 2068
[54] White RD, Ness KF and Robson RE 2002 *Appl Surf. Sci.* **192** 26
[55] Banković A, Marler JP, Šuvakov M, Malović G and Petrović ZLj 2008 *Nuclear Inst. Methods, B* **266** 462
[56] Al-Qaradawi I, Charlton M, Borozan I and Whitehead R. 2000 *J. Phys. B: At. Mol. Opt. Phys.* **33** 2725
[57] Banković A, Dujko S, White RD, Marler JP, Malović G and Petrović ZLj unpublished
[58] Marler JP, Petrović ZLj, Banković A, Dujko S, Šuvakov M, Malović G and Buckman SJ 2009 *Phys. Plasmas* submitted
[59] Vrhovac SB and Petrović ZLj 1996 *Phys. Rev. E* **53** 4012
[60] Banković A, Petrović ZLj, Robson RE, Marler JP, Dujko S and Malović G 2008 *Nucl. Instrum. Methods B* accepted doi: 10.1016/j.nimb.2008.10.034

Journal of Physics: Conference Series **162** (2009) 012003 doi:10.1088/1742-6596/162/1/012003

Positron transport in gases

M Charlton

Department of Physics, Swansea University, Singleton Park, Swansea SA2 8PP UK

M.Charlton@swansea.ac.uk

Abstract. We review the field of positron transport physics in gases, starting with a historical overview. In particular, experiments which have measured positron drift velocities and mobilities in specially constructed cells are highlighted. The relevance of understanding positron transport in current efforts to manipulate trapped clouds of positrons is emphasised. We suggest some ways in which the output of a positron accumulator might be used to develop a new and versatile technique to study positron drift.

1. Introductory remarks and contemporary relevance

Before the advent of low energy, quasi-monoenergetic, positron beams, the only way to extract (experimental) information regarding positron collisions was to use systems in which the positrons emitted from a radioactive source stopped directly in the gas sample under study. Depending upon the target species, and other circumstances, the densities used were typically in the Amagat range (1 Amagat = 2.69×10^{25} m^{-3}) to provide sufficient signal from the positrons stopping and annihilating in the gas, rather than the prompt "background" emanating from those positrons that reached the walls of the confining chamber. Data were usually extracted by monitoring positron annihilations in the gas; see e.g. [1] for a description of the techniques involved.

The kinetic energy of a positron emitted from a radioactive source into a dense gas is moderated into the eV-range within a short (sub-ns) period. The scattering processes involved in the slowing down (ionization and excitation of the gas) have cross sections typically many orders of magnitude larger than that for annihilation with an atomic or molecular electron, such that almost all of the positrons reach the energy range below the ionization energy, $E_{\rm ion}$ of the gas. Of most concern to us here will be those positrons whose energies fall below $E_{\rm Ps} = (E_{\rm ion} - 6.8)$ eV, the so-called positronium formation threshold of the gas. Such positrons cannot capture an electron directly from the target gas to form positronium, the quasi-stable electron-positron bound state (with a ground state binding energy of 6.8 eV), and usually annihilate as so-called "free" positrons, following diffusion and/or drift in the gas.

Experimenters, often with theoretical support, were ingenious in extracting information with the limited techniques at their disposal. To mention just some areas of study: non-equilibrium effects apparent due to annihilation during the slowing down were analysed to extract positron thermalisation times, tests of scattering theory and to probe localisation phenomena at low temperatures; electric fields were applied to gases and the changes with respect to the zero-field case gave crude, but surprisingly accurate, estimates of the elastic scattering cross sections for some of the noble gases; attempts were made to measure positron drift velocities and mobilities in gases and radiation chemical

Journal of Physics: Conference Series **162** (2009) 012003 doi:10.1088/1742-6596/162/1/012003

effects were brought into play to help extract similar information. Further details will be given in the following sections.

The main drawback of all of this work was the lack of control of the experimenter over individual collision events, including impact energy and selectivity of outcome. This had long been known, and the obstacles were finally begun to be overcome with the development of the low energy positron beam, mainly from the early 1970's onwards. Gradually more became known about the details of positron interactions with gas atoms and molecules as total scattering cross sections, and partial cross sections for some channels, were measured with increased confidence and accuracy. Inevitably, interest in what might be termed positron-gas "swarm" techniques began to wane, and there are currently very few groups worldwide still with competence in this area.

Nevertheless, there is renewed interest in positron transport effects in gases. Modern theoretical techniques, which have been applied extensively with great success to electron transport (see this issue), are beginning to be applied to positrons [see e.g. 2-4]. This will perhaps herald a new era for positron transport work, with some qualitatively new phenomena to study [2]. Furthermore, the importance of positron interaction with gases and positron cooling and transport have been highlighted by recent advances in positron trapping and accumulation and their applications [see e.g. 5-8]. This will be discussed further below, since modern positron beam technology could play an important role in reviving interest in the positron transport field.

In the remainder of this article we will describe some of the phenomena involved in positron slowing and transport, including highlight results from the archive. We will briefly discuss the previous attempts to extract positron drift speeds using positron lifetime and related techniques, and describe why such experiments present formidable challenges (and hence why so few were done). Modern techniques which use positron transport will be introduced, with the aim of highlighting synergies with the resurgent theoretical interest in such phenomena. This will lead to a discussion of possible new opportunities and directions in the field.

2. Basic positron phenomenology

Our aim in this section is not to be comprehensive, but rather to highlight relevant aspects from the positron field. It seems logical to follow the positron as it loses its last few electron-Volts of kinetic energy, into equilibrium with its surroundings and interaction and annihilation. The treatment will be mainly experimental/empirical in nature, but reference will be given as appropriate to theoretical work. We start, though, with a brief description of a typical timing experiment.

2.1. The positron lifetime technique

In this method, individual events are collated into a lifetime spectrum by recording the time difference between the emission of a positron from a radioactive source and its subsequent annihilation in the gas. The former is usually done by registering the nuclear gamma-ray which accompanies the beta-decay of ^{22}Na, whilst the annihilation gamma-ray provides the stop for the timing sequence. The gamma-rays are detected using a pair of scintillation counters arranged appropriately on the exterior of the chamber which contains the gas and the source. Standard timing electronics is used to record the lifetime spectra; see [1, 9, 10] for more details. Occasionally, a method in which the beta-particle was timed directly it left the source, using a thin scintillator technique, has also been used [1, 9].

We are only concerned with the free positron component of the lifetime spectrum, to which we assign a time spectrum, $S_f(t)$. The instantaneous annihilation rate of the free positrons, $\lambda_f(t)$, is given by

$$\lambda_f(t) = S_f(t)/\int_t^{\infty} S_f(t')\mathrm{d}t' = \omega\rho Z_{\mathrm{eff}}(t), \qquad (1)$$

where $\omega\rho$ is the Dirac annihilation rate for a free electron gas with number density, ρ, and $Z_{\mathrm{eff}}(t)$ is the time-dependent effective number of electrons available to the positron for annihilation per gas atom or molecule. This parameter is related to the velocity, or energy dependent Z_{eff}, viz $Z_{eff}(E)$, given a time dependent positron energy distribution $f(E,t)$ as,

$$Z_{eff}(t) = \int^{\infty}_{0} Z_{eff}(E)\, f(E,t)\mathrm{d}t / \int^{\infty}_{0} f(E,t)\mathrm{d}t. \quad (2)$$

In equilibrium, Z_{eff} is a constant, averaged over the energy distribution of the positrons, such that the measured annihilation rate is given by

$$\lambda_f = \omega\rho Z_{eff} = 0.201\rho Z_{eff}\,(\mu s^{-1}), \quad (3)$$

where the gas density is in units of Amagat. Thus, an equilibrium free positron lifetime spectrum will consist of single exponential component, possibly preceded by a non-exponential region if the timescales for slowing down and annihilation are comparable. A classic example is the spectrum shown in figure 1 [9], which is for room temperature argon gas at a density of 6.3 Amagat. Note in particular curve (c) from figure 1 which displays the exponential decay at long times, but has a pronounced "shoulder" on approaching $t = 0$ as a result of annihilation whilst slowing down. Such behaviour is typical of the noble gases, and is particularly striking in argon, krypton and xenon where the variation of $Z_{eff}(E)$ leads to dramatic effects upon $Z_{eff}(t)$. The latter, computed from (1), is shown in figure 2 for the same gas and conditions as the example in figure 1.

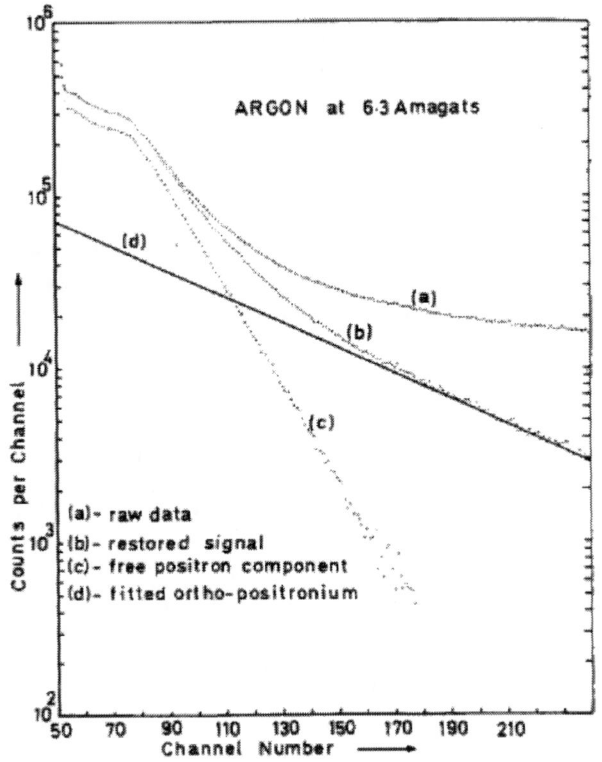

Figure1. Positron lifetime spectrum for argon gas at room temperature and at a density of 6.3 Amagat. The spectrum is broken down into components, the most relevant for the present discussion being that for the free positrons. The effect of non-thermal annihilation (the shoulder) is evident at short times.

Journal of Physics: Conference Series **162** (2009) 012003 doi:10.1088/1742-6596/162/1/012003

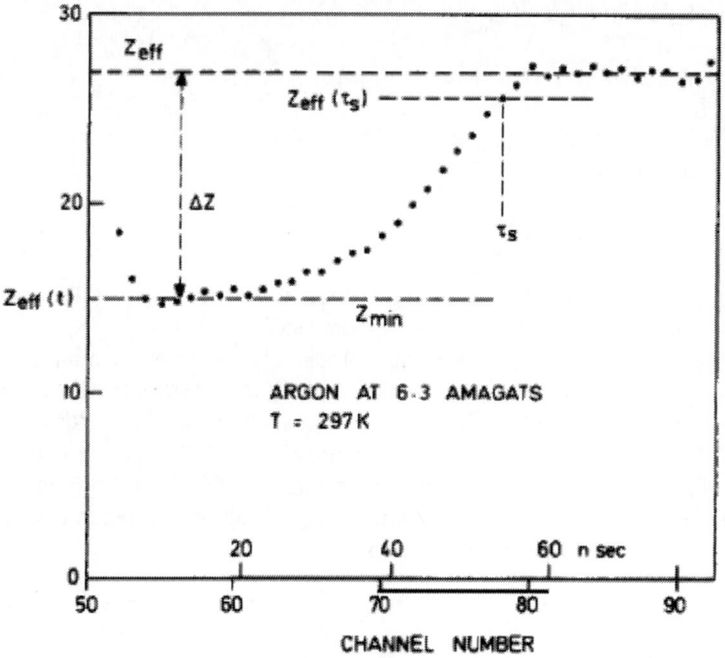

Figure 2. Behaviour of $Z_{eff}(t)$ extracted from the lifetime spectrum illustrated in figure 1. See text for details.

2.2. Positron slowing down

The instantaneous Z_{eff}, $Z_{eff}(t)$, displayed in figure 2, can be characterised in terms of a shoulder width, τ_s, which is related to the thermalisation time of the positrons. It is inversely proportional to gas density (such that values for $\tau_s\rho$ in ns-Amagat are usually quoted) and, following early work [11], can be defined using

$$Z_{eff}(\tau_s) = (Z_{eff} - 0.1\Delta Z), \qquad (4)$$

where $\Delta Z = (Z_{eff} - Z_{min})$, as illustrated in figure 2.

The behavior of $Z_{eff}(t)$ across the shoulder region in the noble gases has, historically, been the subject of much investigation. Indeed, before the advent of positron beams analysis of this region was one of the only ways in which information regarding positron scattering cross sections could be derived. This included an intriguing study [12], comparing the shoulders for argon gas at room temperature and at a much lower temperature (~ 100 K), which deduced that the momentum transfer cross section possessed a Ramsauer minimum at low energies. This was later directly observed in the total scattering cross section using a low energy beam [13]. Overall, little has changed regarding theory-versus-experiment for positron-noble gas shoulders since the situation was last reviewed, and readers are referred to that literature [1, 10] and references therein. However, with an array of more accurate scattering cross sections becoming available, it may be worthwhile to revisit this topic.

It is more difficult to observe the effects of positron slowing down in pure molecular gases, where the presence of rotational and vibrational channels down to sub-eV energies usually means that timescales are sufficiently short that non-thermal effects are not directly visible in lifetime spectra. An exception is N_2, and this gas has been studied in detail [14]. This molecule is also important in positron accumulation physics, as will be described below.

Information on thermalisation times of positrons in molecular gases can be gleaned by studying the behaviour of the shoulder in argon gas upon controlled addition of small quantities of impurities, as was first shown by Paul and Leung [11]. The most recent study using that technique reported thermalisation times for a range of species [15], noting ultra-fast thermalisation in SF_6 for example. The importance of this observation will also be detailed below in connection with its use for cooling in positron accumulators.

The classic early discussion of positron slowing down in gases was that for argon given by Tao, Green and Celitans [16]. It is still worth reading this paper for the clarity of its treatment. A broadly similar approach was taken by Jacobsen [17] who computed estimates for the density-normalised thermalisation distance, $R\rho$, for positrons in various gases. The motivation here was to account for density-dependent positronium formation in dense molecular gases, in which combination of a positron with an electron created during its slowing down in the gas could occur. This will be discussed further in the next section.

Jacobsen [17] found an approximate, but convenient, expression for $R\rho$ as,

$$R\rho = 2 \times 10^5(\tau_s\rho/\sigma)^{1/2} \quad \text{Å-Amagat}, \tag{5}$$

where $\tau_s\rho$ is the density-normalised thermalisation time (in units of ns-Amagat), which could be taken from experiments such as those described above, and σ is an average low energy scattering cross section (in Å^2). The difficulty lay in estimating a value for the latter, but using the same procedure to estimate thermalisation times in the noble gases he found agreement with experiment to around a factor of 2-3, which gave confidence that estimates based upon the slowing down model can be used as a reasonable guide. In principle, with accurate low energy (typically sub-eV) scattering and annihilation cross sections hopefully becoming available for a range of gases of interest, it should be possible to model accurately the approach to thermalisation and derive values for density-normalised time and length scales.

2.3. Electric fields, positronium and radiation spurs

From the early years of positron-gas physics, researchers have applied electric fields, $\boldsymbol{\varepsilon}$, (magnitude, ε) across their chambers in an effort to learn more about positron scattering. The first notable work was that of Marder *et al.* [18] who studied the behaviour of the fraction of positrons which formed positronium as $\boldsymbol{\varepsilon}$ was increased. In most cases they found that the fraction increased with field as the mean energy of the positrons became high enough for them to capture an atomic or molecular electron in collision to form positronium. In simple terms a charge, e, and mass, m, in a field $\boldsymbol{\varepsilon}$ will undergo an acceleration, $\mathbf{a} = e\boldsymbol{\varepsilon}/m$ and hence the change in kinetic energy between collisions is

$$m(\mathbf{v} + \mathbf{a}t)^2/2 - mv^2/2 = e\,\boldsymbol{\varepsilon}\cdot\mathbf{l} + (e\varepsilon l)^2/2mv^2, \tag{6}$$

where $\mathbf{v}$ is the velocity at the beginning of the free path of length, $\mathbf{l}$. The first term on the right hand side will sum to zero over many collisions, so that the positron gains on average a kinetic energy of $(e\varepsilon l)^2/2mv^2$ from the field. Equating this to the mean energy loss per (elastic) collision with an object of mass M of m^2v^2/M it can be seen that the positron kinetic energy will tend towards an equilibrium of the order of

$$(M/8m)^{1/2}\varepsilon l = (M/8m)^{1/2}(\varepsilon/\rho)/\sigma \quad \text{(eV)}. \tag{7}$$

It is easy to see that this can be quite a substantial gain, with $\sigma \sim 10^{-20}$ m^2 and density-normalised fields (ε/ρ) around 1 Td (= 10^{-21} Vm^2 ~ 270 Vcm^{-1}Amagat^{-1}). A detailed Boltzmann equation approach to derive elastic scattering cross sections for helium, neon and argon was developed by Teutsch and Hughes [19] in an accompanying paper to [18] (see also the discussion in [16]). The main result of these investigations is that the positronium yield in the noble gases increases with electric field, typically to saturation. The situation is summarized in figure 3 for argon gas [20].

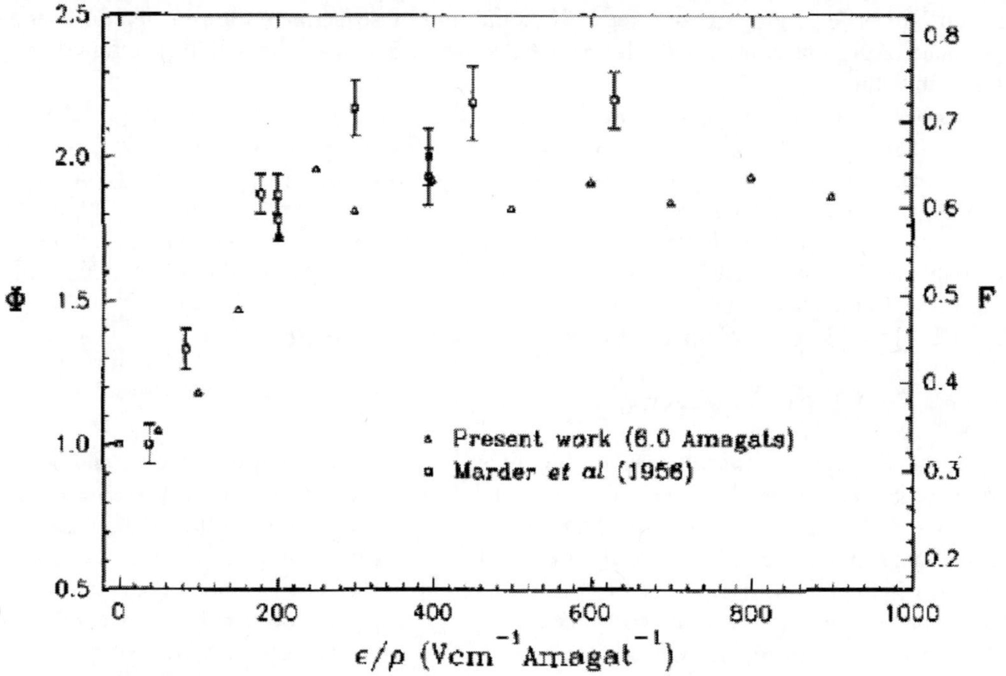

Figure 3. Variation of the yield of positronium atoms (relative to zero field, left scale; absolute fraction, right scale) versus applied electric field for argon gas. The points marked present work are from [20], whilst Marder *et al.* is the work in [18].

However, this situation does not always pertain with molecular gases. Even in the earliest studies on molecular gases an anomaly at a relatively low density of around 2 Amagat of SF_6 was found in which the positronium fraction was found to fall with field [18, 21], before rising as expected at higher fields [21]. Similar phenomena have been seen in a number of other gases [22-24] and other media [25, 26]. This is now understood in terms of the spur model of positronium formation [27, 28] in which positronium can be formed by a positron combining with an electron that it has created in one of its final ionizing collisions. The applied electric field tends to separate the positron and electron and will, at least at low fields, reduce the positronium yield.

At zero externally applied electric field a semi-quantitative model for the fraction of positrons forming positronium, F, [28] yields the expression,

$$F = F_o + (1 - F_o)\exp(-\lambda_f \tau_{Ps})[1 - \exp(-r_c/R)]. \tag{8}$$

Here F_o is the positronium fraction at "zero" density (i.e. that formed directly in collision via the so-called Ore mechanism) and R is the thermalisation distance discussed above. The second factor in (8) is the contribution from the positron spur, and is expected to be density and temperature dependent, and there have been several observations of such effects [1, 10, 29]. The parameter r_c $(= e^2/4\pi\varepsilon_d kT)$ is the critical Onsager radius at which the potential energy of a charged pair in a medium of dielectric constant, ε_d, becomes equal to the thermal energy, kT. The factor $\exp(-\lambda_f \tau_{Ps})$ is the probability that the positron has not annihilated before the pair can recombine with a characteristic time, τ_{Ps}, given by [28],

$$\tau_{Ps} = 4\pi\varepsilon_d R^3/3e\mu_+, \tag{9}$$

with μ_+ the positron mobility for the gas at a particular temperature and density. Thus, the importance of transport physics in positronium phenomena (via R and μ_+) should, at least qualitatively, be clear. Positronium formation via the spur mechanism is a ubiquitous phenomenon in, for instance, dense gases, liquids and molecular solids [27, 30]. Contemporary relevance may be the issue of positron transport leading to positronium formation in bio-materials, particularly of relevance to medicine.

3. Positron drift experiments

As mentioned in section 1, there have been attempts, albeit very limited in number and scope, to directly measure positron drift velocities, **w**, (speed, *w)* in gases. Two main approaches have been used, and we will summarise them here and give some general remarks. However, we note at the outset that, at least to date, the main experimental difficulty has been in getting suitable drift lengths and times to compute **w** using systems in which the positrons are derived directly from a radioactive source in a similar fashion to the lifetime experiments described above.

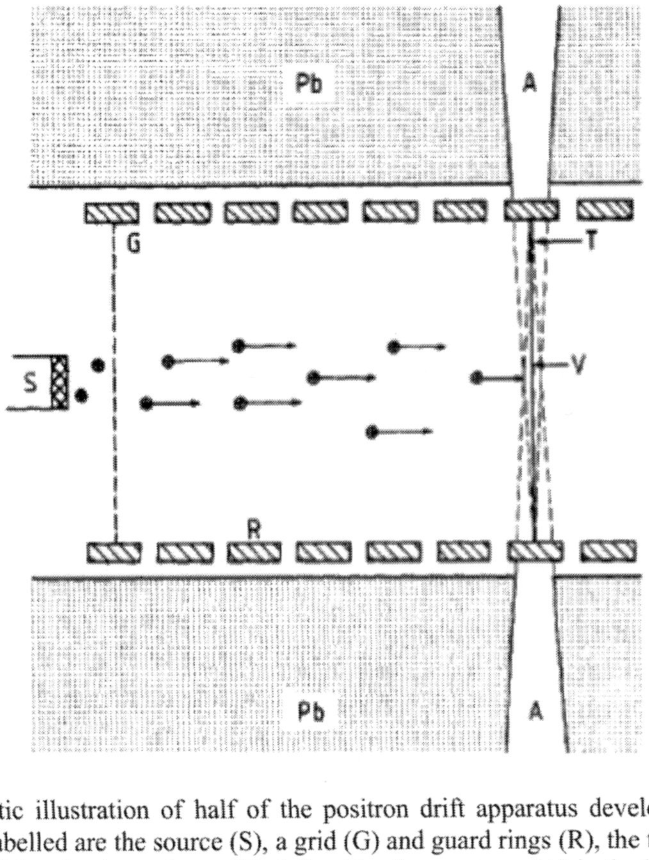

Figure 4. Schematic illustration of half of the positron drift apparatus developed by Paul and co-workers [31-33] Labelled are the source (S), a grid (G) and guard rings (R), the thin annihilation target (T) and a section of the viewing volume (V) defined by the apertures (A) in the lead wall.

3.1. Paul's method

This is the technique developed by Paul and co-workers [31-33], as adapted from the pioneering work of Rodionov *et al.* [34, 35]. Paul's system is shown schematically in figure 4. Positrons from two ^{58}Co sources (only one of which is shown in figure 4) were confined along the axis of a drift system, of total length around 4 cm, by a 0.12 T magnetic field. A small fraction of the positrons thermalised in the gas, which was typically at pressures of 100 Torr or lower ($\rho \sim 0.1$ Amagat). Drift was terminated by annihilation on a thin foil (to reduce annihilation of unmoderated positrons from the sources) in the centre of the cell such that the maximum drift distance, l_m, was about 2 cm. The foil was viewed by gamma-ray detectors in coincidence, via a lead collimation system, to reduce counts from other regions and the sources

The coincidence data of Böse *et al.* [32] for H_2 at pressures in the range 10-200 Torr are shown in figure 5. It can be shown that the fraction, F_d, of positrons that drift into the foil after stopping in the gas [31] is given by

$$F_d = (w/\lambda_f l_d)[1 - \exp(-\lambda_f l_d/w)] \qquad (10)$$

such that fits to the data shown in figure 5 yielded the parameter $w/\lambda_f l_d$. The latter, expressed as w/Z_{eff} is plotted in figure 6, along with w derived using the known annihilation rate in H_2 from the accepted Z_{eff} value, though only at zero electric field.

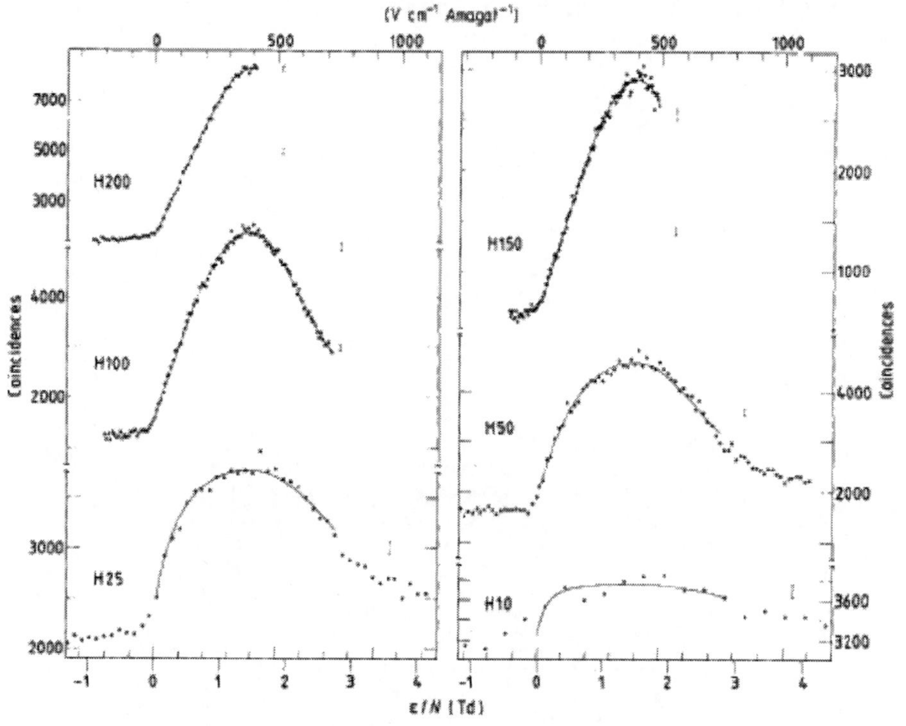

Figure 5. Relative coincidence curves versus applied electric field (here density-normalised as ε/N, rather than ε/ρ) for the detection of positrons drifting into the foil target in Paul's apparatus. The data are for H_2 gas at various pressures (as marked by "H" numbers) between 10 and 200 Torr.

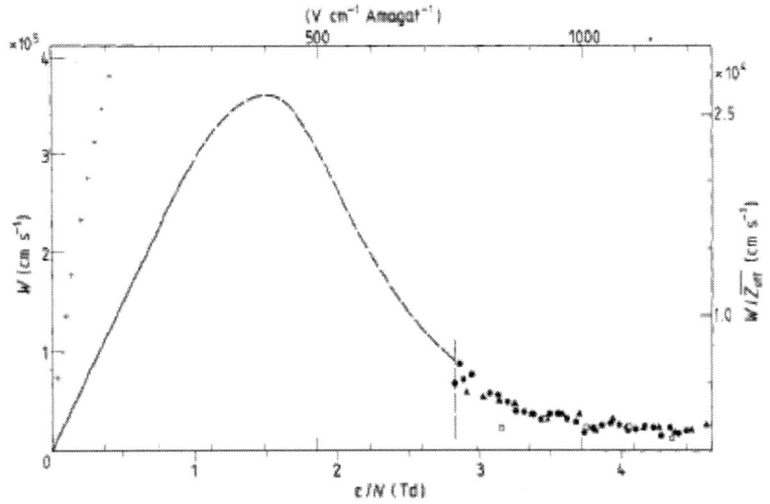

Figure 6. The parameter w/Z_{eff}, which is proportional to $w/\lambda_f l_d$, and the positron drift speed w (see text) for H_2 as derived from the coincidence curves shown in figure 5 plotted versus density-normalised electric field. The crosses are drift speeds for electrons in similar conditions.

The form of these curves has been explained in terms of positronium formation, which is also responsible for the downturns in the coincidence curves shown in figure 5. If the positrons have chance of being sufficiently heated that their kinetic energies might exceed E_{Ps}, then they are very likely to form positronium and be lost. This is, in effect, a reduction in w, since the positron stops drifting, akin to the effect of negative differential conductivity as explained by Petrović and co-workers [2-4]. However, in this experiment, in which the clock to time the drift is provided by the positron annihilation rate, this seems to be equivalent to an increase in λ_f. In any case, flux is lost, and at high enough ε/ρ, virtually no positrons can drift to the foil.

3.2. Charlton's method

This method used the thin plastic scintillator timing technique, described briefly in section 2.1, together with a roughly 1 cm drift cell, which also formed the gas chamber [36, 37]. A gamma-ray detector was used to monitor positron annihilations in the cell, and with an electric field applied, some positrons annihilated prematurely, following drift to one of the electrodes. Thus, the free positron component was field dependent [38], and in particular was terminated at a time τ_{md}, a maximum drift time. This was due to positrons which had drifted across the entire length, l_d, of the cell. Thus, the drift speed was given by $w = l_d/\tau_{md}$, such that the positron mobility could be found from

$$\mu_+ = l_d/\tau_{md}\varepsilon. \qquad (11)$$

A small sample of molecular gases was studied at room temperature using this technique and examples of lifetime spectra showing the effect of τ_{md} are given in figure 7(a). Extracted drift velocities for O_2 and CO_2 are shown in figures 7(b) and (c). However, in many cases it was difficult to locate τ_{md} , such that the derived mobilities are only (statistically) accurate to about 15%. The timing spectra were dominated by the annihilation of positrons in the metal walls of the gas chamber, and very weak sources (and the efficient β^+ counting technique) had to be used in the fight to preserve signal-to-background. Thus, it would not easily be possible to scale this experiment up in terms of source activity or drift length (which would make τ_{md} even harder to locate). It may, however, be possible to extract further information from this type of experiment by utilizing the entire positron lifetime component [38], rather than just relying on the determination of τ_{md}.

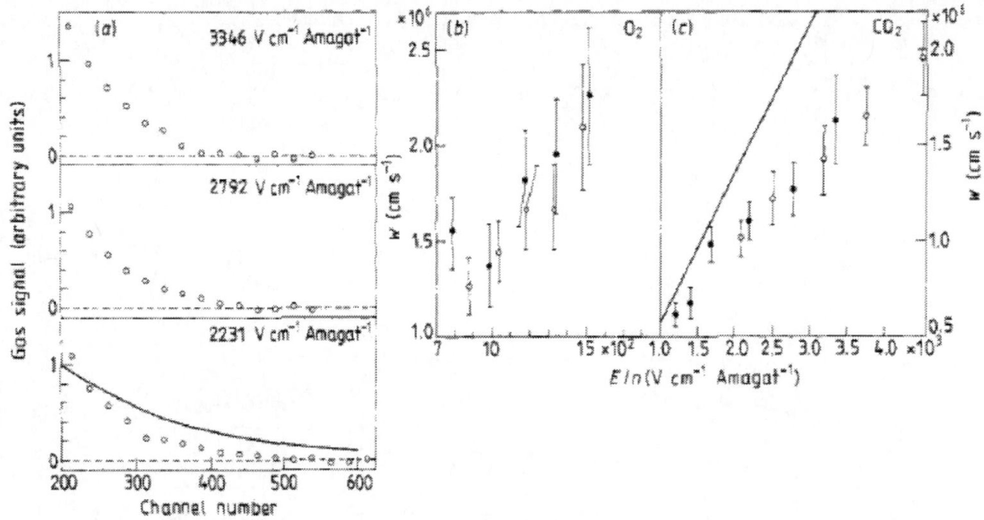

Figure 7. (a) Examples of positron lifetime spectra for CO_2 [36] at various density normalized fields illustrating the effect of the maximum drift time. The line drawn on the lower panel is a schematic of the zero-field spectrum at the density of 0.26 Amagat used in the experiment. (b) Drift velocities for O_2 at various ε/ρ for pressures of 600 Torr (full circles) and 400 Torr (open circles). (c) Drift velocities for CO_2 at various ε/ρ for pressures of 200 Torr (full circles) and 100 Torr (open circles). The full line is for the behaviour of the drift velocities for electrons in the corresponding range of ε/ρ (CO_2 only).

A compendium of mobilities measured using this technique has been presented elsewhere [1], and the interested reader is referred there, and to the primary sources [36, 37] for further information and discussion.

3.3. General remarks

Let us take an example to explore the challenges faced by the two techniques discussed above. Assuming a pressure ~ 100 Torr at room temperature, which corresponds to a density of about $\rho \sim 4 \times 10^{24}$ m^{-3} (or just over 0.1 Amagat) and taking H_2, with $Z_{eff} \sim 14$ as an example, results, using (3), in $\lambda_f \sim 0.03$ μs^{-1}, corresponding to a free positron lifetime of about 30 μs. Thus, for typical positron drift speeds $w \sim 10^5$ cms^{-1}, the drift distance in a lifetime will be around 3 cm. This explains the drift cell dimensions used in both of the experiments described in this section. Indeed for the work of Charlton, which typically operated at higher densities, shorter drift times were necessary to try to keep the signal-to-background manageable in the lifetime spectra.

To date, direct drift velocity data only exist for a few molecular species: this is due to the need to avoid non-thermal effects. The latter, as described in section 2, have been observed in several gases, most notably the nobles (see figure 1 and accompanying discussion). We can parameterise by introducing the density-normalised thermalisation, time, $\tau_{th}\rho$, which is around 14 ns-Amagat for N_2, about 3 ns-Amagat for H_2 and well under 1 ns-Amagat for other molecules [1, 11, 15]. Thus, for molecules at a density of 0.1 Amagat, we find τ_{th} just over 10^{-7} s, which is negligible in comparison to expected drift times.

However, a better figure of merit here is $X = \tau_{th}\rho / \tau_f\rho = 2 \times 10^{-4}(\tau_{th}\rho)Z_{eff}$, which is the ratio of the thermalisation time and the positron lifetime. We have $X << 1$, which is the desired situation, for all molecular gases, but things are not so straightforward for the nobles, as table1 shows.

Table 1. The figure of merit X (see text) for the nobles and selected molecular species: $\tau_{th}\rho$ and $\tau_{f}\rho$ are in units of ns-Amagat.

Gas			$\tau_{th}\rho$			$\tau_{f}\rho$			X
He			1700			1270			1.3
Ne			2300			830			2.8
Ar			360			190			1.9
Kr			325			75			4.3
Xe			180			12			15
H_2			3			340			0.009
N_2			14			160			0.09
CO_2			0.1			94			0.001
SF_6			0.1			51			0.002

Thus, great care will be needed to perform and analyze the results of drift experiments on pure noble gases.

It has recently been suggested [39] that it may be beneficial to borrow from positron beam technology and incorporate a robust metal moderator (see section 4.1) into the drift cell to produce eV-energy positrons from a localized source. The surface of, for instance, an annealed polycrystalline tungsten moderator seems to be very inert. They have been routinely used in gassy environments, and can be left at atmospheric pressure for long periods (after annealing), and still found to operate on insertion into a beamline. We can estimate the flux of moderated positrons and compare this to that stopped directly in the gas. Thus, if we assume a moderator efficiency of $\varepsilon_m \sim 10^{-4}$, which is typical for a polycrystalline sample, how does this compare to the efficiency, ε_g , of positron stopping directly in the gas?

The β^+ implantation coefficient, μ_{imp} is given by $\mu_{imp} = 1.7\rho/E_m^{1.14}$ (m^{-1}) (e.g. [40]), where ρ is the medium density, now in kgm^{-3}, and E_m is the endpoint energy of the spectrum of the particular isotope in MeV. For ^{22}Na, $E_m \sim 0.545$ MeV, so upon converting the density into Amagat and using mass number, A, we find $\mu_{imp} \sim 0.15A\rho$ (m^{-1}).

To estimate ε_g for positrons which stop in a distance, d, say, we use a simple one-dimensional approach, which will underestimate the true fraction. It is easy to show (see e.g. [1]) that here $\varepsilon_g = [1 - \exp(\mu_{imp}d)]$, which is approximately $\mu_{imp}d$, since typically $1/\mu_{imp} >> d$. Taking our H_2 example from above, $A = 2$ and $\rho = 0.1$ Amagat, we find, with $d = 2$ cm (roughly the Paul cell drift length), $\varepsilon_g \approx 6 \times 10^{-4}$. Thus, even in this unfavorable case of low A, ε_g is still comparable to, or bigger than, ε_m. For SF_6, at the other end of the scale, ε_g will be of order 4-5%. Thus, using a moderator in a gas cell appears to offer few benefits.

4. Trapping, accumulating and manipulating positrons – new synergies for positron transport

In this section we will briefly discuss how positrons can be trapped, accumulated and manipulated in special systems, and how positron transport plays a crucial role in the underlying physical processes. Although there are several techniques by which positrons can be accumulated in high vacuum (see e.g. [41], where these methods are reviewed), here we will describe just one, the buffer gas accumulator. This is the most efficient positron accumulator and was pioneered by Surko and co-workers [42-44]. In order to appreciate the operation of this type of instrument it is first necessary to provide a brief introduction to positron beams. The interested reader can find much more information on this latter topic elsewhere; see e.g. [1, 45, 46].

4.1. Positron beam primer

The use of materials to moderate the kinetic energy of positrons emitted from radioactive sources has a long history. It has been found that certain materials, when appropriately treated, can emit positrons at low energy (typically of order eV) into vacuum. The mechanisms responsible for this effect are well understood [45]. These either involve emission after thermalisation in the material from a surface with a negative work function for positrons (e.g. metal moderators, such as tungsten discussed above), or emission at epithermal energies from materials with a positive positron work function. The prime example of the latter are the solid rare gases and solid neon [47, 48] is the most efficient positron moderator to date, with a conversion efficiency from β^+ particles to "slow" positrons in vacuum of around 1%. Thus, with ^{22}Na sources typically available with GBq activities, it is possible to obtain positron beams with intensities of several millions per second.

An example of a positron beamline is shown in figure 8, and it is the apparatus, with a 2-stage accumulator, which is currently in use at Swansea [49]. The detailed practicalities need not concern us here, but the energy of the beam can be altered, typically in the range 50-100 eV in this case, by applying a bias voltage directly to the neon moderator, which is plated onto a cold finger arrangement held in vacuum. The beam is guided along the axis of the system using magnetic fields provided by an array of coils and solenoids. Such beamlines are used for a variety of physics investigations, mainly in atomic and condensed matter physics [1, 46]. The controlled low energy beam also means that the positrons can be trapped and accumulated, as will be described below.

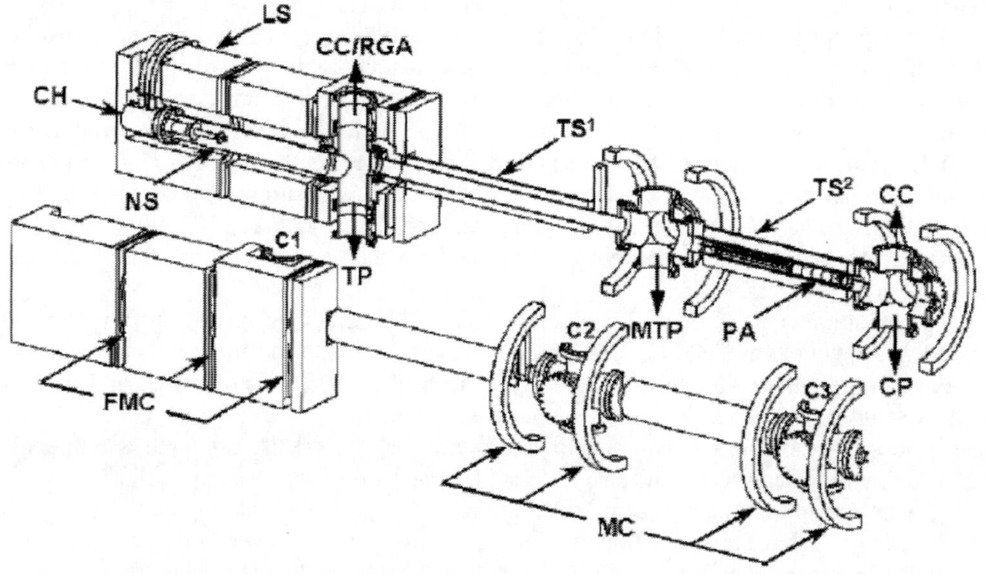

Figure 8. Cutaway diagram of a low energy positron beamline [49] showing the arrangement of source end coldhead (CH) lead shielding (LS) and ^{22}Na source (NS), the coils and solenoids (FMC, MC and TS^1 and TS^2) to confine the beam onto the axis of the system and the two-stage buffer gas positron accumulator (PA). The pumping crosses are labelled C1-3 from left to right and are pumped by turbopumps (TP and MTP) and a cryopump (CP). Vacuum pressure and constitution can be monitored by cold cathode gauges (CC) and a residual gas analyser (RGA).

4.2. Trapping and accumulation of positrons

The low energy positrons are directed into a Penning Malmberg-type trap in which they interact with deliberately introduced nitrogen gas. An example of a 3-stage system, of the type developed by the UCSD group [42] and used at CERN for the formation of cold antihydrogen [6, 7], is illustrated schematically in figure 9. The three stages are formed from different diameter cylindrical electrodes and with nitrogen gas admitted to the centre of the narrow first stage, a pressure gradient is established, with, in the third stage, the pressure being around 10^{-6} mbar.

The on-axis electrical potential is also shown in figure 9 (lower panel) and, following energy loss of the positron as it undergoes an inelastic scattering event (excitation) with the nitrogen gas, forms the axial confinement for the particles. Radial confinement is provided by an axial magnetic field, typically of around 0.15 T in strength. The positrons are trapped after they lose energy in stage 1 (N_2 has a strong positron excitation channel at around 9 eV, which competes effectively with positronium formation [50]) and eventually perform further excitations in stages 2 and 3 before slowing to thermal energies, which is room temperature in the studies to date. Thus, positrons from a quasi-continuous beam are accumulated into the low pressure, third stage, of the trap. Efficiencies can be of order 30% of the incoming beam [44, 49], such that it is possible to accumulate over 1 million positrons per second in this manner.

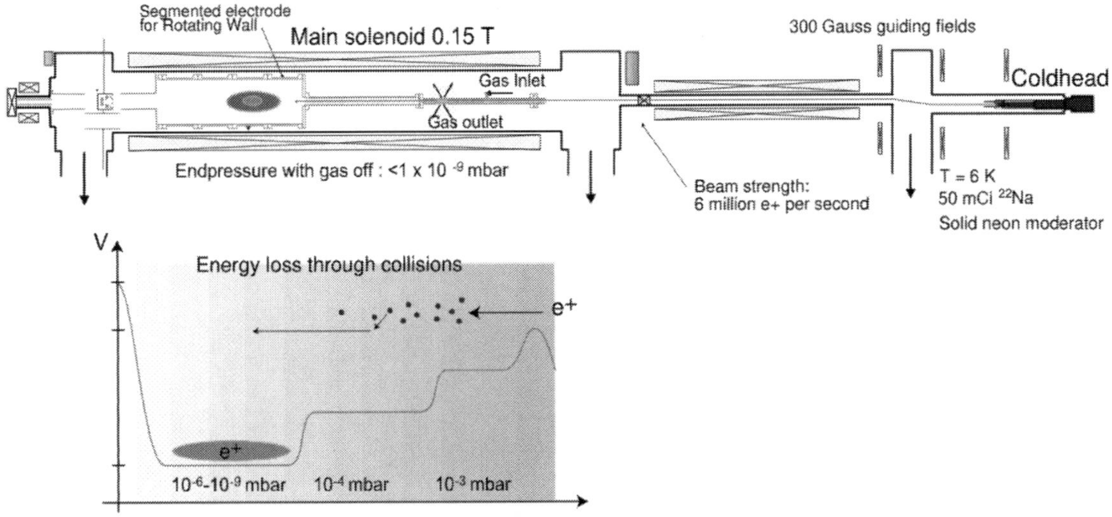

Figure 9. Three-stage positron accumulator [6, 7]. The pressures of the buffer N_2 gas in the three stages are given. The on-axis electrical potential of the Penning Malmberg trap is also illustrated. Radial confinement is provided by an axial magnetic field of 0.15 T in this instrument. One of the electrodes in the third stage is segmented for application of the rotating wall (see text).

Accumulation curves for the number of positrons, $N(t)$, at a given accumulation time are shown in figure 10, and they follow the form,

$$N(t) = N(\infty)[1 - \exp(t/\tau)], \qquad (12)$$

with $N(\infty) = R_+\tau$, where R_+ is the rate of accumulation of the positrons and τ is the positron lifetime in the trap. Note that in the three stage trap it is possible, with care, to obtain positron lifetimes of around 100 s, which is useful for studies which require large instantaneous fluxes of positrons [6-8]. (Note that the third stage can be in a separate vacuum chamber, and be fed repeatedly using a two-stage device, as was the case with Cassidy and Mills [8].) However, many applications can be envisaged for two-stage devices, with accumulation times of a second or less, and there are several such operating devices around the world [8, 49, 51]. Here typically 10^4-10^5 positrons will be accumulated before being ejected for further experimentation.

Figure 10 shows two accumulation cases for the ATHENA/ALPHA three-stage apparatus [52], one of which has a rotating electric field applied to one of the trap electrodes in stage 3, which has been segmented for that purpose. The effect of this will form part of the topic of the following sub-section.

4.3. Manipulation of trapped positrons

We start our discussion with the three-stage trap. Here, there are typically so many positrons accumulated that they form a plasma, a so-called single component, or non-neutral plasma. The behaviour of electron clouds held under similar circumstances to the positrons now forms a long established branch of plasma physics [53]. It was from this field that the rotating electric field, now known as the rotating wall, was developed to counteract plasma expansion [54]. From a comparison of the plots of figure 10 for rotating wall on and off, it is clear that there is a beneficial effect from the wall, which increases the positron lifetime in the gas. (Note that for these measurements, the rotating wall is applied for the entire accumulation period.) From the results of other experiments with this apparatus which examined the behaviour of the positron lifetime as the nitrogen gas pressure is reduced, it is clear that the difference between field on and off is dominated by cross-field transport due to collisions with the nitrogen gas. We have observed similar effects in our two-stage accumulator at Swansea, though on shorter timescales, since the gas pressure is much higher.

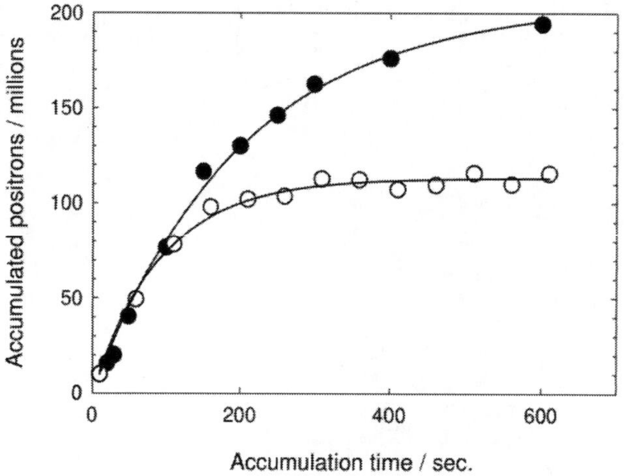

Figure 10. Number of positrons accumulated versus storage time. The open circles are for accumulation without operation of the rotating wall, whilst the latter is switched on for the duration of the accumulation time for the closed circles. See text for details.

The basic physics of the rotating wall can be understood by recalling the behaviour of a non-neutral plasma in a Penning Malmberg trap, which has a uniform axial magnetic field. The radial self electric field due to the plasma, which is proportional to the distance from its centre, r, combined with the axial magnetic field results in an $\mathbf{E} \times \mathbf{B}$ drift velocity which is thus proportional to r. This means that the plasma rotates about its axis with a constant angular frequency, Ω. Applying the rotating wall at a frequency greater than Ω, and in the same sense as the natural rotation, leads to compression of the plasma due to the conservation of angular momentum. Further details of the mechanisms underlying the technique have been explored, primarily by the UCSD group in recent years [55-57].

The application of a rotating electric field to the plasma will cause heating, unless there is a mechanism to introduce cooling. In the ATHENA/ALPHA accumulator, in which the weak rotating wall is applied continuously, the cooling provided by the nitrogen buffer gas is sufficient to prevent excessive heating of the positrons. Systems which employ strong magnetic fields (~ T) in their traps, for instance one of the UCSD traps [58] and the mixing trap region of the ATHENA/ALPHA antihydrogen experiment [7], sufficient cooling is provided for electrons and positrons by the emission of synchrotron radiation. However, if the rotating wall is applied to compress the plasma after accumulation is complete, a strong drive is needed. Given the relatively poor positron cooling capabilities of N_2 (see table 1), an extra gas is needed to achieve compression. SF_6 is often used for this purpose, since it is a strong positron cooler, but has a relatively low Z_{eff} (97 at room temperature [1, 10]); i.e. it has $X \ll 1$.

Thus, it is also important to understand the positron transport properties of SF_6. Optimisation of rotating wall performance is usually carried out experimentally by trial and error. Guidance from theory to help understand limitations to the technique, or perhaps point to directions for improvement would seem to be timely and important.

Further notable work by the Surko group has been done in extracting beams of positrons from their accumulator [59-61]. Such beams have a number of advantages over those typical of emission direct from a moderator. First, the positrons have been cooled to the temperature of the buffer and cooling gases, so the beams are typically of narrow energy width [59]. Furthermore, short time-width pulses containing ~ 10^5 positrons can be obtained using a pulsed extraction technique [62]. In addition, if the extraction is done with carefully adjusted voltage levels, the pulses will be preferentially extracted from the centre of the cloud, which is at the highest potential energy. In this manner narrow beams (so far down to ~ 50 μm in diameter) can be obtained [61]. Once a pulse has been extracted, the plasma quickly relaxes, refilling its centre such that the process can be repeated. Such pulsed micro-beams may prove useful in next generation positron drift experiments, as will be discussed in section 5.

So far, all the discussion has centred on manipulation of positron plasmas. In two-stage accumulators, for instance, it is typically not possible to collect positrons at a sufficiently high density to enter this regime. If rotating wall compression as described above is attempted in such systems there is no effect, perhaps other than positron loss via ejection from the trap.

However, very recently, a new rotating wall regime has been discovered which allows compression of pre-plasma clouds of positrons [63, 64]; i.e. the technique can be applied in the so-called single particle regime, where collective plasma effects can be ignored. As illustrated in figure 11(a), Greaves and Moxom [64] have shown that the central density of a positron cloud can be increased by about two orders of magnitude. Furthermore, figure 11(b) illustrates that the timescales needed for compression are in the ms range [64]. Again the cooler gas SF_6 is needed to observe the effect. Similar compression has recently been achieved in the Swansea two-stage accumulator [65].

The mechanism by which the rotating wall influences the particles and compresses the cloud is not clear at the time of writing. However, it is known that the rotating wall needs to be operated at a frequency at, or close to, the axial bounce frequency of the particles in the trap for compression to occur. Particles which have cooled to near the bottom of the potential well in a Penning Malmberg trap will experience a local harmonic potential. As such, the axial frequency is well defined and is given by,

$$\upsilon_a = [(eV_t/md_t^2)^{1/2}]/2\pi, \qquad (13)$$

where V_t and d_t are trap parameters, respectively the electrical depth and length of the trap. For positrons and electrons (mass, m and charge, e) this frequency is typically in the MHz range. By systematically varying trap depth to change υ_a, Greaves and Moxom [64] were able to show that the effect was caused by coupling to the positrons at their axial frequency and they speculated that an effect known as bounce resonance transport was responsible. Here, the combination of the axial bounce and azimuthal drift motions of particles causes them to move into resonance with an applied asymmetry. Some work has been done on this for electron plasmas [66-68], but further work is urgently needed, since the technique seems to have the promise to revolutionise positron beam physics by allowing cold, narrow beams to be available using relatively simple apparatus.

5. Possibilities for beam-based positron drift experiments

The first suggestion that positron drift experiments could be done using a low energy positron beam seems to have been made by Charlton and Jacobsen [69] in an article discussing applications of pulsed, accelerator-based beams. Unfortunately few of these beams were built, and even fewer were used for atomic physics work, and to the best of our knowledge, positron drift measurements have never been attempted using a positron beam as a source.

The idea of the proposed measurement was simple. Since the magnetically-confined pulsed beams envisaged at that time contained about 10^5-10^6 positrons per pulse, over a diameter of typically 1 cm, a 0.1 mm aperture into a cell containing gas at a pressure of several 10's of Torr of gas was envisaged, both to define the gas, and attenuate the beam. Charlton and Jacobsen [69] discuss various other technicalities of the experiment (including non-thermalisation effects in nobles), but the principle outlined was straightforward. The experiment was to rely on timing between the machine pulse and a particle detector located after the exit of the cell. Anticipating later work, [69] noted that the output of the gas cell would form a pulsed, cooled beam.

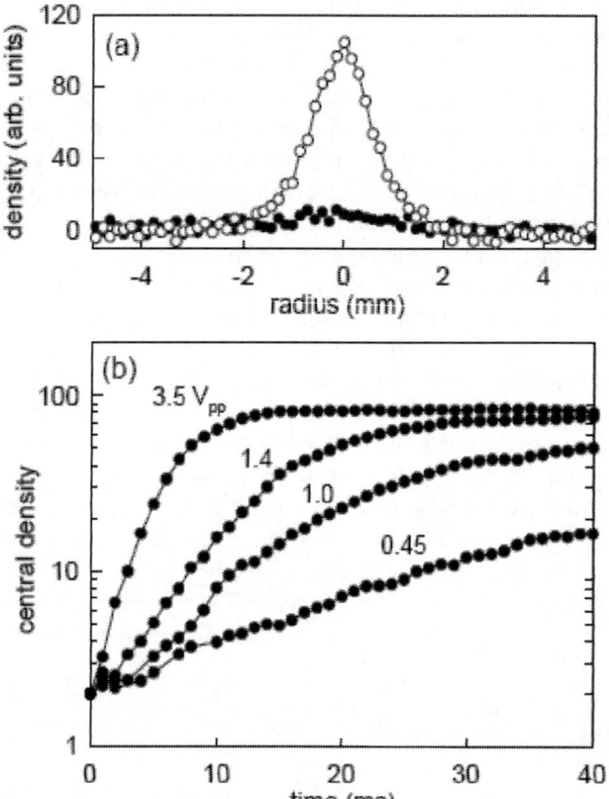

Figure 11. (a) Illustration of compression of positron clouds in the single particle regime as evidenced from the marked increase in central density (open circles versus the closed) of the positrons extracted from an accumulator [64]. (b) The timescales for compression of a cloud of positrons illustrated by the variation of the central density of the cloud at various times after the start of the rotating wall for various peak-to-peak applied voltage amplitudes, V_{pp}.

Whilst it is disappointing that some much early enthusiasm in the positron field was unfulfilled when intense beams, by and large, failed to appear, we have, as outlined in section 4.3, fresh cause for optimism. The new techniques which can be used to create pulsed, micro-beams, which are thus naturally timed (or can be further gated if the precision is insufficient) and which can be injected via a narrow aperture into a gas cell, seem to hold much promise. Indeed, the experiment suggested in [69] would seem, with small modification, to be suitable for use on the output of an accumulator, operating with rotating wall compression and employing the UCSD micro-beam extraction system. It should be recalled that all of the suggested experiments require confinement and transport of the beam using an axial magnetic field.

There may also be a further alternative, if a micro-beam and associated apparatus is not available. This involves taking a pulsed output from a two-stage accumulator and injecting it, through a thin foil, into a gas cell. In this case the beam would have to be injected into a Faraday cage-type system containing the cell such that the beam energy was raised to say 20 keV. Such acceleration systems are in routine use in positron labs around the world, mostly for depth-dependent condensed matter studies (see e.g. [46]), and there are plans to incorporate such a feature into a beamline at the ANU, Canberra [70], where there is a distinguished tradition of interest in swarm physics.

First we will explore positron implantation to see what thickness of foil might be needed to allow injection into the cell. Fortunately, there have been many studies of positron stopping in materials over the years and implantation profiles, $P(z)$, are well understood. Without going into technicalities, the profiles can be reasonably accurately described by a Makhovian of the form [45, 46],

$$P(z) = (mz^{m-1}/z_o^{m}) \exp(-[z/z_o]^m), \tag{14}$$

with m a constant, the so-called shape parameter, and equal to 1.9. (Note here the different use of m from earlier in this article.) Here z_o is given by

$$z_o = \langle z \rangle / \Gamma(1/m + 1), \tag{15}$$

where $\langle z \rangle$ is the mean implantation depth and Γ is the gamma function and is of order unity here. Thus, for our purposes, $z_o \sim \langle z \rangle$, which can be written in the form $\langle z \rangle = A_{mat}E^n$, where A_{mat} is a material-dependent constant and E is the positron impact energy. It turns out, empirically, that this can be approximated by,

$$\langle z \rangle = 40E^{1.6}/\rho \qquad \text{(nm)}, \tag{16}$$

with E in units of keV and ρ in gcm^{-3}. Inserting 20 keV as a kinetic energy and using $\rho = 2.3$ gcm^{-3} (for carbon) we find $\langle z \rangle \sim 2$ μm. Whilst this might appear to be thin, vacuum separation has been reported using very thin supported carbon foils [71]. Conductances as low as 10^{-3} ls^{-1} per cm^2 were reported for foils as thin as 0.1 μm. Presumably thicker foils will have even lower conductances, such that adequate pressure differences between the cell and the surrounding chamber can be achieved. Indeed, depending upon the pressure required in the cell, issues concerning foil fidelity would need to be addressed first, though comments in [71] would appear to offer hope.

In the situation in which the positrons are injected through a foil it is likely that many will annihilate therein, resulting in a large prompt signal. However a down stream annihilation detector, which can be used to monitor the drift time, can probably be selectively sensitised by being switched "on" using a pulsed bias voltage only after the injection peak is over.

6. Conclusions

In this article we have reviewed positron transport in gases, a topic which has been studied since the earliest days of positron annihilation physics. In particular, we have detailed the attempts which have been made to extract drift velocities and mobilities from investigations of positron behaviour in small gas cells under the influence of applied electric fields. We have identified new areas where understanding positron transport is of current relevance for positron trapping, accumulation and manipulation in positron beam systems. We have suggested how trapped positrons might be used in future to make positron drift velocity measurements using new, and potentially versatile, techniques. New theoretical studies of positron transport in gases have also provided a fresh and timely impetus for experiment in this area.

Acknowledgements
The author is grateful to Professor Zoran Petrović for the invitation to contribute this article and to Professor Petrović, Professor Stephen Buckman and Dr. James Sullivan for their interest in this work and for several stimulating discussions. Thanks are also due to Dr Finn Jacobsen who fostered my early interest in the field described in this article, and to Ana Banković who helped to prepare the article for publication. I am grateful to the EPSRC (UK) for their support of my work with positrons, particularly as applied to antihydrogen physics.

References

[1] Charlton M and Humberston J 2001 *Positron Physics* (Cambridge University Press)
[2] Šuvakov M, Petrović ZLj, Marler JP, Buckman SJ, Robson RE and Malović G 2008 *New J. Phys.* **10** 053504
[3] Banković A, Marler JP, Šuvakov M, Malović G and Petrović ZLj 2008 *Nuc. Inst. Meth. in Phys. Res. B* **266** 462
[4] Banković A, Petrović ZLj, Robson RE, Marler JP, Dujko S and Malović G 2008 *submitted*
[5] Greaves RG and Surko CM 2004 *Phys. Plasmas* **11** 2333
[6] Amoretti M *et al.* 2002 *Nature* **419** 456
[7] Jørgensen LV *et al.* 2005 *Phys. Rev. Lett.* **95** 025002
[8] Cassidy DB and Mills Jr AP 2007 *Nature* **449** 195
[9] Griffith TC and Heyland GR 1978 *Phys. Rep.* **39** 169
[10] Charlton M 1985 *Rep. Prog. Phys.* **48** 737
[11] Paul DAL and Leung CY 1968 *Can. J. Phys.* **46** 2779
[12] Canter KF and Heyland GR 1974 *Appl. Phys.* **5** 231
[13] Kauppila WE, Stein TS and Jesion G 1976 *Phys. Rev. Lett.* **36** 580
[14] Coleman PG, Griffith TC and Heyland GR 1981 *J. Phys. B: At. Mol. Opt. Phys.* **14** 2509
[15] Al-Qaradawi I, Charlton M, Borozan I and Whitehead R 2000 *J. Phys. B: At. Mol. Opt. Phys.* **33** 2725
[16] Tao SJ, Green JH and Celitans GJ 1963 *Proc. Phys. Soc.* **81** 1091
[17] Jacobsen FM 1984 in *Positron Scattering in Gases* (Plenum) 85
[18] Marder S, Hughes VW, Wu CS and Bennett W 1956 *Phys. Rev.* **103** 1258
[19] Teutsch WB and Hughes VW 1956 *Phys. Rev.* **103** 1266
[20] Charlton M, Cullen RA, Raza J and Watts MST 1992 *J. Phys. B: At. Mol. Opt. Phys.* **25** 4351
[21] Obenshain FE and Page LA 1962 *Phys. Rev.* **125** 573
[22] Charlton M and Curry PJ 1985 *Il Nuovo Cimento D* **6** 17
[23] Sharma SC, Hyatt SD, Ward MH and Dark CA 1985 *J. Phys. B: At. Mol. Opt. Phys.* **18** 3245
[24] Jacobsen FM, Charlton M and Laricchia G 1986, *J. Phys. B: At. Mol. Opt. Phys.* **19** L111
[25] Pepe I, Paul DAL, Deutsch J and Prieels R 1993 *J. Phys. B: At. Mol. Opt. Phys.* **26** L373
[26] Bisi A, Gambarini G and Zappa L 1983 *Il Nuovo Cimento D* **2** 1465
[27] Mogensen O 1974 *J. Chem. Phys.* **60** 998
[28] Jacobsen FM 1986 *Chem. Phys.* **109** 455
[29] Heyland GR, Charlton M, Griffith TC and Clark G 1985 *Chem. Phys.* **95** 157
[30] Mogensen OE 1995 *Positron Annihilation in Chemistry* (Springer)
[31] Paul DAL and Tsai J-S 1979 *Can. J. Phys.* **57** 1667
[32] Böse N, Paul DAL and Tsai J-S 1981 *J. Phys. B: At. Mol. Opt. Phys.* **14** L227
[33] Paul DAL and Böse N 1982 in *Electron and Ion Swarms* (Pergamon Press) 65
[34] Rodionov SN, Sannikov BP and Solodov EP 1969 *JETP Lett.* **10** 325
[35] Rodionov SN, Sannikov BP and Solodov EP 1971 *Phys. Lett. A* **35** 297
[36] Charlton M 1985 *J. Phys. B: At. Mol. Opt. Phys.* **18** L667
[37] Charlton M and Laricchia G 1986 in *Positron (Electron)-Gas Scattering* (World Scientific) 73
[38] Davies SA, Curry PJ, Charlton M and Jacobsen FM 1986 in *Positron (Electron)-Gas Scattering* (World Scientific) 317

[39] Petrović ZLj 2007 private communication
[40] Beling CD and Charlton M 1987 *Cont. Phys.* **28** 241
[41] Holzscheiter MH, Charlton M and Nieto MM 2004 *Phys. Rep.* **402** 1
[42] Surko CM, Passner A, Leventhal M and Wysoki FJ 1988 *Phys. Rev. Lett.* **61** 1831
[43] Greaves RG, Tinkle MD and Surko CM 1994 *Phys. Plasmas* **1** 1439
[44] Murphy TJ and Surko CM 1992 *Phys. Rev. A* **46** 5696
[45] Schultz PJ and Lynn KG 1988 *Rev. Mod. Phys.* **60** 701
[46] *Physics and Applications of Positron Beams*, 2000 (ed. Coleman PG) (World Scientific)
[47] Mills Jr AP and Gullikson EM 1986 *Appl. Phys. Lett.* **49** 1121
[48] Khatri R, Charlton M, Sferlazzo P, Lynn KG, Mills Jr. AP and Roellig LO 1990 Appl. Phys. Lett. 57 2374
[49] Clarke J, van der Werf DP, Griffiths B, Beddows DCS, Charlton M, Telle HH and Watkeys PR 2006 *Rev. Sci. Inst.* **77** 063302
[50] Sullivan, JP, Marler JP, Gilbert SJ, Buckman SJ and Surko CM 2001 *Phys. Rev. Lett.* **87** 073201
[51] see eg. Sullivan JP, Makochekanwa C, Jones A, Caradonna P and Buckman SJ 2008 *J. Phys. B: At. Mol. Opt. Phys.* **41** 081001
[52] Jørgensen LV, Collier MJT, Fine KS, Watson TL, van der Werf DP and Charlton M 2001 *Materials Science Forum* **363-365** 634
[53] see e.g. the AIP conference series *Non-Neutral Plasma Physics* (AIP, New York)
[54] Anderegg F, Hollmann EM and Driscoll CF 1998 *Phys. Rev. Lett.* **81** 4875
[55] Greaves RG and Surko CM 2000 *Phys. Rev. Lett.* **85** 1883
[56] Danielson JR and Surko CM 2005 *Phys. Rev. Lett.* **94** 035001
[57] Danielson JR, Surko CM and O'Neil TM 2007 *Phys. Rev. Lett.* **99** 135005
[58] Danielson JR, Schmidt P, Sullivan JR and Surko CM 2003 in *Non-Neutral Plasma Physics V* AIP Conf. Proc. No. 692 (AIP, New York) 149
[59] Danielson JR and Surko CM 2006 *Phys. Plasmas* **13** 055706
[60] Danielson JR, Weber TR and Surko CM 2007 *Appl. Phys. Lett.* **90** 081503
[61] Weber TR, Danielson JR and Surko CM 2008 *Phys. Plasmas* **15** 012106
[62] Gilbert SJ, Kurz C, Greaves RG and Surko CM 1997 *Appl. Phys. Lett.* **70** 1944
[63] Cassidy DB, Deng SHM, Greaves RG and Mills Jr AP 2006 *Rev. Sci. Inst.* **77** 073106
[64] Greaves RG and Moxom J 2008 *Phys. Plasmas* **15** 072304
[65] Baker CJ, Isaac CA, Kerrigan SJ, Watkeys PR and Van der Werf DP, 2008 private communication
[66] Eggleston DL and O'Neil TM 1999 *Phys. Plasmas* **6** 2699
[67] Eggleston DL and Carrillo B 2002 *Phys. Plasmas* **9** 786
[68] Eggleston DL and Carrillo B 2003 *Phys. Plasmas* **10** 1308
[69] Charlton M and Jacobsen FM 1987 *Appl. Phys. A* **43** 235
[70] Buckman SJ and Sullivan JP 2007 private communication
[71] Lagomarsino V, Manuzio G, Pozzo A and Testera G 1991 *Nuc. Inst. Meth. in Phys. Res. A* **307** 309

Journal of Physics: Conference Series **162** (2009) 012004 doi:10.1088/1742-6596/162/1/012004

Cross sections and transport properties of negative ions in rare gases

Jasmina V. Jovanović[1,2] and Z. Lj. Petrović[1]

[1]Institute of Physics, P.O.B. 68, Zemun, Belgrade, Serbia

[2]Faculty of Mechanical Engineering, Kraljice Marije 16, Belgrade, Serbia

E-mail: jjovanovic@mas.bg.ac.yu

Abstract.

We have used a combination of a simple semi-analytic theory - Momentum Transfer Theory (MTT) and exact Monte Carlo (MC) simulations to develop momentum transfer cross sections of negative ions in collisions with noble gases based on the available data for reduced mobility at 300K as a function of E/N. At very low energies, we extrapolated obtained cross sections using Langevin's cross section and supplemented it by the total detachment cross section that was used from the threshold around 6 eV up to 100 eV. Other possible reactive processes have not been taken into account. A good agreement for the mean energy and diffusion coefficients is an independent proof of the validity of the cross sections that were derived for the negative ion mobility data.

1. Introduction

The purpose of this work is to provide plasma modelers with cross section data and transport coefficients that may be used in simulations of plasmas containing negative ions. The negative ions play an important role in laser and electrical discharges and in radiation chemistry in the atmosphere. They are present in numerous technological plasmas including plasma etching chemistries. Negative ions determine kinetics of electronegative plasmas [1] and their presence may change critically the nature of plasmas (examples of obvious changes can be seen in [2]). Ion-Ion plasmas [3] have very distinct characteristics. Numerous applications relay on electronegative reactive gases including plasma etching [4], atmospheric plasmas for biomedical application [5], and many more. Furthermore, negative ions may be used directly, for example to reduce the charging [6] of high aspect ratio structures or to be converted to fast neutrals with the aim to achieve fast neutral etching [7]. Modeling of such processes requires either cross sections or transport data or both.

The procedure used here to determine the cross sections is to apply first the momentum transfer theory (MTT) which is fast albeit with limited accuracy to fit the cross sections until the transport data fit the experimental results. In the next step further, but always slight adjustments are made to have a good agreement between experimental data and Monte Carlo (MC) calculations which provide exact results. Monte Carlo simulation code is based on the time integration technique [8, 9]. It was developed to include both isotropic and anisotropic scattering for all processes, non-conservative collisions with proper calculation of both bulk and flux transport coefficients [10], proper calculation of the collision frequency for thermal collisions and it was tested with exceptional results against all known benchmarks. Normally, the disagreement between MTT and MC (and thereby experimental) results was within 11% for all gases. We have extrapolated data to somewhat higher energies based on behavior of similar ions in similar gases and by the addition of reactive processes so a relatively complete set was derived which can be used in modeling of plasmas by both hybrid, particle in cell (PIC) and fluid codes [1, 11].

It is important to note that so far the data presented here and other similar results were interpreted by using interaction potentials [12] which while producing results of high accuracy is not directly applicable in plasma modeling. Plasma modeling requires data of a lesser accuracy (10%-30%), as compared to the derivation of accurate interaction potentials, and the data are required either as drift velocities and diffusion coefficients or as sets of cross sections. Thus we have set out to provide sets of data that may be used in modeling of reactive plasmas. As both hybrid and PIC models involve MC simulations for collisions and as applications to plasma etching devices requires control of negative ions in the afterglow or for neutralization [7, 13] the data for cross sections of negative ions found in argon-fluorocarbon mixtures are of great importance.

We have applied a simple form of MTT [14] based on elastic collisions as the first step in order to develop Br^-, F^-, Cl^-/He, Ne, Ar, Kr, Xe elastic momentum transfer cross section based on the available data for reduced mobility [15, 18]. The analytic equations allow quick calculations of the transport data and their comparisons with the experimental data. Having in mind the mean energy of ions and the overall energy range for any particular E/N and with some experience we modify the cross sections to obtain better and better agreement with experiment. The MTT is well developed to take into account reactive processes [19] but that extension is quite complex so we have applied a simple form of the theory based on elastic collisions since the influence of reactive processes starts at much higher energies for the chosen gases.

In the following sections we present details of determination of the elastic momentum transfer cross sections, transport data and discussion of the observed discrepancies between the results of MTT and MC.

2. Cross sections

The MTT consists of a specific simplification of the Boltzmann equation collision operator and a specific procedure to determine approximate distribution functions.

$$[\frac{\partial}{\partial t} + \vec{v}\frac{\partial}{\partial \vec{r}} + \frac{e}{m}\vec{E}(\vec{r},t)\frac{\partial}{\partial \vec{v}}]f^{mix}(\vec{r},\vec{v},t) = \Im^{mix}(f^{mix}, f_1^{mix}, ...f_\ell^{mix}) \tag{1}$$

Considering ions with number density n in neutral gas in equilibrium at temperature T_0 and assuming that density gradients are weak, we have the following simple approximate balance equations for mobility and mean energy of ions [15, 20, 21]:

$$K = \frac{q}{m\nu_m} = \frac{W}{E}, \tag{2}$$

$$\epsilon = \frac{1}{2}m_0 W^2 + \frac{3}{2}k_B T_0, \tag{3}$$

where q and m are the swarm charge and negative ion mass, respectively, m_0 is the mass of neutral molecule, W is the drift velocity, E is the electric field, k_B is the Boltzmann constant and ν_m is the total momentum transfer collision frequency in the center of mass reference frame given by

$$\nu_m = N\sqrt{\frac{2\varepsilon}{\mu}}\sigma_m \tag{4}$$

where σ_m denotes total momentum transfer cross section, ε is mean energy of ions in the center of mass reference frame and μ is the reduced mass $\mu = \frac{mm_0}{m+m_0}$.

The most important scattering process in the low energy region (up to few eV) will be elastic scattering. Rare gas atoms do not form negative ions and in these cases the charge transfer is not possible so the main reactive process would be detachment. The equations (2-4) are easily solvable and the mobility data [15, 20] could provide cross sections up to 2−3 eV. Even the range where the detachment cross sections become significant was not covered by the earlier cross section data so extrapolation was used. The momentum transfer cross section is supplemented by detachment cross section [22, 23, 24] that was used from the threshold around 6 eV up to 100 eV.

In order to make proper calculations for thermal energies, we allowed the momentum transfer cross sections to converge towards Langevin's cross section [25],

$$Q_P = 2.21\pi\sqrt{\frac{e^2\alpha_D}{2\varepsilon}} \tag{5}$$

Values of α_D (dipole polarizability) [26] for gases used in this paper are given in Table 1 where ε is the relative energy and a_0 is the Bohr radius.

Table 1. Values of the dipole polarizability [26] for some gases of interest.

molecule	polarizability [a_0^3]
He	1.39
Ne	2.68
Ar	11.08
Kr	16.7
Xe	27.1

The cross section sets for Cl^-/He,Ne,Ar,Kr,Xe have been presented in our previous paper [8] so in this paper, in Figures 1 - 5 we presented sets of cross sections for Br^-, F^-/Ar, Kr, Xe. Figure 5 is just an example of the corrections that have been made. The major correction was made in thermal energy region. At higher energies the correction is mostly the consequence of the detachment which starts from the thresholds around 6 eV.

3. Transport data

We have done all the calculations with an assumption of isotropic scattering for elastic and detachment collisions. The MTT results were obtained in the same range of E/N as experimental results and therefore the influence of the detachment process was not included in MTT calculations.The MC calculations were performed over a range of E/N from 1 Td to 1000 Td ($1\mathrm{Td}=10^{-21}\mathrm{Vm}^2$), that is somewhat wider than that covered by the measurements.

3.1. Reduced mobility

Result for drift velocities and mean energies for Br^- in Xe is shown in Figure 6 but these results we have for all mentioned combination of gases. MC perfectly agrees with the experimental results, but there is a significant discrepancy with the MTT results.

The mean energy does not show a significant deviation from the thermal value up to 40 Td. Beyond that point the mean energy increases rapidly, reaching about 6eV at 1 kTd (in all cases).

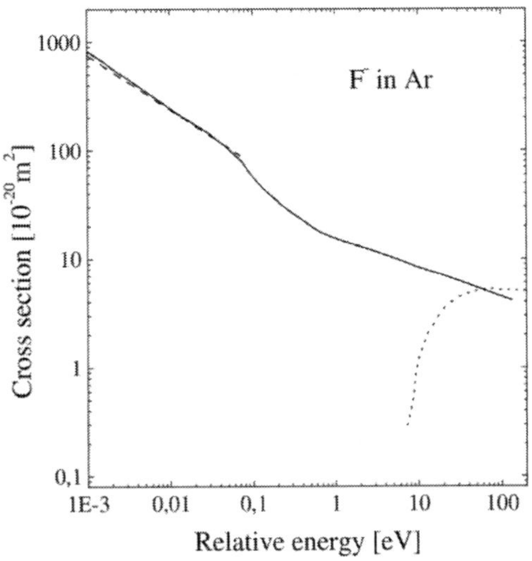

Figure 1. Cross sections for F^- in Ar. Solid line: momentum transfer cross section; dashed line: Langevin cross section; dotted line, detachment cross section from reference [22].

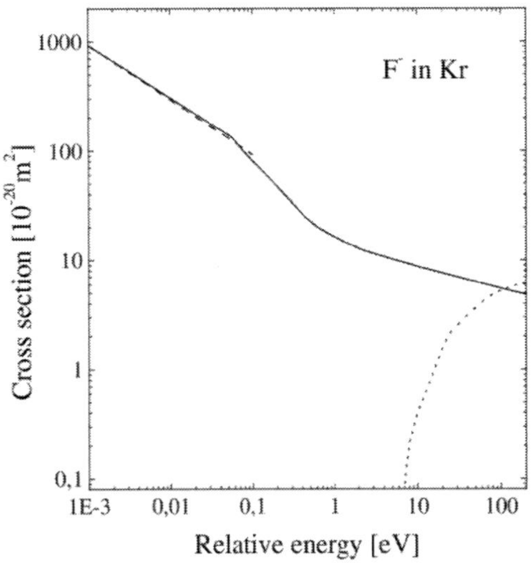

Figure 2. Cross sections for F^- in Kr. Solid line: momentum transfer cross section; dashed line: Langevin cross section; dotted line, detachment cross section from reference [22].

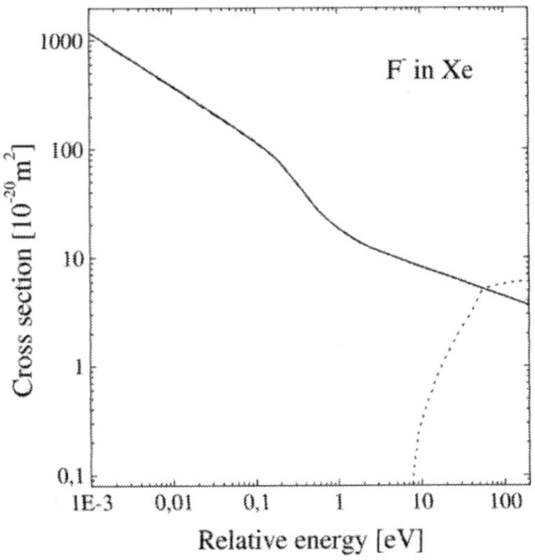

Figure 3. Cross sections for F^- in Xe. Solid line: momentum transfer cross section; dashed line: Langevin cross section; dotted line, detachment cross section from reference [23].

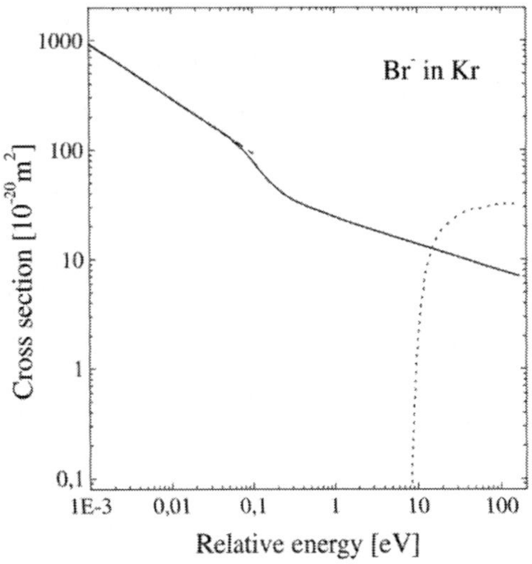

Figure 4. Cross sections for Br^- in Kr. Solid line: momentum transfer cross section; dashed line: Langevin cross section; dotted line, detachment cross section from reference [24].

The data for reduced mobilities of F^- and Br^-ions in rare gases are shown in Figures8-9. The relation between the reduced mobility K_0 and the mobility K of ions is

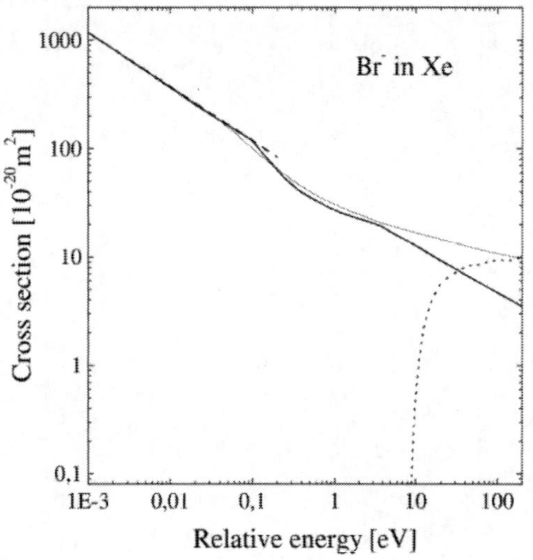

Figure 5. Cross sections for Br^- in Xe. Solid line: momentum transfer cross section; dashed line: Langevin cross section; dotted line, detachment cross section from reference [24]; thin solid line: originally obtained momentum transfer cross section.

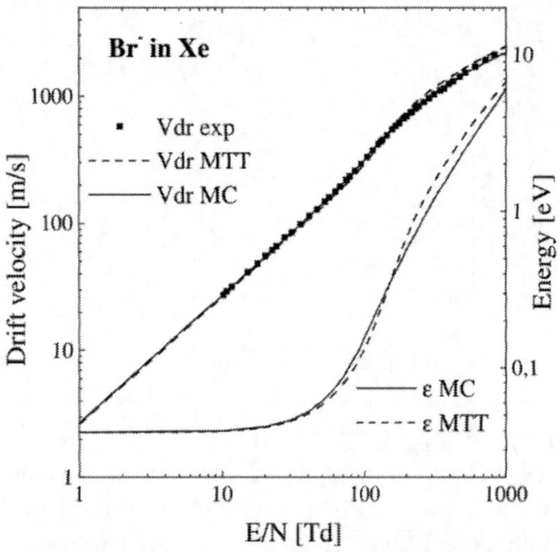

Figure 6. The drift velocity and mean energy for Br^- in Xe at T=300K. Squares are experimental results taken from Lamm et al. [17].

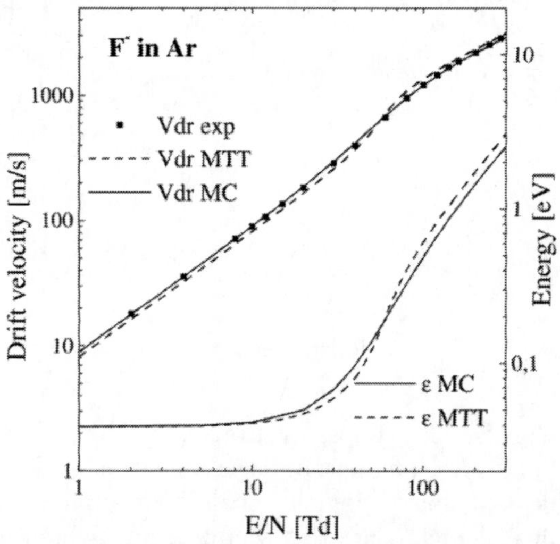

Figure 7. The drift velocity and mean energy for F^- in Ar at T=300K. Squares are experimental results taken from Ellis et al. [15].

$$K_O = \frac{NK}{N_0} \tag{6}$$

where $N_0 = 2.69 \times 10^{25}\mathtt{m}^{-3}$ is gas density at 300K and 760Torr.

MC results are in excellent agreement with the experimental results whose total experimental error does not exceed ±3%, but MTT results deviate between 10% for Br^- and F^- in Xe and 13% in the case of Br^- in Kr. These deviations cannot be related to the ratio ion-neutral gas masses or some systematic error. Thus it is only due to the limited accuracy of MTT that is exact only for the constant collision frequency model.

It can be seen from the equation (2) that the mobility coefficient is proportional to the inverse of ν_m and there are three regions of field to be considered. At low fields mobility is approximately constant due to the influence of the attractive part of the long-range interaction potential arising from the polarization exerted by the ion upon the neutral. At high fields the mobility decreases rapidly due to the dominant repulsive part of the interaction potential arising from the short-range forces. At intermediate fields there is a peak in the mobility curve, corresponding to the enhanced transparency of the gas at intermediate energies, since attractive and repulsive forces cancel out to some extent. These three different regions of the mobility can be related to the corresponding three different portions of the elastic cross sections as displayed in Figures 1-5.

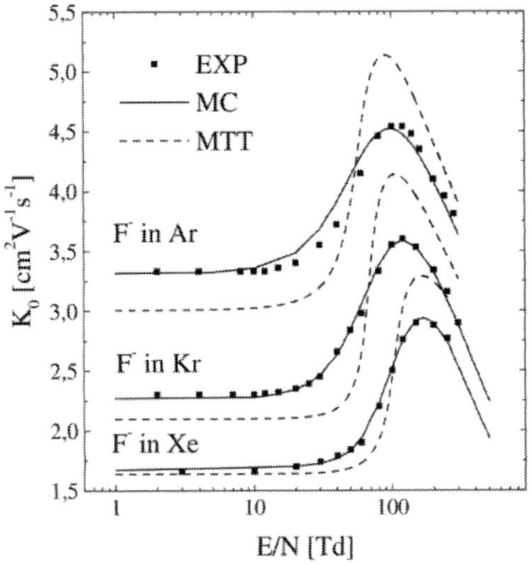

Figure 8. The reduced mobility K_0 for F^- at T=300K. Squares are experimental results taken from Ellis et al. [15].

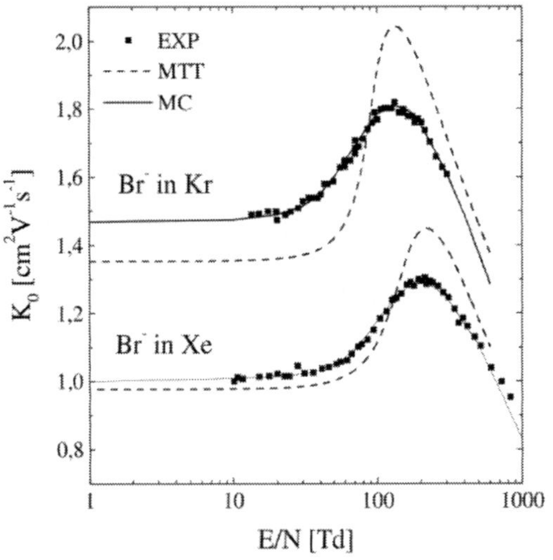

Figure 9. The reduced mobility K_0 for Br^- at T=300K. Squares are experimental results taken from Lamm et al. [17].

3.2. Diffusion coefficients

Diffusion coefficient is a tensor having components that refer to the directions parallel and perpendicular to the electric field named longitudinal and transverse diffusion coefficients respectively. In MTT equations anisotropy of temperature tensor was taken into account [27, 28, 29] and longitudinal D_L and transverse D_T diffusion coefficients were calculated from Generalized Einstein Relations obtained by Robson [27, 28].

$$D_L = \frac{k_B T_L}{q} K(1 + (1 + \Delta)\frac{\mathtt{dln}K}{\mathtt{dln}E}) \tag{7}$$

Journal of Physics: Conference Series **162** (2009) 012004 doi:10.1088/1742-6596/162/1/012004

$$D_T = \frac{k_B T_T}{q} K, \qquad \Delta = \frac{Q}{2K_B T_L W} \tag{8}$$

In the case of constant mean free time where collision frequency ν_m is constant, the ion temperatures and heat conductivity Q are given exactly by the expressions [27, 28, 29]:

$$\begin{aligned} Q &= B m_0 W^3 \\ kT_T &= kT_0 + A_T m_0 W^2 \\ kT_L &= kT_0 + A_L m_0 W^2 \end{aligned} \tag{9}$$

where

$$\begin{aligned} A_T &= \frac{(1+\frac{m}{m_0})\frac{\nu_v}{3\nu_m}}{\frac{2m}{m_0}+\frac{\nu_v}{\nu_m}}, A_L = \frac{\frac{2m}{m_0}(1-\frac{\nu_v}{3\nu_m})+\frac{\nu_v}{3\nu_m}}{\frac{2m}{m_0}+\frac{\nu_v}{\nu_m}}, \\ B &= \frac{(1+\frac{m}{m_0})^2(A_L+\frac{1}{2}+\frac{3m}{2m_0})}{1+\frac{3m^2}{m_0^2}+\frac{4m\nu_v}{3m_0\nu_m}} - (A_L + \frac{1}{2} + \frac{m}{2m_0}) \end{aligned} \tag{10}$$

and ν_v is the collision frequency for viscosity.

Longitudinal and transverse diffusion coefficients are shown in Figures 10-13. The only available experimental results are the experimental data of Ellis et al. [30] for longitudinal diffusion of F^-,Br^-/Kr,Xe. There are no experimental data for the longitudinal diffusion available for F- ions in Ar. The agreement of the results for longitudinal diffusion is an independent proof of the validity of obtained cross sections sets since MC gives an excellent agreement with the experimental results whose uncertainty is 8% at low E/N increasing to 20% at the higher values, but MTT results are again of poor quality. There are two main candidates for the explanation of the discrepancies between the experimental data and MTT results: the approximate nature of the theory and the assumption of isotropic scattering. The differences at high fields are mostly the consequence of the detachment which is not adequately represented in the theory. Additional sources of information such as differential cross sections or transverse diffusion coefficients could be used to improve further the set by adding anisotropies [31] which may be important at higher mean energies where scattering is more likely to occur in forward direction. Nevertheless, we may still claim that a simple form of MTT may provide a reasonable first approximation for diffusion coefficients and further improvement by other but more demanding techniques.

The transverse diffusion coefficients are not available from experimental papers and the obtained MC results may be a useful source of data but one should be aware that further information on anisotropy of scattering may affect the results though much less than in the case of resonant charge exchange dominated transport [31].

3.3. Rate coefficients

In addition to transport coefficients in fluid plasma models one needs the rate coefficients for all relevant processes. The net rate coefficients of elastic scattering and detachment are shown in Figure 14-16. It is obvious that a large rate of detachment as a non-conservative process above 300 Td will affect the real space (bulk) and the velocity space (flux) definitions of transport coefficients. This is shown for Cl^-/Ar in Figure 17. Similar results are found for all other ions in other gases.

4. Summary

We have determined a set of cross sections for F- ion in Ar, Kr and Xe based on the available experimental transport data in a wide range of reduced electric field E/N of 1 Td - 1000 Td

Journal of Physics: Conference Series **162** (2009) 012004 doi:10.1088/1742-6596/162/1/012004

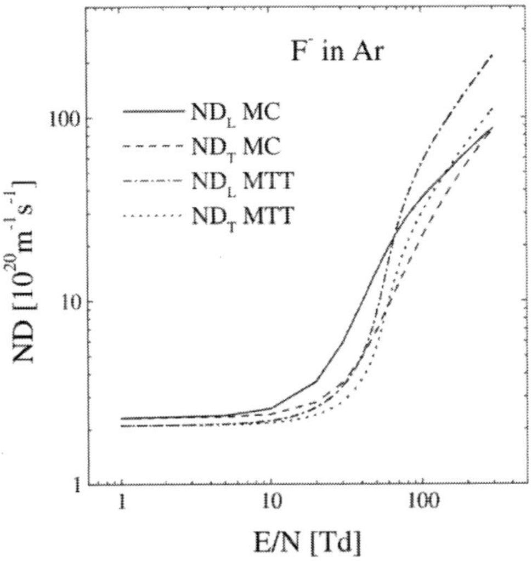

Figure 10. The diffusion coefficients for F^- in Ar at T=300K.

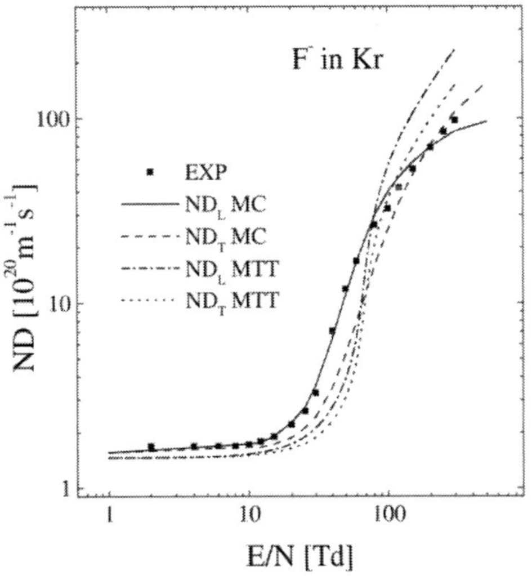

Figure 11. The diffusion coefficients for F^- in Kr at T=300K. Squares are experimental results of Ellis et al. [30].

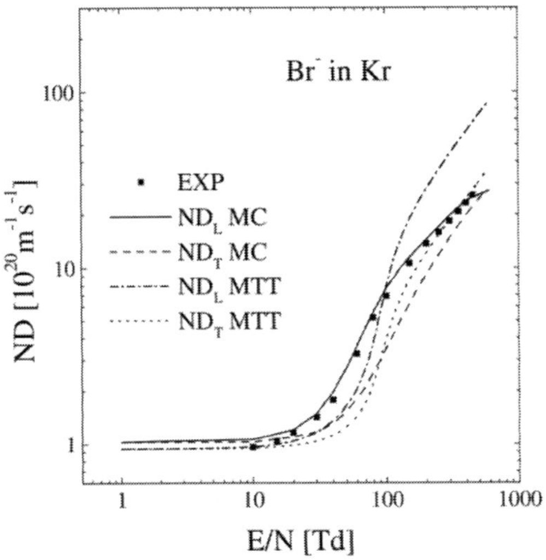

Figure 12. The diffusion coefficients for Br^- in Kr at T=300K. Squares are experimental results of Ellis et al. [30].

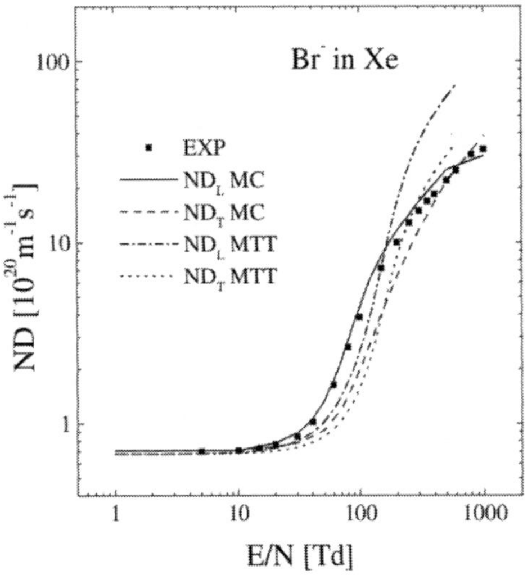

Figure 13. The diffusion coefficients for Br^- in Xe at T=300K. Squares are experimental results of Ellis et al. [30].

by employing a standard swarm technique. The procedure used here to determine the cross sections is to apply first the MTT which is fast albeit with limited accuracy to fit the cross sections until the transport data fit the experimental results. In the next step further, but always slight, adjustments are made to have a good agreement between experimental data and MC calculations which provide exact results. Normally, the disagreement between MTT and MC (and thereby experimental) results was within 13% for all gases.

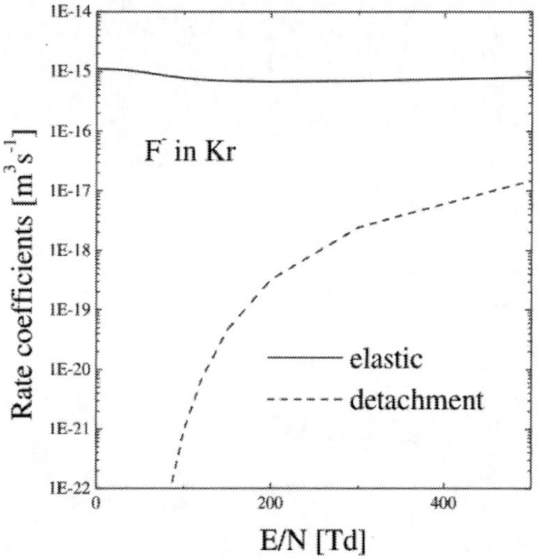

Figure 14. Net rates for elastic scattering and detachment for F^- ions in Kr at T=300K.

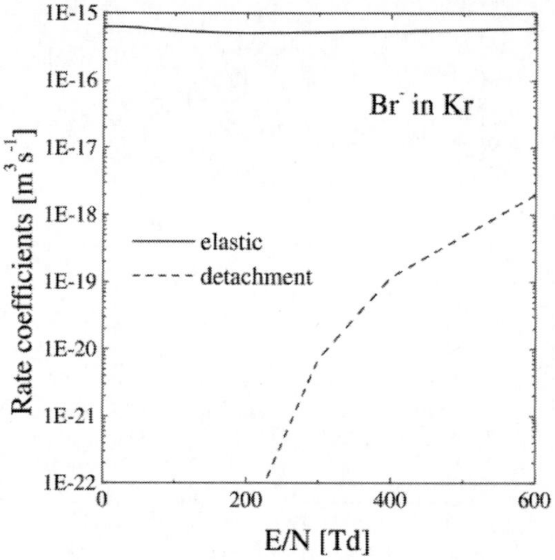

Figure 15. Net rates for elastic scattering and detachment for Br^- ions in Kr at T=300K.

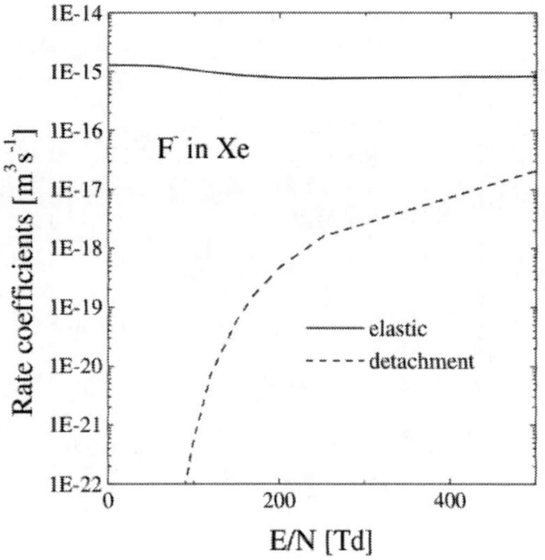

Figure 16. Net rates for elastic scattering and detachment for F^- ions in Xe at T=300K.

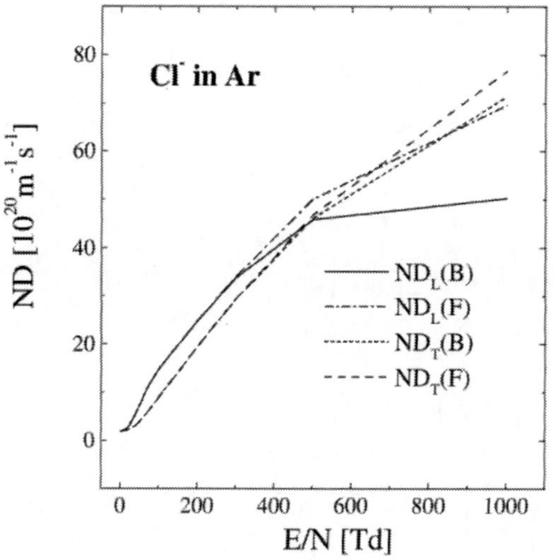

Figure 17. The effect of detachment as non-conservative process on the components of the diffusion tensor of Cl^- ions in Ar. The real space components are denoted by B (bulk) and the velocity space components by F (flux).

The issue of the transport data that may be used in plasma models has been pursued recently [9] as the basis to produce comprehensive plasma models [32]. Fundamental interest in the nature of kinetic phenomena still leads to new insights [33] and in many cases application of inadequate transport data in plasma models may lead to inaccuracies [10] and even failure to represent all relevant physics [34]. In case of negative ions it is important to represent properly the effect of detachment on the transport coefficients as detachment is a nonconservative process.

The cross sections provided here should be employed with the assumption of isotropic scattering and are very accurate in the region covered by the experiments.

5. Acknowledgments

This work has been supported by MNTRS 141025 project and FP6 INCO project 026328.

[1] Lieberman M A and Lichtenberg A J 2005 *Principles of Plasma Discharge and Materials Processing* (Wiley Hoboken, New Jersey).
[2] Stefanovic I, Kovacevic E, Berndt J and Winter J 2003 *New J. Phys.* **5** 39.
[3] Economou D 2007 *Appl. Surf. Sci.* **253** 6672.
[4] Nakano N, Shimura N, Petrovic Z Lj and Makabe T 1994 *Phys. Rev. E* **49** 4455.
[5] Pintassilgo D, Kutasi K and Loureiro J 2007 *Plasma Sources Sci. Technol.* **16** S115.
[6] Matsui J, Nakano N, Petrović Z Lj and Makabe T 2001 *Appl. Phys. Lett.* **78** 883.
[7] Samukawa S, Sakamoto K and Ichiki K 2002 *J. Vac. Sci. Technol.* A **20** 1566.
[8] Petrović Z Lj, Jovanović J V, Stojanović V, Raspopović Z M and Ristivojević Z 2008 *Eur. Phys. J. D* **48** 87.
[9] Petrović Z Lj, Raspopović Z M , Stojanović V D, Jovanović J V, Malović G, Makabe T and de Urquijo J 2007 *Appl. Surf. Sci.* **253** 6619.
[10] Robson R E, White R D and Petrović Z Lj 2005 *Rev. Modern Phys.* **77**(4)1303; Petrović Z Lj, Suvakov M, Nikitović Z, Dujko S, Sasić O, Jovanović J V, Malović G and Stojanović V 2007 *Plasma Sources Sci. Technol.* **16** S1.
[11] Makabe T and Petrović Z Lj 2006 *Plasma Electronics* (New York: Taylor and Francis).
[12] Buchachenko A A, Krems R V, Szczsniak M M, Xiao Y-D, Viehland L A and Chaasfski G 2001 *J. Chem. Phys.* **114** 9919.
[13] Hamaoka F, Yagisawa T, Makabe T 2007 *J. Phys: Conf. Series* **86** 012018; Kitajima T, Takeo Y, Petrović Z Lj and Makabe T 2000 *Appl. Phys. Lett.* **77** 489.
[14] Robson R E 2006 *Introductory Transport Theory for Charged Particles in Gases* (Singapore: World Scientific Publishing).
[15] Ellis H W, McDaniel E W, Albritton D L, Viehland L A, Lin S L and Mason E A 1978 *At. Data Nucl. Data Tables* **22** 179.
[16] Thackston M G, Eisele F L, Pope W M, Ellis H W, Gatland I R and McDaniel E W 1979 *J. Chem. Phys.* **70**, 3996.
[17] Lamm D R, Chelf R D, Twist J R , Holleman F B, Thackston M G, Eisele F L, Pope W M, Gatland I R and McDaniel E W 1983 *J. Chem. Phys.* **79** 1965.
[18] Thackston M G, Eisele F L, Pope W M, Ellis H W, McDaniel E W and Gatland I R 1980 *J. Chem. Phys.* **73** 3183.
[19] Vrhovac S B and Petrović Z Lj 1996 *Phys. Rev E* **53** 4012; Vrhovac S B, Petrović Z Lj, Viehland L A and Santhanam T S 1999 *J. Chem. Phys.* **110** 2423.
[20] Viehland L A and Mason E A 1995 *At. Data Nucl. Data Tables* **60** 37.
[21] Viehland L A and Kirkpatrick C C 1996 *Chem. Phys.* **202** 285.
[22] Huq M S, Doverspike L D, Champion R L and Esaulov V A 1982 *J. Phys. B: At. Mol. Phys.* **15** 951.
[23] Champion R L and Doverspike L D 1976 *Phys. Rev. A* **13** 609.
[24] Smith B T, Edwards III W R, Doverspike L D and Champion R L 1978 *Phys. Rev. A* **18** 945 .
[25] Langevin P 1905 *Ann. Chim. Phys.* **8** 245.
[26] Mason E A and McDaniel E W 1988 *Transport Properties of Ions in Gases* (New York: Wiley-Interscience).
[27] Robson R E 1994 *Aust. J. Phys.* **47** 279.
[28] Robson R E 1976 *J. Phys. B: At. Mol Phys.* **9** L337.
[29] Skullerud H R 1973 *J. Phys. B: At. Mol Phys.* **6** 728.
[30] Ellis H W, Thackston M G, Mc Daniel E W and Mason E A 1984 *At. Data Nucl. Data Tables* **31** 113.
[31] Jovanović J V, Vrhovac S B and Petrović Z Lj 2002 *Eur. Phys. J. D* **21** 335.
[32] Rauf S and Kushner M J 1997 *J. Appl. Phys.* **82** 2805; Yagisawa T, Shimada T and Makabe T 2005 *J. Vac. Sci. Technol. B* **23** 2212.
[33] White R D, Ness K F and Robson R E 2002 *Appl. Surf. Sci.* **192** 26.
[34] Tadokoro M, Hirata H, Nakano N, Petrović Z Lj and Makabe T 1998 *Phys. Rev. E* **57**, R43; Vasenkov A V and Kushner M J 2003 *J. Appl. Phys.* **94**, 5522.

2nd Int. Workshop on Non-equilibrium Processes in Plasmas and Environmental Science IOP Publishing
Journal of Physics: Conference Series **162** (2009) 012005 doi:10.1088/1742-6596/162/1/012005

Nonequilibrium calculations of the role of electron impact in the production of NO and its emissions

Laurence Campbell and Michael J. Brunger

ARC Centre for Antimatter-Matter Studies, School of Chemistry, Physics and Earth Sciences, Flinders University, GPO Box 2100, Adelaide SA 5001, Australia

E-mail: `laurence.campbell@flinders.edu.au`

Abstract. We review our recent work on nonequilibrium modelling of the density of nitric oxide and its infrared emissions in the Earth's upper atmosphere. The aim of these studies was to investigate the contribution of electron impact excitation to the NO density and the sensitivity of this process to the electron impact cross sections. The results are compared with satellite measurements of NO densities in equatorial and auroral high-latitude conditions and with rocket measurements of infrared emissions in auroral conditions. Particular findings are that electron impact excitation of N_2 makes a significant contribution to the NO density at altitudes around 105 km and to auroral infrared emissions for the $(1 \rightarrow 0)$ ground-state emission from NO. The sensitivity of the NO fundamental emissions to various measured and theoretical integral cross sections is investigated and found to be significant.

1. Introduction

Electron impact (photoelectron and auroral) in the Earth's upper atmosphere plays a major part in the production of nitric oxide [1] and in infrared emissions from this species [2]. The main production mechanism is electron impact ionisation and dissociation of atmospheric species, which produces $N(^2D)$ (excited N) that then reacts with O_2 to produce NO. A minor source is electron impact excitation of $N_2[A^3\Sigma_u^+]$, which reacts with O atoms to produce NO and $N(^2D)$ [3]. This $N(^2D)$ reacts with further O_2 to produce even more NO, enhancing the effect of this excitation source. The NO produced by $N(^2D) + O_2$ is vibrationally excited, leading to chemiluminescent infrared emissions. Recently direct electron impact excitation has also been shown (in calculations) to contribute to infrared emission by NO [4].

In order to investigate the importance of the $N_2[A^3\Sigma_u^+]$ source of NO, a model which predicts NO density in the atmosphere was developed [5, 6]. Above 130 km the NO density reaches a chemical equilibrium value that varies with solar insolation and auroral input. NO is transported downwards by molecular diffusion at altitudes above about 110 km and by eddy diffusion below this height. Vertical winds produced by auroral heating may also redistribute NO vertically. As a result of these transport processes the NO density does not reach a chemical equilibrium value at altitudes below 130 km, so a nonequilibrium calculation is required to predict the NO density in the region of its peak at around 110 km.

The model has been applied to predict the NO density profile in the upper atmosphere at the equator, and at 65° N with moderate auroral input [6]. It was run with and without the $N_2[A^3\Sigma_u^+]$ + O reaction and in both the equatorial and high-latitude cases it was found that the contribution of this reaction is significant, and that the predicted densities are closer to the observed values when the $N_2[A^3\Sigma_u^+]$ contribution is included.

The model was then used to calculate the background NO density as part of an investigation of infrared emissions from NO [7]. Auroral electron impact produces such emissions principally by chemiluminescent emission from NO* (vibrationally excited NO) that is produced by the $N + O_2$ reaction [8]. However, measurements [9, 10] of electron impact excitation of the first three vibrational levels of NO made possible calculations of the contribution of electron impact to the infrared emissions. It was found that the contribution is significant relative to chemiluminescence for the $\nu' = 1$ level, while the smaller contribution for the other levels may nevertheless alter the spectrum from that of chemiluminescence in a way that could be exploited in remote sensing [4]. The nonequilibrium model to calculate NO density allowed a detailed calculation [7] of these electron-driven infrared emissions for the particular conditions of a rocket measurement [8]. It allowed both a calculation of the background NO density for the time and place, and of the increase in the density due to the auroral input. While there was a discrepancy between the absolute value of the predicted emissions and the experiment, it was found that the shape of the predicted spectrum of NO fundamental emissions matched the measurements better when direct electron excitation was included. This simulation of the rocket measurements was recently repeated [11] with new measurements [12] and theoretical calculations [13] of electron impact vibrational excitation of NO. In both cases the agreement with the measurements is improved for the $\nu' = 1$ level, particularly using the new theoretical cross sections. However, the new values lead to an overestimation for $\nu' = 2$.

2. Computational methods

Two computational methods are used in this work. A time-step simulation, in which the change in density of each species is calculated for a small time step, is used for slower reactions. For species which reach equilibrium quickly, a statistical equilibrium calculation is used.

In the time-step calculation the incremental change in density $\Delta[z]$ for each species z is calculated at successive small time steps of Δt as $\Delta[z] = R[x][y]\Delta t$, where R is the rate for the reaction $x + y \to z$. To implement this computationally, all density changes in the time step for each species z are summed, so that positive and negative increments cancel before the densities are adjusted for all species at the end of the time step.

Alternatively, for fast reactions, the positive and negative contributions to $\Delta[z]$, $\Delta[z_p]$ and $\Delta[z_n]$, are summed separately. The next value of $[z]$ is calculated by solving the continuity equation, in which the gain $\Delta[z_p]$ is assumed equal to the new value of the density $[z]'$ multiplied by the loss rate in the current interval of $\Delta[z_n]/[z]$, i.e.

$$[z]' = \frac{\Delta[z_p][z]}{\Delta[z_n]} \tag{1}$$

Equation (1) is applied for all species being considered, and iteration of this process continues until the populations of all species reach "statistical equilibrium" [14].

In some cases, such as the determination of the population densities of the excited states of N_2 in the atmosphere, produced by photoelectrons and auroral electrons, there is a wide range of excitation probabilities and radiative transition probabilities. This large range, together with the large number of different pathways, makes a time-step calculation impractical and the statistical equilibrium calculation is used. Equation (1) is applied for all electronic states being considered, and iteration of this process continues until the populations of all states reach equilibrium. The population densities of the excited states of N_2 are then imported into the time-step calculation.

There is a wide range of different rates of chemical reactions in the upper atmosphere. While a time-step calculation at 1-s intervals is suitable for most reactions, those that are too fast

Journal of Physics: Conference Series **162** (2009) 012005 doi:10.1088/1742-6596/162/1/012005

for this interval must be identified and the equilibrium calculation performed. The issue of combining the two approaches has been described previously [15].

3. Atmospheric model

The current atmospheric model has its origin in a study of the role of excitation of the $N_2[A^3\Sigma_u^+]$ state in setting the electron density in the upper atmosphere [5]. Inputs required by the original model from other work were:

- densities of N_2, O_2, and O atoms from the MSISE90 [16] model,
- an established model of photoionisation and photodissociation by sunlight [17], and
- measured NO densities [18, 19].

A flux distribution for auroral electrons was developed [5] by combining various measurements and model results.

The photoionisation and photodissociation produced by sunlight, plus ionisation and dissociation produced by photoelectrons and auroral electrons, are calculated for particular solar and auroral conditions. The ions and electrons produced take part in a series of other reactions, as outlined by Barth [1]. These processes were modelled in a time-step calculation which allows for nonequilibrium conditions, such as solar input varying with time of day and short-duration aurora. Using a time-step simulation also allows the inclusion of molecular diffusion downwards at higher altitudes and eddy diffusion, also downwards, at lower altitudes. Simultaneous calculations of the populations of the excited state and ground state vibrational levels of N_2 allowed the importance of the role of the $N_2[A^3\Sigma_u^+]$ state to be calculated. It was found that the contribution of the $N_2[A^3\Sigma_u^+]$ state to the population levels of the ground state was not as important as the higher electron temperature produced by auroral heating. Nevertheless the model produced a range of values for atmospheric parameters that may be compared with other models or measurements.

The model was then used to calculate profiles of electron density in the atmosphere, using two different sets of reaction rates for the reaction

$$O^+(^4S) + N_2(\nu'') \rightarrow NO^+ + N(^4S) \qquad \text{(R1)}$$

which is critical in setting the electron density due to the high recombination rate of NO^+ with electrons. The predictions in each case were compared with those of the International Reference Ionosphere (IRI) [16]. It was shown [20] that the $\nu'' = 0$ rate of St. Maurice and Torr [21], enhanced for $\nu'' > 0$ by factors determined by Schmeltekopf *et al.* [22], led to predictions of electron densities in better agreement with the IRI than predicted using the reaction rates of Van Zandt and O'Malley [23].

The model was then modified [15] to calculate the NO densities in the upper atmosphere. It was applied to the specific nonequilibrium case of a 4.5-day simulation of sunlight plus moderate auroral activity, with 1 h of strong auroral activity at the end. This simulation predicted a value of the NO density at 106 km that was consistent with the maximum density measured by the SNOE (Student Nitric Oxide Explorer) satellite [24]. While this consistency was not definitive, as the duration of the strong aurora was arbitrarily chosen, it justified further development of the model for the recent work reviewed in the next section.

4. Recent Applications

The model described above was applied to predict atmospheric NO densities [6] in situations where averages of satellite measurements were available. As part of this study it was necessary to include a new reaction rate and consider the effect of vertical winds.

It was then applied to make predictions of NO infrared emissions [7] for the specific conditions of a rocket measurement. This required both the nonequilibrium calculation described above to predict the background NO density for the location, then another nonequilibrium calculation of the increase in NO due to the extra auroral activity at the time of the rocket flight.

Finally the predictions of the rocket measurements were repeated [11] for recent measurements and recent theoretical calculations of the integral cross sections for electron impact vibrational excitation of NO, to investigate the sensitivity of the emissions to the cross sections.

4.1. Nonequilibrium calculations of NO density

Average values of NO density in the upper atmosphere at 11:00 LT have been determined [24] from measurements by the SNOE satellite for several intervals around the spring and autumn equinoxes, for equatorial conditions and for moderate geomagnetic conditions at 65° N geomagnetic latitude.

The model described previously [15] was used, but updated with some recent reaction rates for,

$$N(^2D) + O_2 \rightarrow NO + O \ [25] \quad \text{(R2)}$$

and

$$N(^4S) + O_2 \rightarrow NO + O \ [26]. \quad \text{(R3)}$$

As reported previously [15] molecular and eddy diffusion of NO and $N(^4S)$ were included. Details of the parameters used for these transport processes were detailed later [6], including the extension to the equatorial case.

As the atmospheric model was based on one by Barth, a comparison is made in figure 1 of the current calculations with previous predictions by Barth [1]. The quantity compared is the rate of NO production by the reactions $N(^4S) + O_2$ and $N(^2D) + O_2$. There is substantial similarity, giving credibility to the current model.

Also plotted in figure 1 are the production rates of NO due to the reactions

$$N_2[A^3\Sigma_u^+] + O \rightarrow NO + N(^2D) \quad \text{(R4)}$$

and

$$N_2[X^1\Sigma_g^+](\nu' > 11) + O \rightarrow NO + N\ . \quad \text{(R5)}$$

While these reactions contribute less than one tenth of the N-atom contribution, reaction (R3) is nevertheless significant because it produces excited N atoms, which can then react with O_2 to produce even more NO. The model was run with and without the excited N_2 to quantitatively investigate the effect of reactions (R4) and (R5).

The nonequilibrium nature of the model is illustrated in figure 2. It is run for 101 hours, starting at 6:00. The calculated NO densities are shown (horizontal axis) as a function of altitude at 11:00 on each successive day. While equilibrium is reached quickly above 150 km, it takes two days for the NO density to build up to a near constant value at around 110 km. This is a non-physical example because the NO density does not start at zero. However, it illustrates the time constant of the processes acting. (An early estimate of NO densities by Mitra [27] is included to illustrate the deficiency of an equilibrium calculation for NO density).

Figure 3 shows the relative importance of the contributions of excited N_2 and the enhanced rate of reaction (R2), which is greater than in the earlier modelling. While the full model gives densities in fair agreement with the measurements (except around 120 km), the calculated NO densities get smaller as the model is run progressively without the contributions of $N_2[A^3\Sigma_u^+]$, $N_2[X^1\Sigma_g^+](\nu' > 11)$ and the enhancement of the rate for $N(^2D) + O_2$. The contribution of vibrationally excited N_2 is small, particularly at the lower altitudes, while the contribution of $N_2[A^3\Sigma_u^+]$ is more significant than the enhancement of (R2) in the region of the peak NO density. The enhancement of the $N(^2D)$ rate is relatively more significant at higher altitudes.

Similar results are shown in figure 4 for the auroral region. SNOE measurements are shown for high geomagnetic activity at 65° N geomagnetic latitude and for the highest density measured. The simulation in this case is run for 77 hours, which is a reasonable duration for high geomagnetic activity, starting at 6:00. The model is run for auroral fluxes of 0.5 erg $cm^{-2}s^{-1}$ and 1.0 erg $cm^{-2}s^{-1}$, the latter being the value chosen by Barth [1]. The calculated NO densities are

Journal of Physics: Conference Series **162** (2009) 012005 doi:10.1088/1742-6596/162/1/012005

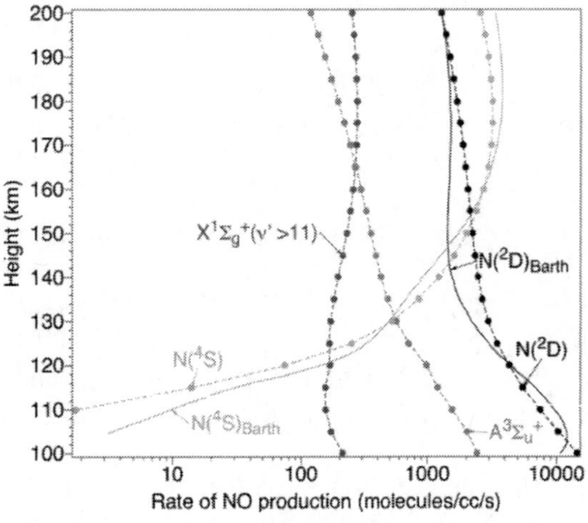

Figure 1. Comparison of current calculations (•) with results of Barth [1](———) for production rates of NO due to reactions involving species as labelled.

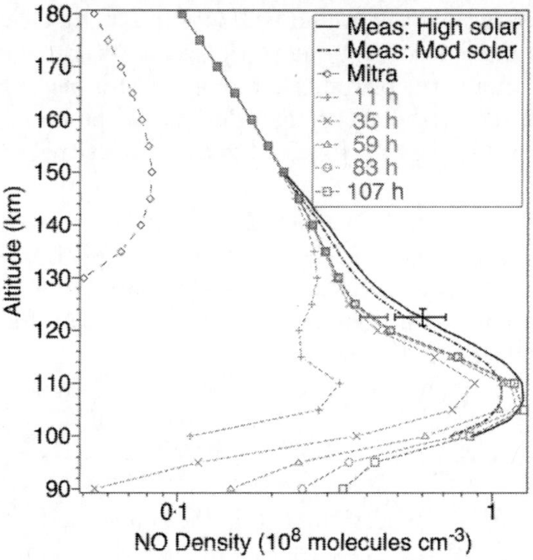

Figure 2. A comparison of the calculated NO density at 11 am on successive days of the simulation with the SNOE measurements for high (———) and moderate (—·—·) solar activity and with an empirical model (◇). Error bars at 122.5 km show the estimated measurement uncertainty and the altitude range for the high solar activity case. The single horizontal error bar shows the estimated uncertainty due to solar input at 122.5 km to the calculated density for 107 hours.

close to the average values for the lower auroral flux, while agreement with the highest density at 106 km is given by using the higher auroral flux. As for the equatorial case, the calculated NO densities are substantially less (by a factor of ~3) in the region of peak density when the $N_2[A^3\Sigma_u^+]$ contribution is excluded.

In figure 4 a simulation of Barth's predictions [1] was made by running the current model (without the excited N_2 contribution and the two updated reaction rates) for the conditions of Barth's calculations. The predicted NO densities agree well with those of Barth, giving credibility to the current model and the main conclusion that excited N_2 must be included in order to produce reasonable agreement with the measurements.

As part of this study the role of vertical winds was investigated, as such winds can be produced by auroral heating. To investigate whether vertical winds can affect the calculated NO densities, a measured wind record [28] at 120 km for auroral conditions was applied in the simulation. This record (shown in figure 5) is for one night, so it is repeated on successive nights and the daytime values are interpolated as shown in figure 5. This gives a predominately downward wind prior to the SNOE measurement time of 11:00. As the wind measurements are only for one height, the vertical wind values at heights above 120 km were found by scaling [7] based on other observations in which vertical winds in auroral conditions were measured at two heights simultaneously [29]. Below 120 km, assuming conservation of mass, the speed is scaled as the reciprocal of the atmospheric density. The speeds were divided by 20, given that the

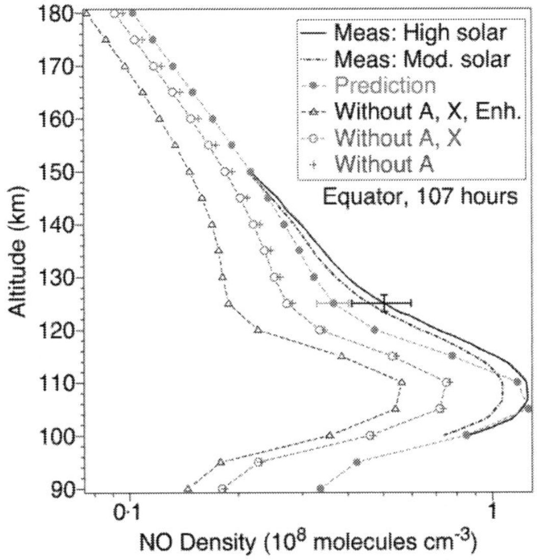

Figure 3. Measured (for high and moderate solar activity) and calculated NO densities for equatorial conditions at 11:00. Calculated densites are for the full model (•) and with progressive removal of the contribution of $N_2[A^3\Sigma_u^+]$ (+), $N_2[X^1\Sigma_g^+](\nu' > 11)$ (∘) and the enhancement of the $N(^2D) + O$ rate (△).

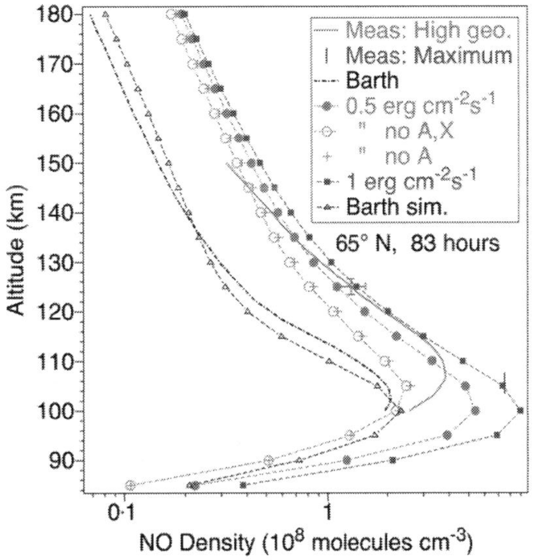

Figure 4. Measurements of NO densities at 11:00 for high geomagnetic activity at 65° N (——) and of the highest density observed (|). Calculated NO densities for 1 erg $cm^{-2}s^{-1}$ auroral input (■) and for 0.5 erg $cm^{-2}s^{-1}$ with the full model(•), then without $N_2[A^3\Sigma_u^+]$ (+) and finally without $N_2[X^1\Sigma_g^+](\nu' > 11)$ (∘). Results of Barth [1] (–·–·–) and a simulation of Barth's calculation (△).

auroral flux of 0.5 erg $cm^{-2}s^{-1}$ that we assume is likely to be much less than that at the time of the wind measurements. Wind is included in the model by calculating, in each time step, the change in density in each height interval due to the amount of NO that is added or removed at the top and bottom of the height interval by the wind.

In figure 6 the calculated NO densities are shown for the application of the "downward" vertical wind just described, and for an "upward" wind case implemented by taking the inverse of all the wind speeds. As expected, the downward wind transfers NO from higher to lower levels while the upward wind transfers it the other way. However, the effect is not so large as to change the overall shape of the density profile with height, and in practice it is to be expected to be smaller due to averaging of the effects of upward and downward vertical winds over the time and space of the SNOE measurements. The addition of the effect of vertical winds illustrates how the nonequilibrium nature of the current model allows the effect of other parameters to be easily included and studied.

4.2. Nonequilibrium calculations of NO infrared emissions

Nonequilibrium calculations are required to predict NO infrared emissions during auroral conditions [15]. The NO density increases with time due to NO production by auroral electron impact. Also the temperature and density of the thermal electrons in the atmosphere increase with auroral input, so the electron-impact induced vibrational excitation of the NO increases [15].

The model described above has been modified to enable prediction of infrared emissions from

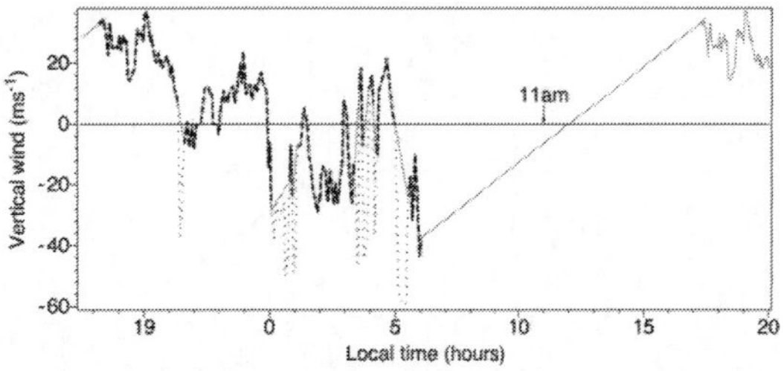

Figure 5. Measurements of vertical neutral wind (positive upward) [28] with low (– – –) and high (· · ·) error and the fit (———) to these measurements used in this work. The time of the NO measurements (11:00 LT) is indicated.

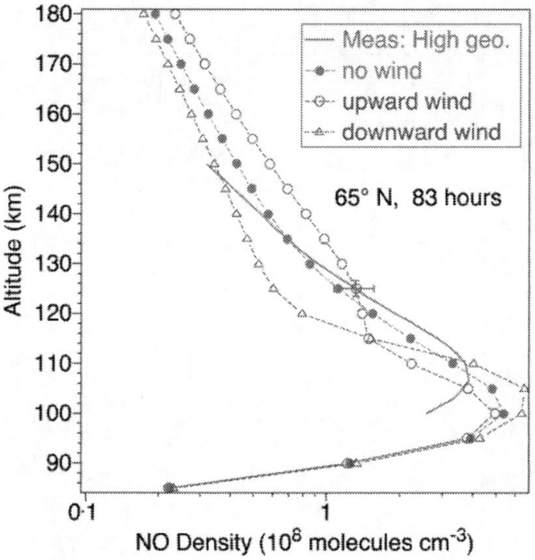

Figure 6. A comparison of calculated NO densities with measurements (———) for the cases of no vertical wind (•), upward wind before the measurement time (○) and downward wind before the measurement time (△).

NO. The previously accepted sources [2] for this emission are nightglow, produced by radiative decay following the collisional excitation:

$$NO(\nu''=0) + O \rightarrow NO(\nu'=1) + O \qquad (R6)$$

and the chemiluminescent reactions,

$$N(^2D) + O_2 \rightarrow NO(\nu') + O \quad (\nu' = 1-12) \qquad (R2')$$

$$N(^4S) + O_2 \rightarrow NO(\nu') + O. \quad (\nu' = 1-9) \qquad (R3')$$

Recent measurements [22] of the vibrational excitation cross sections for NO at low impact energy, i.e.

$$NO(\nu''{=}0) + e^- \rightarrow NO(\nu') + e^- \quad (\nu' = 1,2,3) \qquad (R7)$$

have enabled calculations of the infrared emission due to electron impact to be made. This was first investigated with an equilibrium auroral model [4]. The development of the nonequilibrium atmospheric model has allowed simulations to be performed for specific conditions. In particular, it allowed the calculation of both the background NO density and the increase due to auroral input [7] for the conditions of a particular rocket measurement [8].

Figure 7 shows the calculated NO densities after a 71-h simulation for the time and place of the rocket measurement, again including the calculation with and without the excited N_2 contributions and with the smaller reaction rate for (R2). For reference the SNOE measurements are included.

The model was then run for another 3 minutes with much stronger auroral activity added and the infrared emissions calculated. Due to the speed of the processes involved (radiative decay and collisional quenching of the vibrationally excited NO) a time step of 0.001 s was required. An aurora of strength 5 kR (kRayleighs [30]) was applied, as this was the quoted strength (from ground-based observations of the 557.7 nm line) at the time of the rocket measurements. The calculated fundamental emissions, i.e. $(1 \rightarrow 0)$, $(2 \rightarrow 1)$ ..., are compared in figure 8 with the

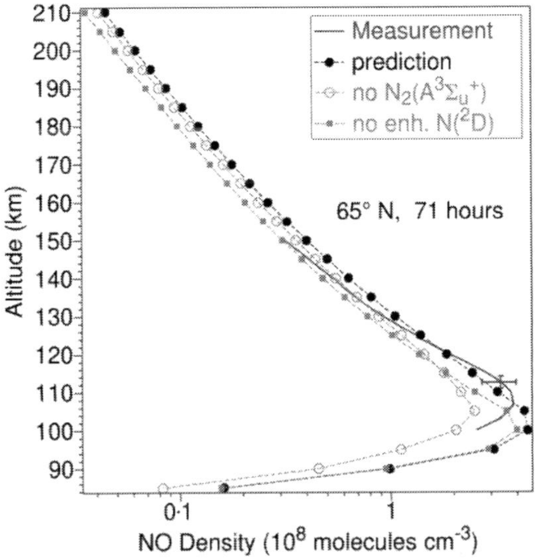

Figure 7. Calculated NO densities at 23:00 hours for the full calculation (•) and for the same calculations without the reaction $N_2(A^3\Sigma_u^+) + O \rightarrow NO + N(^2D)$ (∘) or without the enhancement of the $N(^2D) + O_2$ rate (▪).

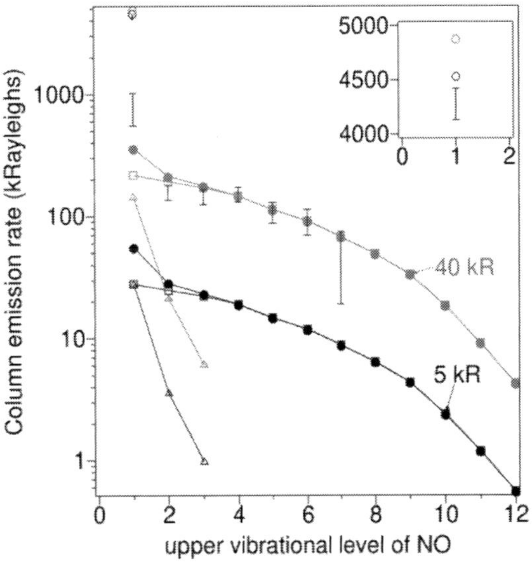

Figure 8. Predicted emission rates for the NO ($\nu' \rightarrow \nu' - 1$) emission due to electron impact (△), chemiluminescence (□), the sum of these two processes (•) and nightglow (∘), and the rates deduced from the measurements (I). The nightglow values are magnified in the inset. Calculated values are given for a 5-kR aurora (lower values) and a 40-kR aurora (higher values).

values of these emissions deduced [8] from the measurements. The $(1 \rightarrow 0)$ emission was divided into nightglow and auroral components, with the much larger one being the nightglow.

The calculated brightness of the nightglow, chemiluminescent emissions and emissions due to electron impact are shown for the current model. There is reasonable agreement for the nightglow value around 4500 kR, indicating that the calculated NO density for the conditions of the measurements is valid. There is a large discrepancy between calculations and the measurements of the auroral emissions. However, the rocket was launched when the aurora had a strength of 50 kR, and when a value of 40 kR is used in the current calculation, it produces agreement within error for all except the $(1 \rightarrow 0)$ emission. So it may be that the quoted reduction of the auroral strength from 50 to 5 kR during the time of the rocket measurements is not appropriate to the region in which the infrared emissions occurred.

The electron impact contribution to the emissions is much smaller than that of the chemiluminescent reactions, except for the $(1 \rightarrow 0)$ emission. The addition of electron impact improves the agreement of the calculated $(1 \rightarrow 0)$ emission with the measured value, and improves the overall agreement for the shape of the spectrum.

4.3. Sensitivity of NO emissions to cross sections

The most recent use of the nonequilibrium model was to investigate the sensitivity of the calculated NO emissions to the cross sections for electron impact vibrational excitation of NO. Figure 9 shows three sets of integral cross sections for electron impact excitation of the $\nu' = 1$ level of NO. In addition to the cross sections [4] used in the "current"' model above, there are excitation functions [12] scaled by 4π and theoretical calculations [13].

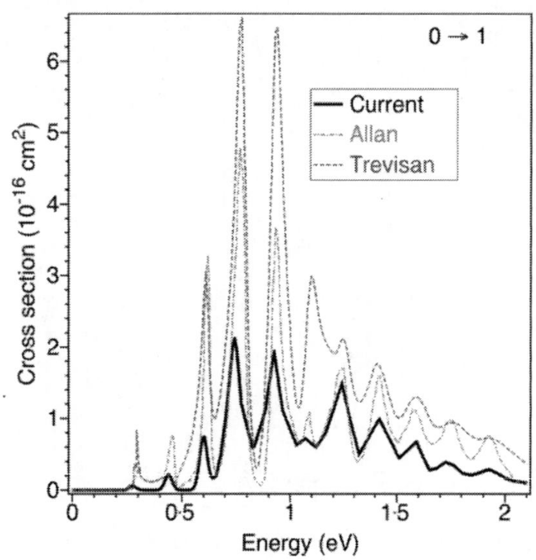

Figure 9. Integral cross sections for electron impact vibrational excitation ($0 \to 1$) of NO from measurements [4] (———), 4π-scaled excitation functions [12] (– · – · –) and theory [13] (- - -).

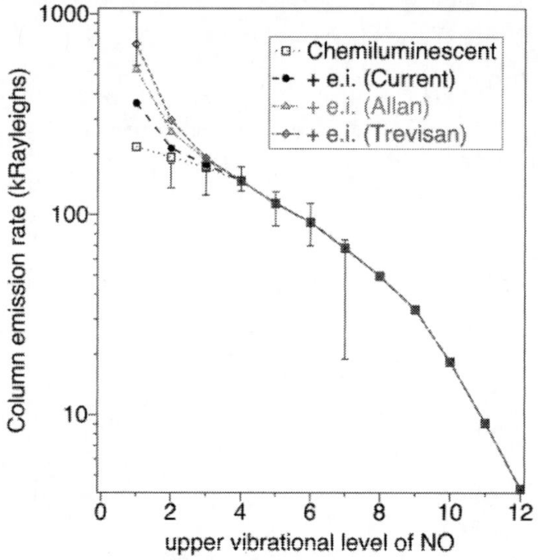

Figure 10. Measured [8] NO fundamental infrared emissions (I) as a function of the vibrational level of NO. Also shown are current predictions for chemiluminescent reactions only (··□··), and for chemiluminescent reactions plus electron impact calculated for the current cross sections (• – – •), 4π-scaled excitation functions [12] (△-·-△) and theoretical calculations [13] (◇ - - - ◇).

The comparison of calculations (for a 40-kR aurora) with the rocket measurements is shown in figure 10. It can be seen that use of the scaled excitation functions improves the agreement for the ($1 \to 0$) emission, while the use of the theoretical cross sections brings the calculated ($1 \to 0$) emission into agreement with the measurements. However, for the ($2 \to 1$) emission, the use of the scaled excitation functions and of the theoretical cross sections makes the calculated brightness too large [as compared to the higher fundamental lines, such as ($3 \to 2$]. A possible explanation is that the separation of the nightglow and auroral components [8] of the ($1 \to 0$) emission was done incorrectly (possibly because the role of electron impact was not considered). Alternatively the relative rates of the chemiluminescent reactions are not correct, or the strength of the $\nu' = 2$ excitation is overestimated in the newer cross sections. The latter might be resolved by definitive low-energy cross-section measurements.

5. Conclusions

A nonequilibrium atmospheric model has been shown to predict NO densities and NO infrared emissions in good agreement with other models and with particular measurements. In each case the agreement with measurements is substantially improved when electron impact excitation is considered. At the height of the peak NO density around 110 km, inclusion of the $N_2[A^3\Sigma_u^+]$ + O reaction increases the predicted NO density by a factor of ~2 at the equator and a factor of ~3 at high latitudes. An additional nonequilibrium calculation of processes driven by a limited-duration aurora gives predictions of NO infrared emissions in good agreement with rocket measurements of the spectral shape, except for the ($1 \to 0$) emission. Addition of electron impact vibrational excitation of NO increases the ($1 \to 0$) emission, thus improving the agreement of

Journal of Physics: Conference Series **162** (2009) 012005 doi:10.1088/1742-6596/162/1/012005

the shape of the spectrum with the measurements. (There is an unexplained discrepancy for the absolute values.) Finally the calculations are repeated for more recent scaled excitation functions and theoretical integral cross sections for electron impact vibrational excitation of NO. These give, respectively, better and within-error agreement with the peak in the measured spectrum for the $(1 \rightarrow 0)$ emission, although both produce an increased discrepancy for the $(2 \rightarrow 1)$ emission. Improved experimental measurements of the NO electron impact vibrational cross sections are needed to investigate this discrepancy.

Acknowledgments
We thank the Australian Research Council for financial support of this research and the International Science Linkages programme of the Department of Innovation, Industry, Science and Research for funding international collaboration which led to the investigation of the sensitivity to NO electron impact cross sections.

References

[1] Barth C 1992 *Planet. Space Sci.* **40** 315–36
[2] Caledonia G E and Kennealy J P 1982 *Planet. Space Sci.* **30** 1043–56
[3] Swider W 1976 *Geophys. Res. Lett.* **3** 335–7
[4] Campbell L, Brunger M J, Petrovic Z Lj, Jelisavcic M, Panajotovic R and Buckman S J 2004 *Geophys. Res. Lett.* **90** L10103-1–4
[5] Campbell L, Cartwright D C, Brunger M J and Teubner P J O 2006 *J. Geophys. Res.* **111** A09317-1–19
[6] Campbell L, Cartwright D C and Brunger M J 2007 *J. Geophys. Res.* **112** A08303-1–10
[7] Campbell L and Brunger M J 2007 *Geophys. Res. Lett.* **34** L22102-1–5
[8] Espy P J, Harris C R, Steed A J, Ulwick J C, Haycock R H and Straka R 1988 *Planet. Space Sci.* **36** 543–51
[9] Josić L, Wróblewski T, Petrović Z Lj, Mechlińska-Drewko J and Karwasz G P 2001 *Chem. Phys. Lett.* **350** 318–24
[10] Jelisavcic M, Panajotovic R and Buckman S J 2003 *Phys. Rev. Lett* **90** 203201-1–4
[11] Campbell L, Brunger M J and Allan M 2008 *J. Phys. Conf. Ser.* **115** 012003-1–12
[12] Allan M 2005 *J. Phys.* B **38** 603–14
[13] Trevisan C S, Houfek K, Zhang Z, Orel A E, McCurdy C W and Rescigno T N 2005 *Phys. Rev.* A **71** 052714-1–12
[14] Cartwright D C 1978 *J. Geophys. Res.* **83** 517–31
[15] Campbell L and Brunger M J 2007 *J. Phys. Conf. Ser.* **71** 012011-1–10
[16] Space Physics Data Facility, 2002 *http://modelweb.gsfc.nasa.gov/models/*
[17] Richards P G, Fennelly J A and Torr D G 1994 *J. Geophys. Res.* **99** 8981–92
[18] Barth C A and Solomon S C 1999 *Geophys. Res. Lett.* **26** 1259–62
[19] SNOE web site 2002 *http://lasp.colorado.edu/snoe/descr/science.html*
[20] Campbell L, Brunger M J, Cartwright D C and Bolorizadeh M A 2006 *AIP Conf. Proc.* **876** 41–50
[21] St-Maurice J P and Torr D G 1978 *J. Geophys. Res.* **83** 969-77
[22] Schmeltekopf A L, Ferguson E E and Fehsenfeld F C 1968 *J. Chem. Phys.* **48** 2966-73
[23] Van Zandt T E and O'Malley T F 1973 *J. Geophys. Res.* **78** 6818-20
[24] Barth C A, Mankoff K D, Bailey S M and Solomon S C 2003 *J. Geophys. Res.* **108** 1027–37
[25] Duff, J W, Dothe H and Sharma R D 2005 *Geophys. Res. Lett.* **32** L17108-1–4
[26] Pintassilgo C D, Loureiro J and Guerra V 2005 *J. Phys D Appl. Phys.* **38** 417-30
[27] Mitra A P 1968 *J. Atmos. Terr. Phys.* **30** 1055-114
[28] Ishii M 2005 *Adv. Polar Upper Atmos. Res.* **19** 63–70
[29] Price G D, Smith R W and Hernandez G 1995 *J. Atmos. Terr. Phys.* **57** 543–631
[30] Hunten D M, Roach F E and Chamberlain J W 1956 *J. Atmos. Terr. Phys.* **8** 345–6

2nd Int. Workshop on Non-equilibrium Processes in Plasmas and Environmental Science IOP Publishing
Journal of Physics: Conference Series **162** (2009) 012006 doi:10.1088/1742-6596/162/1/012006

Recent developments on *PLASMAKIN*– a software package to model the kinetics in gas discharges

N R Pinhão

ITN - Nuclear and Technological Institute, Estrada Nacional 10, 2685 Sacavém Portugal

E-mail: npinhao@itn.pt

Abstract. *PLASMAKIN* is a user-friendly software package to handle physical and chemical data used in plasma physics modeling and to compute the production and destruction terms in fluid models equations. These terms account for the particle or energy production and loss rates due to gas-phase and gas-surface reactions. The package has been restructured and expanded to (a) allow the simulation of atomic emission spectra taking into account line broadening processes and radiation trapping; (b) include a library to compute the electron kinetics; (c) include a database of species properties and reactions and, (d) include a Python interface to allow access from scripts and integration with other scientific software tools.

1. Introduction

The modeling of non-equilibrium plasmas through fluid equations requires the computation of terms expressing, for the several species involved in the discharge, the changes in density, momentum or energy[1] due to collisional or radiative processes. These terms represent the local 'chemistry' and are, in most of the cases, independent of the transport model or algorithm chosen for the solution of the fluid model equations. In most cases the evaluation of these terms is straightforward. However as the number of species and reactions involved can be very high, a good data management facility is required.

One common approach for the computation of these terms is to write down the equations for each species directly in the program. However this has the disadvantage of linking the program to a specific discharge model, gas composition and set of reactions. The risk of making trivial but probably time consuming errors is also high. With this approach, the 'chemistry modeller' and the 'programmer' have to be the same person.

A better alternative, followed in several commercial [1] and non-commercial packages [2, 3] for chemical kinetics or plasma modeling, is to keep the information on species and reaction in an external file. This file can then be pre-processed by a parser to automatically generate a program code, to be further compiled with a user's main program. With this approach the pre-processing and compilation can be done in an automatic way and the task of selecting the reactions and writing the datafile is separated from coding. However the final program is still specific for a given model and conditions. Any change in reaction values requires a new pre-processing/compiling cycle.

[1] usually the energy conservation equation is only solved for electrons.

A further improvement can be obtained if all processing of species and reactions is handled by a library called from a user's program. This library can provide the information needed by the main program, e.g. species production and destruction terms for continuity equations without requiring any customization or compilation. The properties and processing of species and reactions are hidden from the calling program and the user's program only has to be compiled and linked with the library once. Any change on gas composition or reactions is only reflected on the data input file. Besides the same program can be used for a different gas mixture provided the type of discharge is the same.

This approach has driven the development of the first version of *PLASMAKIN* [4] as a chemical kinetics library. However the study of gas discharge kinetics is not limited to the study of the chemical kinetics.

Another frequent requirement is to obtain electron rate and transport coefficients from the solution of the electron Boltzmann equation (eBE) for a given set of conditions and cross-section data. A common practice is to obtain the required data in tabulated form as a function of the reduced electric field, E/N, or mean electron energy, ε_e. However the solution of the electron Boltzmann equation is a demanding problem requiring high specialization. Fortunately there are also several eBE solvers available either commercially [5] or non-commercially [6]. Both of those codes rely on the classical two-term expansion for the angular dependency of the electron distribution function. The results obtained with this approximation are adequate for most discharges. However it is well known that this approximation is insufficient in high E/N or when the cross sections for inelastic processes become comparable with those for elastic processes [7] and incapable of predicting the values of transport coefficients with the same precision as that achieved in electron swarm experiments.

Finally, a model of a plasma discharge needs to be validated against experimental results. The usual output of a discharge model are the values of species density in space and in time, a voltage or current signal and other diagnostics such as the electron temperature or the relative contribution of each reaction for each species. These are very useful information to understand what happens in the discharge and how it happens. However except for the voltage or current signal, it is not always easy to get experimental access to the quantities obtainable by a model. Often experimental information on values predicted by a model is obtainable indirectly through, e.g. the analysis of the UV-VIS emission of the plasma. The optical spectrum can give information on the density of species, the plasma and electron temperatures or the pressure. It seems reasonable to expect that discharge models would, in turn, also compute the emission spectrum. Besides, a good amount of the information needed to simulate the spectrum is already present in the models: species densities, temperature profiles and emission coefficients. However very few models offer this possibility and among the reasons are the complexity of computing line profiles when radiation trapping is present, the need to consider a large number of levels to obtain realistic spectra, and the treatment of molecular spectra. However a few programs are available through the CPC program library to compute the effect of radiation trapping [8, 9, 10] likewise, a program to simulate the emission spectra of some diatomic molecules is freely available [11].

The present communication reports the recent developments on the OpenSource package *PLASMAKIN* in order to address a few of these concerns.

PLASMAKIN has grown from a chemical kinetics library to a collection of libraries and tools for plasma modeling. It presently includes a database of species and reactions, libraries to compute the chemical kinetics or the electron kinetics, the simulation of the emission spectra of atomic gases and a binding module to allow access of the libraries' routines from Python scripts or command line.

The next chapter introduces the overall structure of the package. The new capacities added to *PLASMAKIN* are detailed in chapters 3 through 6. Finally, we draw some conclusions.

2. Package Structure

The package has been split into several Fortran modules allowing a better organization of the code and facilitating future development and introduction of new facilities. These modules perform the following tasks:

(i) Set data structures and common routines for all *PLASMAKIN* modules (`pkCommon`);
(ii) read and process data either from text datafiles or a database (`pkDatasql`);
(iii) define methods for inquiring and setting species and reaction properties (`pkBase`);
(iv) compute the chemical kinetics, the power losses in collisions and the atomic emission spectrum (`pkChemicalKin`); and
(v) compute the electron kinetics: the electron velocity distribution function (evdf), transport parameters and electron collision rate coefficients (`pkElectronKin`).

The relationship between these modules is shown in figure 1.

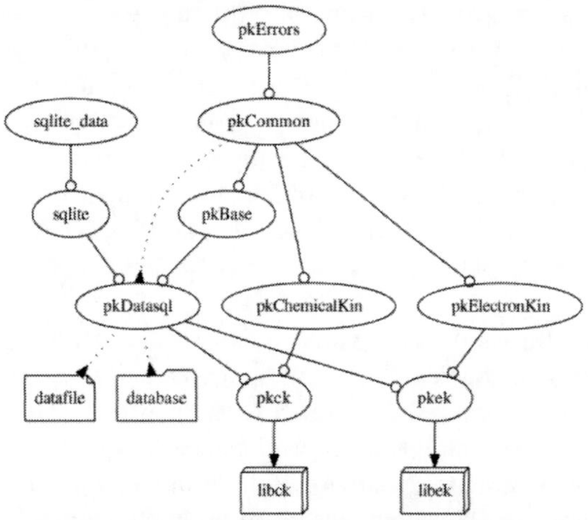

Figure 1. Structure of modules, data sources and libraries in the *PLASMAKIN* package. (Not shown in the figure, the package also includes a Python module to allow access of routines of both libraries from Python scripts or command line.)

The compilation of these modules produces two libraries (`libck` and `libek`) responsible for dealing with the chemical kinetics and the electron kinetics, respectively.

The compilation of the libraries is managed by the GNU `autotools` system that takes care of checking the necessary requirements and adapt the compilation options to the capacities of the local operating system. To obtain a working installation, only three instructions are required: *configure*, *make* and *make install* (as root). To check if everything is well installed the instruction *make check* runs several tests and compares the output with preset values.

The package also includes a sample database of species and reactions and a Python module providing bindings for the libraries. The installation of these tools is optional and requires support to *sqlite* [2] and a Python interpreter, respectively.

3. Data types and data input

3.1. Species in Local Thermodynamic Equilibrium

The type of species supported in *PLASMAKIN* has been extended to include levels in local thermodynamic equilibrium (LTE) or partial LTE. The population of these levels, n_k, is defined relative to a reference level n_o by a Boltzmann factor:

$$n_m = n_o \frac{g_m}{g_o} \exp\left(-\frac{\varepsilon_m - \varepsilon_o}{k_B T_e}\right) \tag{1}$$

[2] *sqlite* is a public domain software library that implements a SQL engine that does not require a server and accesses database files on a local disk. It is widely used and can be found in all operating systems.

where g and ε are the statistical weight and energy of the levels, and respectively, T_e the electron temperature.

The LTE levels don't need to be tracked in a program using *PLASMAKIN* but contribute to other levels through radiative decay and are taken into account in the simulation of emission spectra.

The population of LTE levels is computed by the library from T_e and the density of the reference level. If the later is a cascade level (e.g. produced by electron excitation and rapidly decaying to a lower excited level), the density n_o in equation (1) is replaced by

$$n_o = \frac{\sum_{ij} k_{ij}^o n_i n_j}{\sum_m A_{om}} \tag{2}$$

where the sums include all reactions producing level o and all radiative transitions from level o, respectively. If any of the species involved in (2) is also a cascade level, its density is determined again from (2).

3.2. Database

The way data is input in *PLASMAKIN* has been extended in order to reduce the work in preparing the datafile.

Until now all the information on species and reactions was written in an text file. This method of input it is still valid but from now on is also possible to keep most of the data in a database and limit the use of a text datafile to the following: the indication of the database, the database tables used, to impose restrictions on data and to include additional species or reactions missing in the database. With this approach, it is possible for example to keep well established results in the database and use the text datafile only to test the influence of new data on a discharge model.

4. Simulation of atomic emission spectra

The spectral radiation intensity, I_ν, emitted by a plasma and observed at a boundary x_0, is given by

$$I_\nu(x_0) = \int_0^{x_0} \epsilon_\nu(x) \exp\left[-\tau(\nu, x, x_0)\right] dx \tag{3}$$

where $(0, x_0)$ are the boundaries of the plasma along the direction of observation, $\epsilon_\nu(x)$ is the spectral emission coefficient of the plasma along the direction of observation and $\tau(\nu, x, x_0)$ is the optical path between x and x_0,

$$\tau(\nu, x, x_0) = \int_x^{x_0} k(\nu, x')dx' \tag{4}$$

and $k(\nu, x)$ the spectral absorption coefficient. In (3) we neglect any contribution from radiation outside the plasma.

In order to solve equation (3) and simulate the emission spectra we have considered the following assumptions:

(i) We neglected the stimulated emission in the plasma;

(ii) we assume that the emission and absorption line profiles are the same, $P_e(\nu) = P_a(\nu) \equiv P(\nu)$;

(iii) we assume complete frequency redistribution to compute the line profiles;

(iv) we limit ourself to two discharge geometries: plane parallel and cylindrical. In the last case the direction of observation is perpendicular to the cylinder axis and crosses the cylinder at a height Y from the axis. Additionally, in this case we assume that the discharge has radial symmetry; and

(v) we presently limit the simulation of spectra to atomic emission lines.

4.1. Optically thin lines

With the above assumptions, and considering for now that $\tau \ll 1$, equation (3) can be written

$$I_\nu(R) = \zeta \frac{h\nu}{4\pi} \sum A_{ij} \int_Y^R n_i(r) P_{ij}(\nu, r) \eta_Y(r) dr \tag{5}$$

where the sum is on all radiative transitions between levels i and j, R is the cylinder radius or the outer boundary and $\eta_Y(r)$ a geometric factor. For plane parallel geometry $\zeta = 1, Y = 0, \eta_0(r) = 1$ while for cylindrical geometry $\zeta = 2$, $\eta_Y(r) = r/\sqrt{r^2 - Y^2}$. The factor $\zeta = 2$ for cylindrical geometry accounts for photons emitted along the line of sight from points with the same value of r but symmetrically positioned regarding the plane perpendicular to the line of sight and passing by the axis of the cylinder.

The normalized line profile, $P(\nu, r)$ depends, apart from the gas temperature and the atom and electron concentrations, upon the constants that characterize the interaction between the radiating atom and the neighboring particles, and upon the broadening mechanism.

Three line profiles are of importance:

(i) A Gaussian distribution due to Doppler broadening of the lines and given by

$$P_D(x) = \frac{1}{\pi^{1/2}} \exp\left(-x^2\right), \; x = \frac{\nu - \nu_0}{W_D} \tag{6}$$

with ν_o the line central frequency and $W_D = \frac{\nu_0}{c}\sqrt{\frac{2k_BT}{m}}$, the Doppler width.

(ii) A Lorentz distribution from natural broadening or pressure broadening in the impact approximation,

$$P_L(x) = \frac{1}{\pi} \frac{1}{1 + x^2}, \; x = \frac{\nu - \nu_0 - \Delta}{W_L} \tag{7}$$

with Δ a line shift and W_L a Lorentz width characteristic for each type of pressure broadening process. Three types of broadening processes need to be considered:

atom–impact resulting in a van der Walls interaction from a Lennard–Jones potential with a characteristic constant C_6;

electron– and ion–impact resulting in Stark broadening, calculated from the Sahal-Brechot theory [12];

resonance broadening resulting from resonance transfer of excitation from collisions of two identical atoms, one of which is excited.

Expressions for the line width and line shift for each type of broadening can be found in [13, 14] and are summarized in Table 1.

(iii) A line profile from pressure broadening computed in the quasi–static approximation and valid for the far wings,

$$P_I(\nu) = \frac{3}{k\Delta\nu} \left(\frac{\Delta\nu}{\nu - \nu_0}\right)^{\frac{3+k}{k}} \exp\left[-\left(\frac{\Delta\nu}{\nu - \nu_0}\right)^{\frac{3}{k}}\right] \tag{8}$$

with $\Delta\nu = C_k r_0^{-k}$ and $r_0 = \left(\frac{4}{3}\pi n_0\right)^{-1/3}$ equal to the mean distance between particles. For Stark effect $k = 4$ and for Van der Waals interaction $k = 6$.

Table 1. Line width and line shift for different types of pressure broadening in the impact approximation.

	Type of broadening		
	e- and ion-impact	Atom-impact	Resonance
$2W_L$	$(C_eT^{\gamma_e} + C_iT^{\gamma_i})\,n_e$	$8.08C_6^{2/5}\bar{v}^{3/5}n_0$	$K_r\frac{\pi e^2}{m_e\nu_0}fn_0$
Δ	$\left(B_eT^{\beta_e} + B_iT^{\beta_i}\right)n_e$	$0.7W_L$	0

The actual line profile results from the convolution of the line profiles for all broadening mechanism. Stormberg [15] and Damelincourt *et. al.* [14] have shown that the convolution of the different profiles can be approximated by the sum of a Voigt profile and a correction term from the quasi–static profile, $P(x) = P_V(x) + C(x)$ with,

$$P_V(x) = \frac{1}{\pi^{1/2}}H(a,x), \quad H(a,x) = \frac{a}{\pi}\int_{-\infty}^{+\infty}\frac{\exp(-y^2)}{a^2+(x-y)^2}dy \tag{9}$$

where x is the normalized Doppler frequency in equation (6) and $a = W_L/W_D$. Here W_L is the sum of the Lorentz widths for all broadening processes. The Voigt integral is integrated using the polynomial approximation of Humlicek [16] which gives an accuracy of 10^{-4}.

In actual measurements the spectrum obtained depends on the instrument resolution, δ. To take this into account, the spectrum is computed in intervals of width δ where equation (3) is integrated with a sufficiently small resolution.

$$\bar{I}_k = \frac{1}{\delta}\int_{\nu_k}^{\nu_k+\delta} I_R(\nu)d\nu, \quad k = 1,\ldots,N_\lambda \tag{10}$$

To obtain the spectrum the user must call a routine supplying a wavelength range, the instrument resolution, δ, the grid points, $\vec{X}$, along the line of sight or the radius, respectively for plane parallel or cylindrical geometries, the density of species along $\vec{X}$, and as options, the constants listed in table 1 for pressure broadening processes for each relevant species, gas and electron temperature profiles, $T_g(\vec{X})$, $T_e(\vec{X})$, respectively, and Y.

Figures 2 – 4 show the result of three tests on the routine.

Figures 2 and 3 show the simulation of an emission spectra from neon in the range 550 – 900 nm, with the population of levels in LTE with an electron temperature of 5 eV and obtained with two different instrument resolutions. All transitions from the $2p^22p^53s$, $2p^22p^53p$, $2p^22p^54s$, and $2p^22p^53d$ configurations have been included. A comparison of figure 2 with a Saha-LTE spectrum from the NIST Atomic Spectra Database [17], obtained in the same conditions, shows a very good agreement with just a few lines from transitions involving higher configurations missing in figure 2.

In another example using the same neon data in LTE, we have considered an electron temperature radial profile given by a Bessel function, $T_e(r) = T_e(0)J_0(j_{0,1}\cdot r/R)$ where $j_{0,1}$ is the first root of J_0. Figure 4 shows the ratio between the normalized spectra obtained with $T_e(0) = 10\,\text{eV}$ and $T_e(0) = 5\,\text{eV}$, respectively.

4.2. Optically thick lines

In this case, for a plane–parallel geometry, equation (3) is written

$$I_\nu(x_0) = \frac{h\nu}{4\pi}\sum A_{ij}\int_0^{x_0} n_i(x)P_{ij}(\nu,x)\Phi_{j,0}(\nu;x,x_0)dx \tag{11}$$

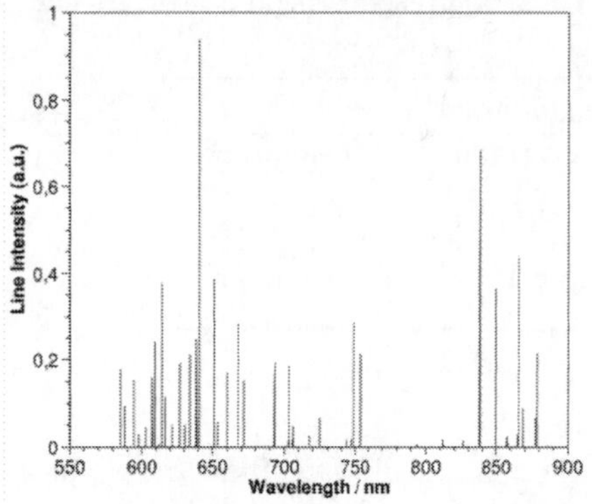

Figure 2. Simulated neon emission spectrum in LTE conditions with $T_e = 5\,\mathrm{eV}$ and an instrument resolution of $\delta = 0.025\,\mathrm{nm}$.

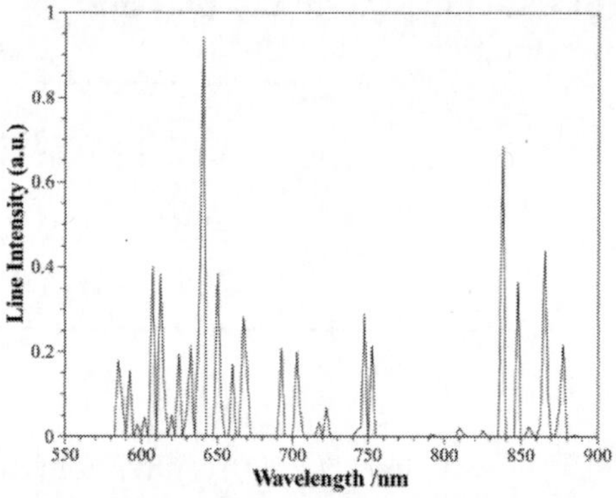

Figure 3. Same conditions as in Figure 2, with an instrument resolution of $\delta = 2.5\,\mathrm{nm}$.

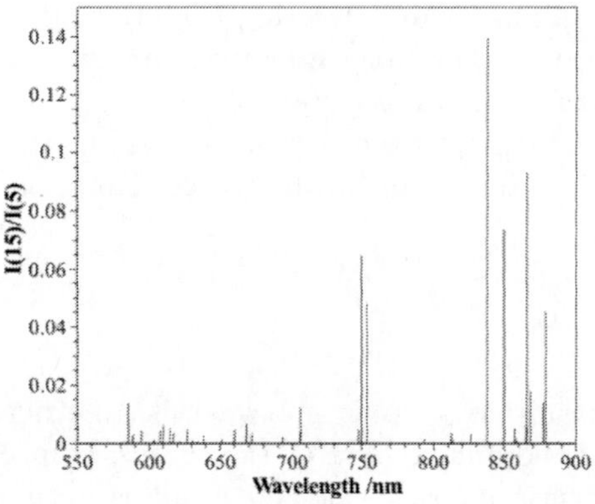

Figure 4. Ratio between the normalized line intensities of neon emission spectra in LTE conditions with a Bessel radial electron temperature profile with, $T_e(0) = 10, \mathrm{eV}$ and $T_e(0) = 5\,\mathrm{eV}$, respectively. $\delta = 0.025\,\mathrm{nm}$.

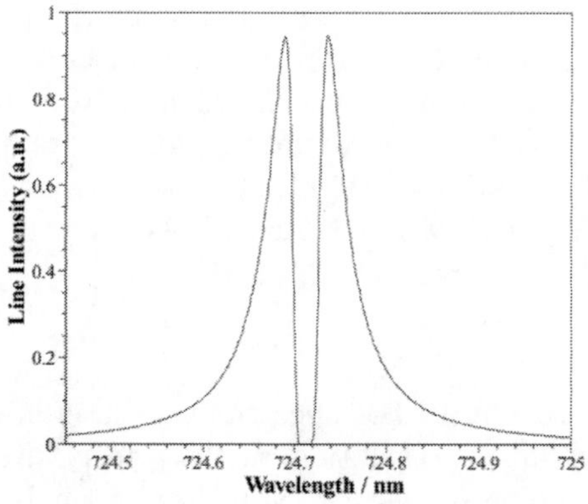

Figure 5. Simulation of a self-reversed line at $p = 1\,\mathrm{atm}$ and a Voigt parameter of $a_V = 0.2$.

with $\Phi_{j,Y}$ the function,

$$\Phi_{j,Y}(\nu; a, b) = \exp\left[-k_{j,0}\int_a^b n_j(x')P_{ji}(\nu, x')\eta_Y(x')dx'\right] \tag{12}$$

where $k_{j,0} = (c^2/8\pi\nu_0^2)(g_j/g_i)A_{ij}$ and the remaining symbols defined as in equation (5).

For cylindrical geometry, we obtain

$$I_\nu(R) = \frac{h\nu}{4\pi}\sum A_{ij}\int_Y^R n_i(r)P_{ij}(\nu, r)\Psi_{j,Y}(\nu; Y, r, R)\eta_Y(r)dr \tag{13}$$

with

$$\Psi_{j,Y}(\nu; Y, r, R) = \Phi_{j,Y}(\nu; Y, r)\Phi_{j,Y}(\nu; Y, R) + \Phi_{j,Y}(\nu; r, R) \tag{14}$$

The function Ψ is equivalent to ζ in equation (5) but takes into account the absorption along the line of sight. The first term accounts for absorption of photons emitted beyond the plane defined by the axis of the cylinder and perpendicular to the line of sight and the last term for photons emitted in front of this plane.

Taking into account that $\Phi_{j,Y}(\nu;Y,R) = \Phi_{j,Y}(\nu;Y,r)\Phi_{j,Y}(\nu;r,R)$, we can write

$$\Psi_{j,Y}(\nu;Y,r,R) = \frac{\Phi_{j,Y}(\nu;Y,R)^2}{\Phi_{j,Y}(\nu;r,R)} + \Phi_{j,Y}(\nu;r,R). \tag{15}$$

As $\Phi_{j,Y}(\nu;r,R) = \Phi_{j,Y}(\nu;r,r+\delta r)\Phi_{j,Y}(\nu;r+\delta r,R)$ and $\Phi_{j,Y}(\nu;R,R) = 1$, the evaluation of Ψ requires the computation of $\Phi_{j,Y}(\nu;r,r+\delta r)$ in each grid interval, starting from $r = R$ and moving backwards to $r = Y$.

Figure 5 shows a line profile obtained for an optically thick line. In this case the broadening values were chosen to evidence self–reversal.

5. The electron kinetics library

A collection of routines to study the electron kinetics under well defined assumptions and compute the electron velocity distribution function, *evdf*, and the transport and rate coefficients for the electrons are now included in the package. The assumptions used apply to a discharge in a constant electric field with or without grown of the electron density.

The main purpose of this library is to obtain values of the transport and rate coefficients for the electrons as a function of E/N or the electron mean energy that can be used on fluid models.

The behaviour of the electrons in a discharge is described by the electron Boltzmann equation

$$\frac{\partial f}{\partial t} + \vec{v}\nabla_r f + \vec{a}\nabla_v f = C(f)\,, \tag{16}$$

for the *evdf*, $f(\vec{r},\vec{v},t)$, where $\vec{r}$, $\vec{v}$ and t are the position, velocity and time, respectively, $\vec{a}$ denotes the acceleration produced by the external field and $C(f)$ is the rate of exchange due to collisions.

The above equation is far too complex to have a general solution and for the time been the algorithm implemented in the library is limited to spatially uniform and constant electric fields. In this case we have $\vec{a} = -(e/m)\vec{E}$.

We further consider that the electrons are moving in the hydrodynamic regime, meaning that although the electron density grows exponentially, the distribution in velocity space and the transport parameters have reached equilibrium values, independent of their initial value. Under these conditions the time derivative term in equation (16) can be replaced by a rate term,

$$\frac{\partial f}{\partial t} = \Omega f \tag{17}$$

where Ω is the macroscopic effective ionization frequency.

The theory of the hydrodynamic regime is well established [18, 19] and only a very brief outline will be given here.

To allow for the influence of electron density gradients in the *evdf*, $f(\vec{r},\vec{v},t)$ is represented as,

$$f(\vec{r},\vec{v},t) = \sum_{k=0}^{\infty} F^{[k]}(\vec{v}) \otimes^k (-\nabla)^k n(\vec{r},t)\ , \tag{18}$$

where $F^{[k]}(\vec{v})$ are velocity-dependent tensors of rank k, $\otimes^k$ is the k-fold inner-product operation and $(-\nabla)^k$ is the k-fold outer product of the gradient operator with itself. Obviously, for a discharge with a spatially homogeneous electron density, equation (18) reduces to the usual representation, $f(\vec{v},t) = F(\vec{v})n(t)$.

Inserting (17) and (18) into Boltzmann equation (16), a hierarchy of equations is obtained allowing to solve for the functions $F^{[k]}$ and transport coefficients $\Omega^{(k)}$. The first three of these coefficients are the macroscopic effective ionization frequency, Ω, the drift velocity, $\vec{W}$ and the diffusion tensor, $\hat{D}$.

The transport parameters measured in time–resolved swarm experiments with $\vec{E}$ along the z–axis, can be written as

$$\begin{aligned}
W &= \int v_z F^{[0]} d\vec{v} + \int \nu_{\mathrm{eff}} F_z^{[1]} d\vec{v} \\
D_L &= \int \tfrac{v_z - W}{\nu+\Omega} v_z F^{[0]} d\vec{v} - \vec{a}\cdot \int \tfrac{v_z}{\nu+\Omega} \nabla_v F_z^{[1]} d\vec{v} + \int \nu_{\mathrm{eff}} F_{zz}^{[2]} d\vec{v} \\
D_T &= \tfrac{1}{2}\left\{ \int \tfrac{v_T^2}{\nu+\Omega} F^{[0]} d\vec{v} - \vec{a}\cdot \int \tfrac{v_z}{\nu+\Omega} \nabla_v F_T^{[1]} d\vec{v} + \int \nu_{\mathrm{eff}} F_T^{[2]} d\vec{v} \right\}
\end{aligned} \tag{19}$$

where D_L and D_T are the longitudinal and transverse components of the diffusion tensor, $\nu_{\mathrm{eff}}(v) = \nu_{\mathrm{ion}}(v) - \nu_{\mathrm{att}}(v)$ is the net ionization frequency, $\nu(v) = \nu_{\mathrm{m}}(v) + \nu_{\mathrm{ex}}(v) + \nu_{\mathrm{eff}}(v)$ is the total collision frequency, and the indices z and T respectively, indicate the longitudinal and transverse components of the velocity $\vec{v}$ or of functions $F^{[k]}$.

The drift velocity W is the center of mass velocity for the electron swarm and, if non-conservative reactions are included, is different from the mean velocity, $\langle v \rangle$, given by the first term on the right side.

The method used to obtain the distribution functions $F^{[k]}$ is an half-range method applied to the density gradients representation developed to explain the results of electron swarm experiments and described in [20]. This method solves the hierarchy of equations for the functions $F^{[k]}$ directly, using a finite elements algorithm on a (v, θ) grid. A comparison between the results obtained using this algorithm and other Boltzmann and Monte Carlo techniques can be found in [7].

The results obtained for $F_n^{[0]}(v)$ can be linked with the usual spherical harmonic expansion terms using the relation

$$F_n^{[0]}(v) = \frac{2n+1}{2} \int F^{[0]}(v,\theta) P_n(\cos\theta) \sin\theta d\theta \tag{20}$$

where $P_n(\cos\theta)$ are Legendre polynomials.

Figure (6), taken from [7], shows a comparison between the first six (6) Legendre expansion coefficients obtained with (20) and a multi–term Legendre expansion using eight (8) terms [21] The agreement improves if the number of intervals in θ is increased.

6. The Python interface

PLASMAKIN now includes a Python interface allowing the use of the library functions from Python scripts.

The development of a Python module serves several purposes:

(i) It allows a broader audience to use the library since it does not require knowledge of Fortran or C;

(ii) it integrates the library with the rich set of Python modules. Of special importance to plasma modeling is the integration with other modules for scientific computing as `Scipy` [22] or `Matplotlib` [23];

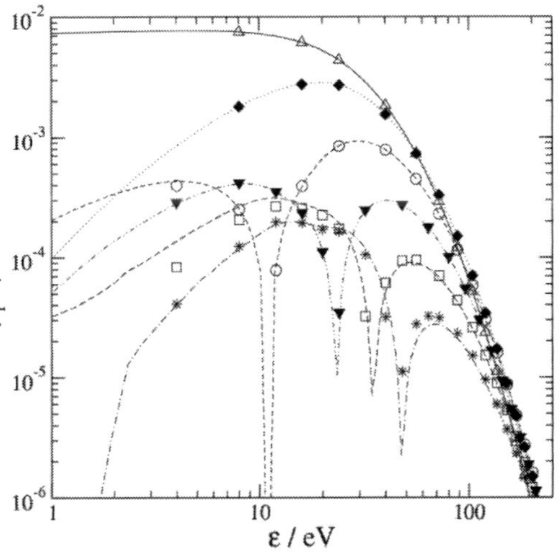

Figure 6. Comparison between the Legendre expansion coefficients $F_n^{[0]}$ for $0 \leq n \leq 5$ calculated using a multi–term Legendre expansion [21] (lines) and the present method at 500 Td in neon: △, ——$n = 0$; ◇, $\cdots\cdots n = 1$; O , – – –$n = 2$; ▽, — · · —$n = 3$; □, — — —$n = 4$; ⋆, — · —$n = 5$. For $n > 1$ the initial values of the expansion coefficients are negative.

(iii) although the execution of a script is slower than a compiled program, for 0– or 1–dimension problems the time penalty of running a Python script is not significant. On the other hand development in Python is usually much faster and simpler than in conventional programming languages;

(iv) it allows the use of Python development tools as `doctest` or `unittest` that have proved useful to track errors in the library itself; and

(v) speeds up the pre–processing or post–processing of data in a comparable way as what is found in commercial products.

Further details on the Python interface can be found in [24].

7. Conclusions

PLASMAKIN is an OpenSource package for plasma physics modeling open to contributions from the plasma physics community.

Presently is able to handle the chemical kinetics, the electron kinetics for constant E/N and provides usefull information as the relative contribution of each process, the power losses in the discharge or the emission spectrum of atomic gases.

It has been extended with a database of species properties and reactions and a Python interface.

Further development of the package has been simplified with a modular organization and the use of the autotools compilation system. Possible directions of development include the simulation of molecular spectra, the treatment of high frequency fields on the electron kinetics, the computation of ion transport properties or the always unfinished task of documentation.

References

[1] Kee R J, Miller J, Rupley F and Meeks E 1996 Chemkin - iii URL `http://www.reactiondesign.org`
[2] Carver G D, Brown P and Wild O 1997 *Comp. Phys. Commun.* **105** 197
[3] Pancheshnyi S, Eismann B, Hagelaar G and Pitchford L 2008 computer code zdplaskin URL `http://www.zdplaskin.univ-tlse.fr/`
[4] Pinhão N R 2001 *Comp. Phys. Commun.* **135** 105–131
[5] Morgan W L and Penetrante B M 1990 *Comp. Phys. Commun.* **58** 127–152
[6] Hagelaar G J M and Pitchford L C 2005 *Plasma Sources Sci. Technol.* **14** 722–733
[7] Pinhão N R, Donkó Z, Loffhagen D, Pinheiro M J and Richley E A 2004 *Plasma Sources Science and Technology* **13** 719–728 URL `http://stacks.iop.org/0963-0252/13/719`

[8] Molisch A F, Oehry B P, Schupita W and Magerl G 1993 *Comp. Phys. Commun.* **74** 81–90

[9] Molisch A F, Oehry B P, Schupita W and Magerl G 1993 *Comp. Phys. Commun.* **77** 255–62

[10] Molisch A F, Oehry B P, Schupita W and Magerl G 1996 *Comp. Phys. Commun.* **93** 127–35

[11] Luque J and Crosley D R 1999 Lifbase: Database and spectral simulation Tech. Rep. MP 99-009 SRI International URL `http://www.sri.com/psd/lifbase/`

[12] Sahal-Brechot S 1974 *Astr. Astrophys.* **35** 319

[13] Karabourniotis D 1986 *Radiative Processes in Discharge Plasmas* (*NATO ASI Series* vol 149) (Plenum Press) chap Self-Reversed Emission Lines in Inhomogeneous Plasmas, pp 171–247

[14] Damelincourt J J, Aubes M, Fragnac P and Karabourniotis D 1983 *J. Appl. Phys.* **54** 3087–3097

[15] Stormberg H P 1980 *J. Appl. Phys.* **51** 1963–69

[16] Humlicek J 1982 *JQSRT* **27** 437–44

[17] Ralchenko Y, Kramida A, Reader J, Saloman E, Sansonetti J, Curry J, Kelleher D, Fuhr J, Podobedova L, Wiese W, Olsen K *et al.* 1979-2008 NIST atomic spectra database URL `http://physics.nist.gov/PhysRefData/ASD/index.html`

[18] Kumar K, Skullerud H R and E R R 1980 *Australian Journal of Physics* **33** 343–448

[19] Sakai Y, Tagashira H and Sakamoto S 1977 *J. Phys. D: Appl. Phys.* 1035

[20] Ségur P, Yousfi P M and Bordage M C 1984 *J. Phys. D: Appl. Phys.* **17** 2199–2214

[21] Loffhagen D and Winkler R 1996 *J. Phys. D: Appl. Phys.* **29** 618–27

[22] Jones E, Oliphant T, Peterson P *et al.* 2001- Scipy: Open source scientific tools for python URL `http://www.scipy.org`

[23] Dale D, Droettboom M, Firing E, Hunter J *et al.* 2008 Matplotlib URL `http://matplotlib.sf.net`

[24] Pinhão N R 2007 *The Python Papers* **2** 35–47 URL `http://pythonpapers.org/`

Hollow Cathode Discharges: Volt-Ampere Characteristics and Space-Time Resolved Structure of the Discharge

D Marić[1], N Škoro[1], G Malović[1], Z Lj Petrović[1], V Mihailov[2] and R Djulgerova[2]

[1]Institute of Physics Belgrade, Pregrevica 118, 11080 Zemun, Belgrade, Serbia
[2]Institute of Solid State Physics of Bulgarian Academy of Sciences, 1784 Sofia, Bulgaria

E-mail: draganam@phy.bg.ac.yu

Abstract. We report measurements of electrical properties and spatial emission profiles of hollow cathode discharge, both for steady state and transient phases during formation of the discharge. A commercial hollow cathode tube filled with neon at 3.5 Torr has been studied. Our aim was to relate discharge anatomy to the voltage-current characteristics in a wide range of the operating conditions. It was particularly interesting to track down in time formation of the discharge structure, which is typical for "hollow cathode effect".

1. Introduction

Hollow cathode discharges were studied almost as long as gas discharges. In addition to applications in spectrometry and ion gas lasers [1,2], plasma processing applications such as ion etching, thin film deposition, surface treatment [3] and analysis of nanostructured surfaces [4] have been recently introduced.

Compared to parallel plate glow discharges with similar dimensions and gas pressure, hollow cathode discharges have lower breakdown voltages and operate at lower voltages for the same current density [5-7]. Enhanced discharge efficiency is due to the fast electrons and ions, which are confined in the potential minimum inside the cathode cavity [8]. When a negative glow regions facing the opposite cathode surfaces overlap, "hollow cathode effect" appears. Hollow cathode effect is the most specific feature of these devices. It is manifested as a large increase in the current density and discharge light intensity followed by a drop of a sustaining voltage. Whether this effect will occur or not depends on geometry and operating conditions in the discharge.

Our aim was to establish a relationship of the discharge structure to the electrical properties in a wide range of discharge currents for steady state conditions, but also in transient stages during formation of the discharge. We were particularly interested in development of the hollow cathode effect. Analysis of formation and maintenance of the hollow cathode discharges is based on comparisons with simpler parallel-plate discharges (e.g. [9-11]). It was possible to identify and explain several critical points in the Volt-Ampere characteristics where the shape of the discharge changes.

We believe that such measurements may be of interest for micro discharges where hollow cathode geometry is often applied even though the conditions for the hollow cathode effect are not always met [12,13]. Thus, one may perhaps identify the regime of operation from the Volt-Ampere characteristics.

Journal of Physics: Conference Series **162** (2009) 012007 doi:10.1088/1742-6596/162/1/012007

2. Experimental setup

We use a Hilger and Watts commercial hollow cathode lamp which is normally applied as a spectral source. The lamp is a sealed glass tube filled with Ne (p=3.5 Torr) containing a Mn cylindrical hollow cathode open at one end and a ring shaped anode. Cathode hole is 3 mm in diameter and 15 mm long. Cathode is placed inside a closely fitting glass tube which prevents ignition of the discharge at the outer wall. Schematic of the electrode system is shown in figure 1.

Our measurements include ICCD imaging of the discharge emission, supported by voltage and current measurements. Details of electrical circuit and technique of measurement are presented in e.g. [9-11]. Our power circuitry is able to impose a current pulse of desired length in addition to a basis low current DC discharge. This technique allows us to avoid long breakdown delay times and it enables us to follow formation of the discharge [11]. Spatial profiles are recorded in the visible range of spectra using an ICCD camera (Andor, iStar DH720-18U-03). Recording of voltage-current signals and camera gating are synchronized with development of the voltage and current pulse.

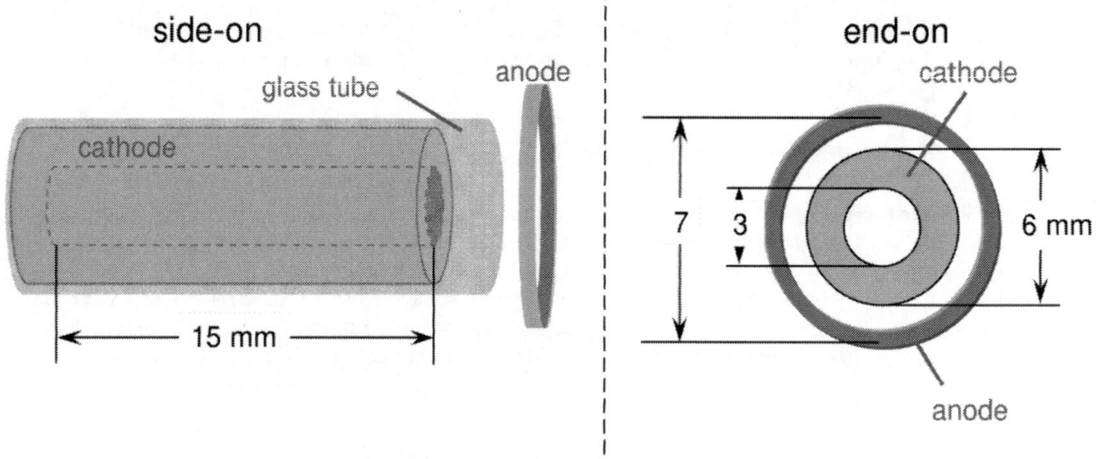

Figure 1. Schematics of the electrode system in side-on (left) and end-on view (right).

3. Results and discussion

3.1. Steady-state measurements

In this paper we show preliminary results that will find its final presentation later [14]. Figure 2 shows voltage-current (V-I) characteristic and current dependence of the peak emission intensity at the centre of the cathode cavity. Peak intensities are taken from end-on images of the discharge structure recorded by ICCD camera. Figure 3 shows 2D images of the discharge at the selected voltage and current conditions, where labels (a)-(f) correspond to those in figure 2.

In the low current limit, the discharge mainly operates outside the cathode cavity i.e. between the outer surface of the cathode and the anode (see figures 2 and 3, label (a)). Behaviour of the V-I characteristics as well as the structure of the discharge in this range of currents resembles discharge in parallel-plate geometry [9-11]. The gap in the descending part of V-I characteristic corresponds to the region of free-running oscillations. With further increase of current the discharge goes into the mode similar to normal-glow discharge between parallel electrodes – the V-I characteristics becomes flat, radial profile of the discharge is more constricted. The emission becomes more intense, but the discharge still operates both inside the cavity and from the outer surface of the cathode (figure 3(b)-(c)). Further on, as a sudden change in the discharge regime takes place; there is a distinct drop in the discharge voltage and an increase in intensity (figure 3(c)-(d)-(e)). At the same time, the discharge becomes significantly constricted and it operates mainly inside the cavity. A large increase in current density followed by the voltage decrease indicates that the discharge switches to a more "efficient" regime typical for the "hollow cathode effect". Further rise in current is accompanied by the discharge expansion within the cavity and emission increment. This is typical for the normal glow in parallel plate geometry. At even higher currents, the voltage-current slope becomes positive. Almost entire cathode cavity is occupied by the discharge (figure 3(f)), which, can be related to the behaviour of abnormal glow discharge in parallel-plate geometry.

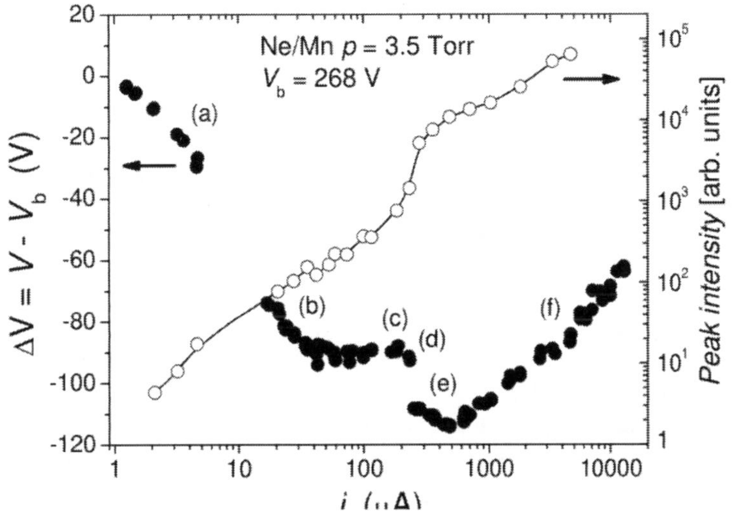

Figure 2. Discharge voltage (solid symbols) and peak emission intensity (open symbols) plotted against discharge current [14]. The discharge voltage is presented as the difference between the operating (V) and the breakdown (V_b) voltage. Labels (a)-(f) correspond to the images shown in figure 3.

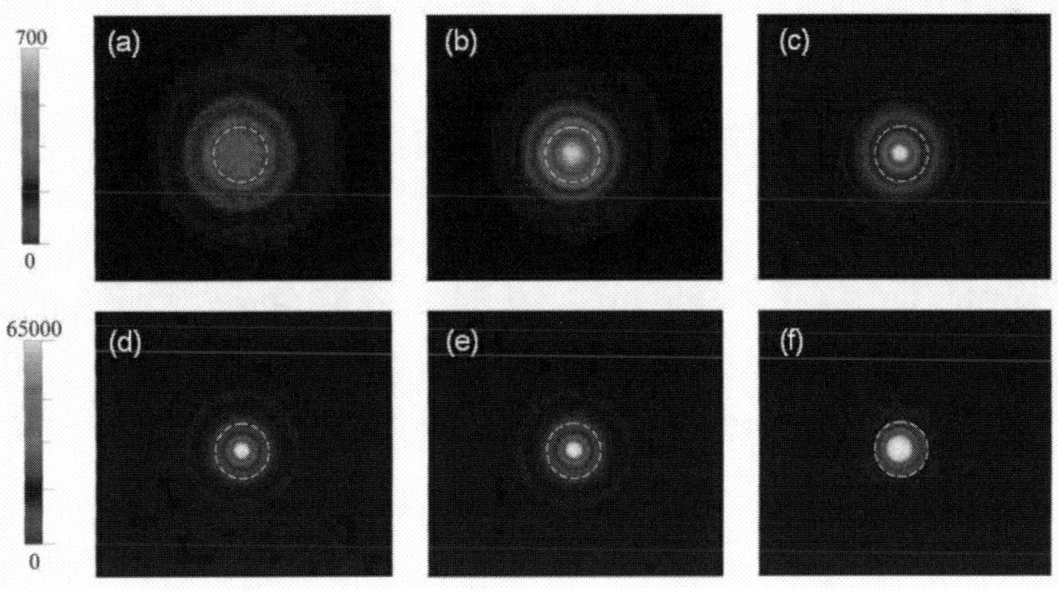

Figure 3. End-on images of the discharge structure for different discharge currents: (a) $i = 3.2$ μA, (b) $i = 28$ μA, (c) $i = 184$ μA, (d) $i = 231$ μA, (e) $i = 278$ μA, (f) $i = 4690$ μA [14]. White dashed lines mark the cathode cavity edge.

3.2. Time resolved measurements

In order to understand better the formation and maintenance of different regimes of the discharge with the hollow-cathode geometry, we tracked down in time the development of the discharge structure. We were able to obtain temporally resolved discharge profiles by using a fast ICCD camera synchronized with the electrical circuit. The technique of measurements that we use can provide perceptible images even for short exposition and low light intensities. Discharge profile formation was traced from the beginning of the current pulse, through the transient behaviour and to the steady state. We have made measurements for several different modes of the discharge marked by labels (a)-(f) in figures 2 and 3. We will present here only several selected regimes while a more detailed presentation will be given in [14].

3.2.1. Townsend-like discharge. Structure of the low-current steady state regime of the discharge is shown in figure 3(a). As it was noted, the discharge operates dominantly from the surface of the cathode that is facing the anode, similar to the case of Townsend discharge in parallel-plate geometry. During the formation of this regime, the discharge structure does not change significantly, while the intensity of the emission closely follows current along the pulse.

3.2.2. Free-running oscillations. As mentioned earlier, gap in the descending part of the *V-I* characteristics (figure 2) belongs to the region of free running oscillations. Oscillatory behaviour of low-pressure discharges in parallel-plate geometry has been studied in a number of papers [11,15]. Oscillations in hollow-cathode geometry have been studied in [13,16]. These studies included detailed analysis of the effect of circuit elements and operating conditions on frequency and damping of oscillations and on oscillation boundaries.

We have traced in time changes in structure of the discharge along development of oscillations. Figure 4(a) shows voltage and current waveforms that occur after application of the voltage pulse. In figure 4(b) we have plotted voltage and current trace in time against steady-state *V-I* characteristics. Such representation gives us a better insight in transient behaviour of the discharge. Images of the discharge structure for selected times of measurements labelled by 1-7 are presented in figure 5.

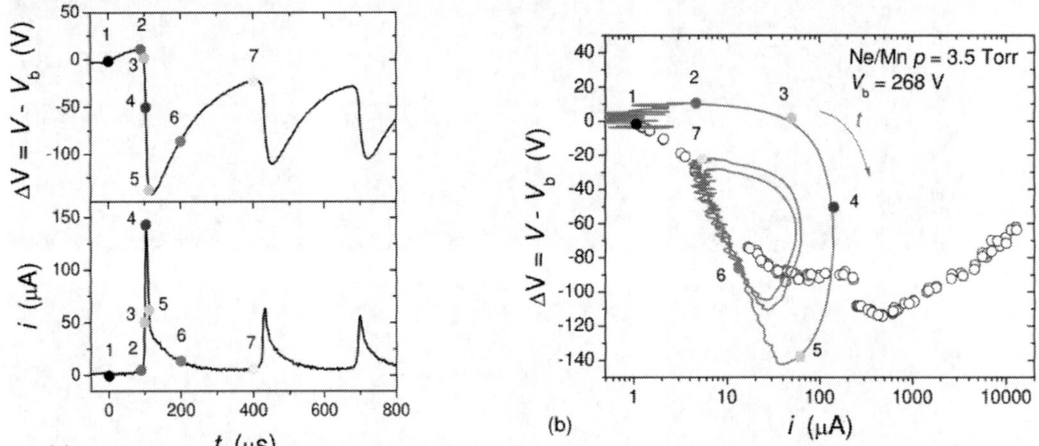

Figure 4. (a)Voltage and current signals throughout free running oscillations [14]. (b) Trajectory of several oscillation periods, plotted against steady state voltage-current characteristics. Points labelled by 1-7 indicate moments of recording of the discharge structure by ICCD camera. The corresponding images are shown in figure 5.

Figure 4 shows that after the initial voltage pulse, discharge goes through a transient phase within the first oscillation period, after which periodically repetitive behaviour takes place. The discharge starts from the initial Townsend-like regime which operates mainly outside the cavity (figure 5, label 1). As the discharge current increases, the emission intensity increases, but the structure of the discharge does not change (label 2). Further increase of the current is followed by a decrease of the voltage. Emission intensity inside the cathode hole seems to increase significantly – label 3 to 5. As the voltage reaches the minimum value, it appears that discharge operates only inside the cathode cavity, which indicates that the hollow-cathode effect has developed. However, the discharge cannot operate at such low voltage for a long time, it keeps self-adjusting – the current decreases, the discharge practically shuts down, until the voltage recovers to the appropriate value (labels 6 and 7). Furthermore, as the discharge cannot reach steady-state conditions, it exhibits repetitive oscillations with a similar temporal behaviour as in first oscillating period. Images recorded by ICCD camera supported by voltage and current measurements clearly show that discharge goes through significant changes in the regime of operation, following the oscillations of voltage and current.

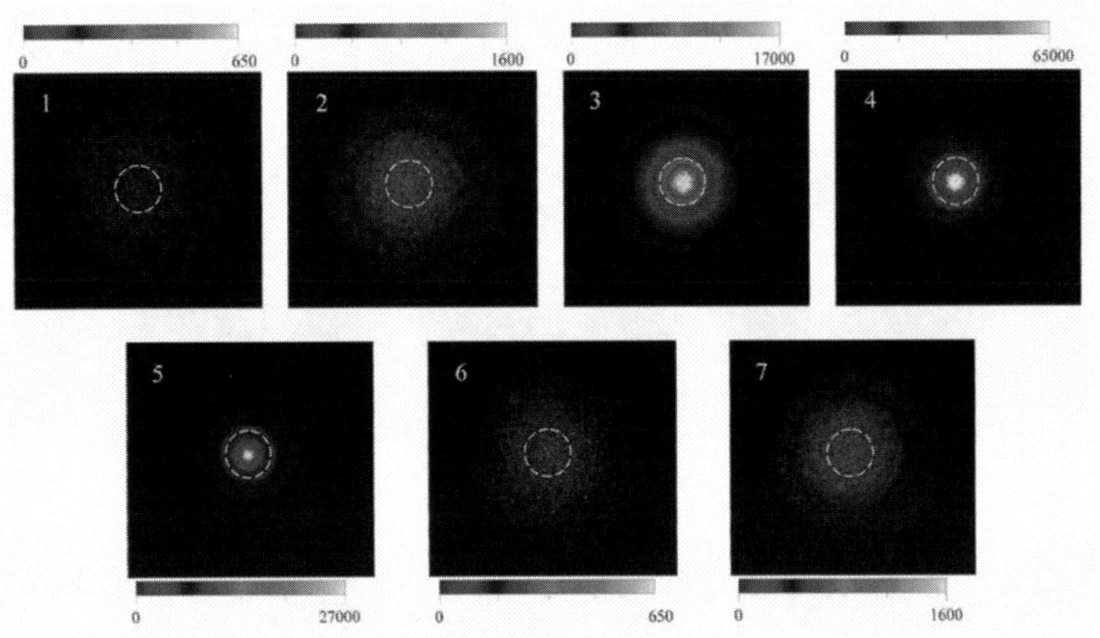

Figure 5 2D end-on images of the discharge structure in selected moments during free-running oscillations [14]. Labels 1-7 indicate time of measurement, marked in figures 4(a) and (b).

3.2.3. Normal glow-like discharge. During formation of the normal glow-like regime of discharge, transient phase is very similar to the previous case, so it will not be discussed here in significant detail. After the application of the voltage pulse, the discharge goes from the Townsend regime, gradually exhibits transition to the regime with hollow-cathode effect and finally relaxes to the steady state which appears to be a combination of the discharge outside and inside the cathode cavity. Spatial structure of the steady-state normal glow-like discharge is shown in figure 3(b)-(c).

3.2.4. Transition from normal glow to hollow cathode discharge. We were able to follow the transition from parallel-plate-like discharge to the regime of discharge with distinct hollow-cathode effect developed, within a single pulse. This corresponds to the gradual change of discharge regime from the point labelled by (d) to the point (e) in the *V-I* characteristic (figure 2). We have shown trajectory of the voltage and current signals in time against the steady-state *V-I* characteristic in figure 6 in order to illustrate more clearly this transition. It was not possible to find an adequate combination of circuit elements to start from the very low basis DC current (1-2 μA) and reach higher current modes. Thus, in this case, our basis current is $i_0 = 30$ μA. After the application of the voltage pulse, the discharge goes through transient phase – from point labelled by 1 to point 5. Further on, it slides from the normal glow-like regime (label 5 to the regime with hollow-cathode effect developed (label 8).

Discharge voltage and current waveforms are shown in figure 7. Transient part at the beginning of the pulse is magnified in figure 7(a) in order to show rapid changes of the amplitude of the waveform. Labels 1-8 in figures 6 and 7 correspond to spatial images in figure 8. At the beginning of the pulse, both discharge voltage and current have a steep rise. At the same time, 2D images show that discharge becomes radially inhomogeneous moving from the cathode edge into the cavity. Profile intensity increment follows current signal peeking at the current maximum. Further changes in voltage and current are mirrored only through profile intensity variations while discharge stays inside the cathode. At the moment when current runs though the minimum, local maximum voltage is achieved with minor changes in the profile intensity. Discharge briefly sets at one current and then slips to another mode accompanied by a slight current increase and significant voltage decrease. Major increase in emission intensity is noticeable.

Although we expected some kind of the discharge instabilities to occur in the region of negative slope of *V-I* characteristics, this transition appears to be quite smooth. However, it is obvious that the discharge cannot operate stably in the region between the two different regimes.

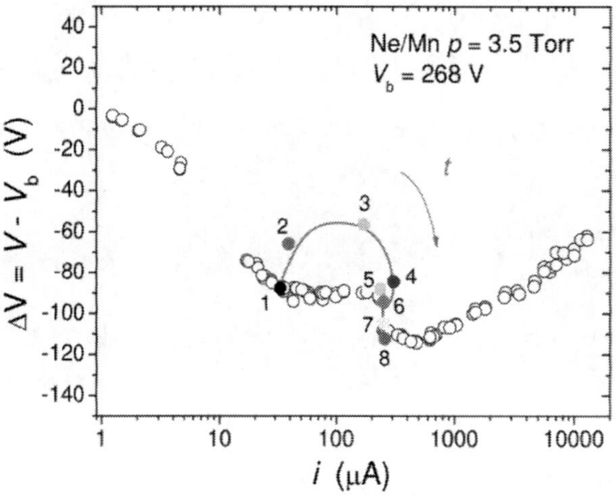

Figure 6. Voltage and current trace in time during transition from normal glow to the regime of discharge with hollow-cathode effect developed, plotted against steady-state voltage-current characteristics [14].

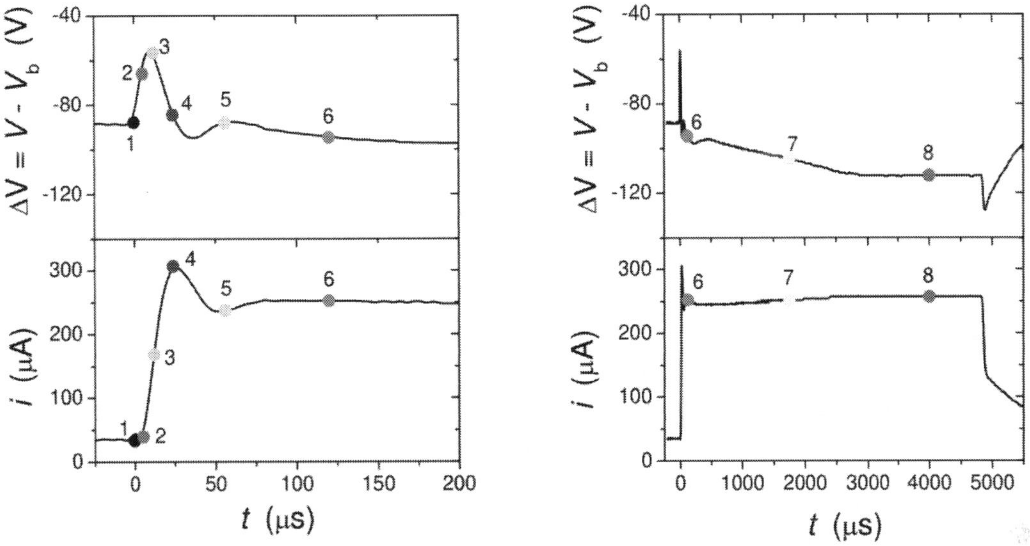

Figure 7. Voltage and current signals after the application of the voltage pulse: (a) beginning of the pulse zoomed in, (b) entire pulse [14].

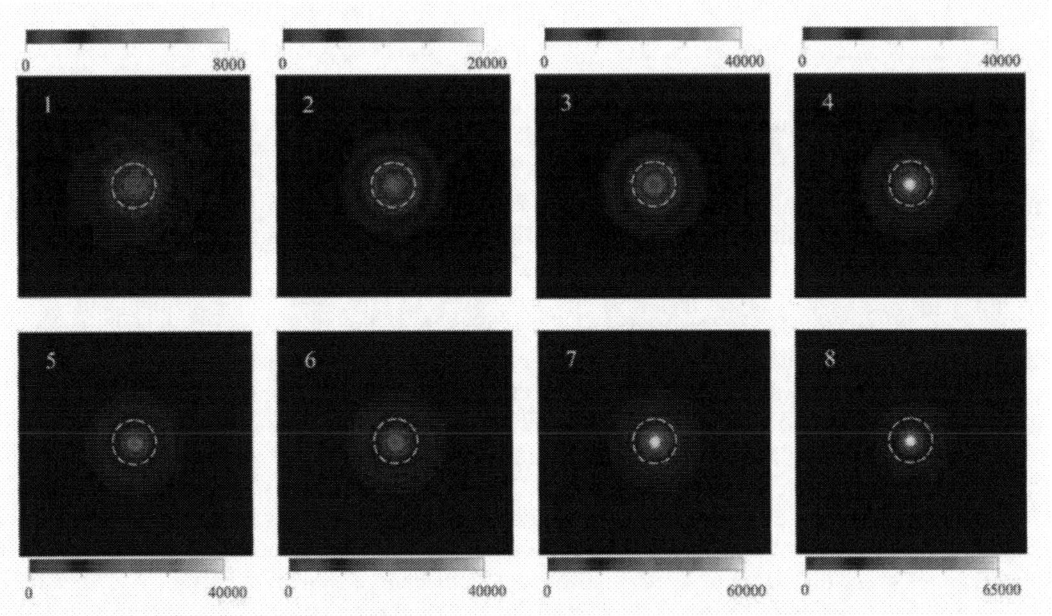

Figure 8 Images of the discharge structure in selected moments during transition from normal glow to hollow-cathode discharge [14]. Labels 1-8 correspond to those in figures 6 and 7.

3.2.5. Hollow cathode regime of the discharge. We will analyze formation of the discharge regime with a fully developed hollow-cathode effect. Voltage and current signals after application of the external voltage pulse are shown in figure 9. We will look more closely to the changes in the profile of emission with time. Figure 10 shows profiles of emission taken along the discharge chamber diameter at selected moments, labelled by 1-6 in figure 9. Corresponding 2D images are shown in figure 10.

We can see that in time interval labelled by 1-3, the discharge profile is radially homogeneous; it operates both inside and outside the cathode cavity. This interval corresponds to the rise of the voltage pulse. As the voltage starts to decrease, the discharge structure changes significantly. At times ranging from 4 to 6, the discharge gradually constricts and finally it operates mainly inside the cavity. Besides that, we can notice that peaks of emission at the edge of the cathode cavity become significant. However, it cannot be clear if these peaks are due to the reflection at the edge of the cavity or due to heavy particle processes close to the cathode surface inside the cavity [16].

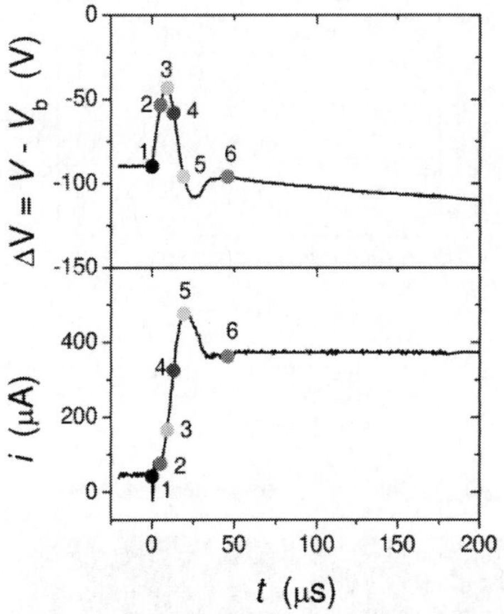

Figure 9. Voltage and current signals during formation of the hollow-cathode effect [14].

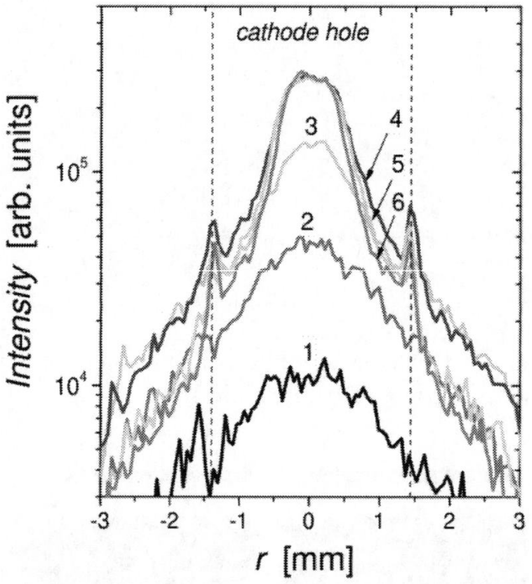

Figure 10. Radial emission profiles of the discharge taken at moments labelled by 1-6 in figure 9 [14].

3.2.6. Abnormal glow discharge. As the formation of abnormal glow discharge is basically the same as the formation of previously described formation of the hollow-cathode discharge, we will not discuss it here. Essentially, the only difference in formation of these two regimes is that in the abnormal glow discharge occupies almost entire cathode cavity, as it can be seen in figure 3(f).

4. Summary

In this paper we have presented studies of spatial structure of the discharge in hollow cathode geometry, supported by voltage and current measurements. Our study included analysis of several distinct regimes of the discharge in steady state as well in discharge formation period. We were able to relate electrical properties to the spatial structure of the discharge. It has been shown that hollow cathode discharge can be represented as a combination of discharges between parallel plates and inside the cavity. Results of measurements revealed gradual change of the parallel-plate-like structure of the discharge to the regime with more efficient operation where hollow-cathode effect is fully developed.

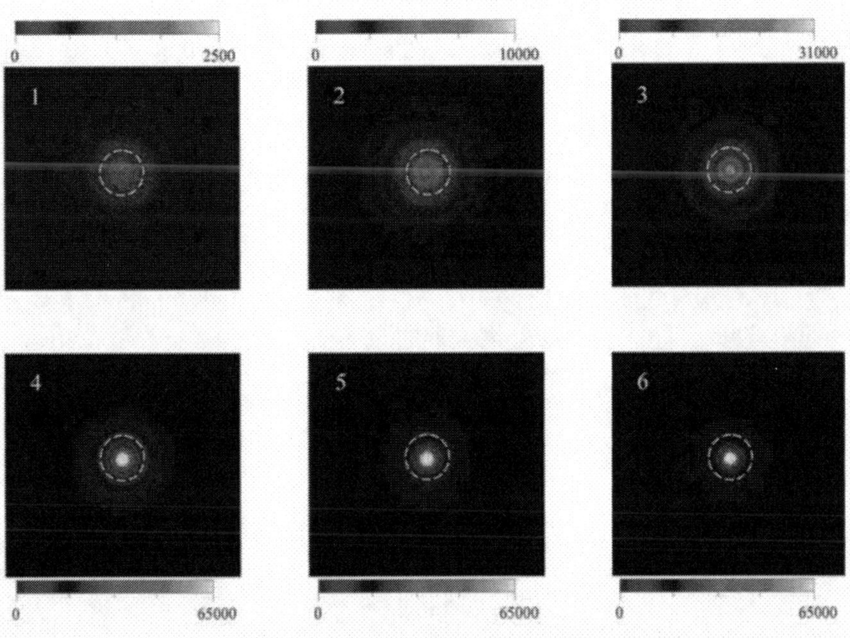

Figure 11. Images of the discharge structure in selected moments during formation of the discharge with hollow-cathode effect fully developed [14]. Labels 1-6 correspond to those in figures 9 and 10.

By complementing these results by a model such as the one described in [17] it would be possible to get an even better insight in the elementary processes that determine the discharge behaviour. For example it would be useful to explore the potential influence of heavy particle processes.

Our aim is to extend this study to different pressures and geometry, which will enable us to study scaling laws relevant for miniaturization of hollow-cathode discharges [18]. Hollow-cathode discharges in micro dimensions are a very attractive research field with vast application possibilities. At the same time there is a possibility that some kinetic phenomena due to non-local nature of charged particle transport may be revealed and connected to the fundamental transport kinetic phenomena [19]

Acknowledgments
This work was funded by the Ministry of Science of the Republic of Serbia number 141025 and by the collaborative program between Serbian Academy of Sciences and Arts and Bulgarian Academy of Sciences.

References
[1] Caroli S 1983 *Prog. Analyt. Atom. Spectosc.* **6** 253
[2] Peard A, Donko Z, Rozsa K, Szalai L and Tobin R C 1994 *IEEE J. Quant. Electron.* **30** 2157
[3] Becker K H, Schoenbach K H and Eden J G 2006 *J. Phys. D: Appl. Phys.* **39** R55
[4] Djulgerova R B, Popova L I, Mihailov V I, Beshkov G D, Szytula A, Gondek L and Petrović Z Lj 2004 *Spectrochim. Acta* B **59** 905
[5] Pillow M E 1981 *Spectrochim. Acta,* **36B** 821
[6] Baguer N, Bogaerts A, Gijbels R 2002 *Spectrochim. Acta Part* B **57** 311
[7] Eichhorn H, Schoenbach K H and Tessnow K. H 1993 *Appl. Phys. Lett.* **63** 2481
[8] Arslanbekov R R, Kudryavtsev A A and Tobin R C 1998 *Plasma Sources Sci. Technol.* **7** 310
[9] Marić D, Kutasi K, Malović G, Donko Z and Petrović Z Lj 2002 *Eur. Phys. J.* D **21** 73
[10] Marić D, Hartmann P, Malović G, Donko Z and Petrović Z Lj 2003 *J. Phys. D: Appl. Phys.* **36** 2639

[11] Marić D, Malović G, and Petrović Z Lj 2007 *Journal of Physics: Conference Series* **86** 012009
[12] Boeuf J P, Pitchford L C and Schoenbach K H 2005 *Appl. Phys. Lett* **86** 071501
[13] Rousseau A and Aubert X 2006 *J. Phys. D: Appl. Phys.* **39** 1619
[14] Marić D, Škoro N, Malović G, Petrović Z Lj, Mihailov V and Djulgerova R, to be published
[15] Petrović Z Lj and Phelps A V 1993 *Phys. Rev.* E **47**(4) 2806
Jelenković B M, Rózsa K and Phelps A V 1993 *Phys. Rev.* E **47**(4) 2816
Phelps A V, Petrović Z Lj and Jelenkovic B M 1993 *Phys. Rev.* E **47**(4) 2825
[16] Donkó Z, 1999 *J. Phys. D: Appl. Phys.* **32** 1657
[17] Kutasi K and Donkó Z 2000 *J. Phys. D: Appl. Phys.* **33** 1081
[18] Petrović Z Lj, Škoro N, Marić D, Mahony C M O, Maguire P D, Radmilović-Rađenović M and Malović G 2008 *J. Phys. D: Appl. Phys.* **41** 194002
[19] Petrović Z Lj, Šuvakov M, Nikitović Ž, Dujko S, Šašić O, Jovanović J, Malović G and Stojanović V 2007 *Plasma Sources Sci. Technol.* **16** S1

An overview of modelling of low-pressure post-discharge systems used for plasma sterilization

K Kutasi[1], C D Pintassilgo[2,3], J Loureiro[2]

[1] Research Institute for Solid State Physics and Optics of the Hungarian Academy of Sciences, POB 49, H-1525 Budapest, Hungary

[2] Instituto de Plasmas e Fusão Nuclear, Instituto Superior Técnico, 1049-001 Lisboa, Portugal

[3]Departamento de Engenharia Física, Faculdade de Engenharia da Universidade do Porto, 4200-465 Porto, Portugal

E-mail: kutasi@mail.kfki.hu

Abstract. We present the modeling of a low pressure post-discharge system used for plasma sterilization, which consists of a flowing microwave discharge, an early-afterglow region developed downstream of the discharge in the same tube, and a large post-discharge reactor. The kinetic model valid for the discharge and early-afterglow region is based on the Boltzmann and rate balance equations, while in the post-discharge reactor a 3-D hydrodynamic model is used. The calculations are conducted in an N_2-O_2 mixture, which has seen to constitute an efficient system for sterilization purpose. The possibilities offered by modeling is presented through a discussion of the effects on relevant species densities and distributions caused by varying: (*i*) the initial gas mixture composition; (*ii*) the length of early-afterglow zone; (*iii*) the gas flow rate and gas pressure.

1. Introduction

In the last decade the use of polymer based heat sensitive tools in the medical praxis or pharmaceutical industry have brought the need of new low temperature sterilization and decontamination methods, which ensure complete inactivation or removal of all possible infectious microorganism: e.g. bacterial spores, viruses, or other potentially harmful biological residuals: e.g. endotoxins and proteins present on the used instruments. For the heat sensitive medical devices the most common sterilization methods in use are based on chemical treatment, which imply toxic active agents, or exposure to ionizing radiation, which may alter the bulk properties of polymers being sterilized.

A new method, which overcomes these drawbacks, implies the use of low-temperature plasmas. Low-temperature plasmas can be created either at atmospheric or low ($p \leq$ 10 Torr) pressures. Numerous experimental studies have been carried out in different discharge configurations in a wide pressure range: i.e. atmospheric [1–15], mTorr [16–21] and Torr regions [22–29]; in various gas mixtures: e.g. N_2 [16, 18, 21], O_2 [13, 16, 18, 21], N_2-O_2 [7, 16, 21, 27, 30], Ar [6, 11, 14], Ar-O_2 [24], Ar-H_2 [19], O_2-H_2O [20], O_2-CF_4 [17], in order to test the sterilizing ability of these plasmas. Either the active discharge regions have been used [1, 8, 11, 16–19], or the afterglow region of discharges [7, 22, 24, 31]. *In this work we focus on the afterglow of low-pressure, Torr regime, discharges.*

Journal of Physics: Conference Series **162** (2009) 012008 doi:10.1088/1742-6596/162/1/012008

Moreau *et al.* [22] have shown that total inactivation of an initial 10^6 spore (*Bacillus Subtilis*) population can be obtained using the flowing afterglow of a microwave discharge sustained in $N_2-2\%O_2$ at reduced gas pressure (typically 1−10 Torr). Moisan *et al.* [31] have shown that in these kind of afterglows the plasma sterilization can be explained by a synergistic effect between $O(^3P)$ atoms and UV photons emitted by $NO(B^2\Pi)$ molecules in the 250−320 nm spectral range (the so-called $NO\beta$ system). More recently, a detailed investigation of the UV spectrum correlated with the bacterial inactivation rate coefficient has shown that the $NO\gamma$ system (generated by $NO(A^2\Sigma^+)$), and not the $NO\beta$ system as stated in earlier publications, is the main UV photon contributor (Moisan *et al.* to be submitted, see also [32]). Philip *et al.* [25] state that the sterilization is the most efficient at the condition when the UV intensity is the highest, in their case at 2 Torr in N_2-0.7%O_2, while Villeger *et al.* [27] have found the best sterilization condition when the O-atoms density is the highest and the UV radiation is still reasonably high, but not necessarily the highest, in their case at 5 Torr in N_2-5%O_2. Nevertheless, the inactivation of spores have been also observed under very little or no UV radiation [33]. Furthermore, it has been shown that spores inactivation can be achieved also at 1-2 Torr in pure N_2 flowing afterglow due to the N atoms [34], which suggests that the N atoms present in an N_2-O_2 afterglow (where similar N densities can exist as in pure N_2) also contribute to sterilization [35]. Villeger *et al.* have shown that the N atoms can contribute to the sterilization through the temperature rise of the substrate induced by their recombination. Consequently, for an efficient sterilization large and homogeneous density distributions of N and O atoms, and NO(A), NO(B) molecules are necessary in a large volume reactor. Therefore, when analyzing the sterilization plasmas in order to define the sterilizing agents and to contribute to the optimization of the system it is important to determine the density distributions of these species in the reactor. The densities of the these species are influenced besides the discharge parameters, by the gas flow rate and gas pressure.

The densities of N and O atoms have been determined experimentally at various discharge conditions - by varying the gas pressure, gas flow rate and initial gas mixture composition - in the early-afterglow of a low-pressure N_2-O_2 post-discharge used for medical sterilization [24, 25, 27, 29, 36, 37]. The UV emission stemming from the $NO(B^2\Pi)$ upper sate that yields photons over the $NO\beta$ bands (250-320 nm) has been registered in reactors (late-afterglow) for the same conditions [24,25,27,29,36]. The absolute N and O atoms densities are usually determined by NO titration in the early afterglow region, and therefore provide only the densities of atoms at the reactor's entrance. Boudam *et al.* [29] have determined the N atoms relative density in several positions in the reactor, mostly in the flow direction, with the Bockel's method [38], which is based on the intensity measurements at 580 nm of the first positive system ($N_2(B,v'=11)$). Here spatially integrated intensities are recorded, due to which a spatially averaged density is defined and thus the spatial inhomogeneity of the N atoms densities is masked. Contrary to this, the discharge modelling allows to obtain the density distributions of the various species present in the reactor, even of those whose densities cannot be measured experimentally. Calculations in post-discharge reactors used fpr surface treatments have been performed by Belmonte *et al.* [39]. Cylindrical reactors were modelled in Ar-N_2(-H_2) [40] and Ar-O_2(-H_2) [41] mixtures, used for nitriding and zirconia thin films deposition, respectively, with the help of the PHOENICS commercial code. The model gave as a result the density distributions of the species in the late post-discharge, however the initial densities at the beginning of the post discharge were not calculated accurately with a self consistent model.

In the sterilization chamber, besides the active species relevant for sterilization, there are also species, which may interact with the sterilized tools and induce damages. Some experimental works have reported the existence of a significant erosion of the spores and a variety of polymers in the afterglow chamber when the samples were located in the vicinity of the discharge tube axis [33]. In contrast, at a few cm away from the axis, no significant erosion has been observed at low O_2 percentages. However, with increasing O_2 percentage the erosion appears, which according to Boudam *et al.* may be attributed to the enhancement of O-atoms density [29]. On the other hand, the strong on-axis erosion is suggested by the same authors [29] to be the action of the metastable state molecules: $N_2(A)$, $N_2(a')$, and nitrogen molecular ions: N_2^+, N_4^+, species which characterize the so-called early afterglow. These facts draw another necessary condition to be satisfied by an optimal sterilizer, the reduction or elimination of the species believed to induce erosion from the reactor. This can be achieved by controlling the O_2 percentage in the initial gas mixture composition, the gas flow rate and the distance between the discharge and reactor. A self-consistent model of the discharge and of the early-afterglow allows the determination of the optimal discharge and geometrical parameters, where the above mentioned conditions are met.

Summarizing, the density distributions and the absolute densities of different species in the reactor can be controlled with: (i) the distance between the discharge and post-discharge reactor, as well as the geometry of this connecting tube, (ii) the discharge characteristics: tube radius, microwave field frequency, input power, (iii) the gas pressure, (iv) the initial gas mixture composition and (v) the gas flow rate. The density distributions of the species in the chamber can be also significantly influenced by the structure of the chamber, namely by the position of the inlets and outlets, which define the gas flow. The self-consistent model of the whole system makes it possible to investigate the effect of all these parameters on the density distributions of different species, as it will be presented in this paper.

The paper is structured as follows. In Section 2 the system set-up, while in the Section 3 the system's model is presented. The results are discussed in Section 4. Finally, the conclusions are drawn in Section 5.

2. System set-up

The system set-up, similar to the one used by the Montréal group in their experimental investigations [32], is shown in Figure 1. The surface-wave discharge is generated with a microwave field of 433 MHz, 915 MHz or 2450 MHz frequency in a tube of 0.6 cm inner diameter, which widens into a 2.6 cm inner diameter tube, that is connected to the reactor. During the calculations the tubes are assumed to be of variable lengths. The main frame of the parallelepipedic 50 L reactor, with dimensions of 60 cm×30 cm×28cm, is made of aluminium and is equipped with fused silica windows, which allows spectroscopic observations. The 2.6×2.6 cm^2 square inlet and outlet are located in the middle of the left side and bottom walls, respectively.

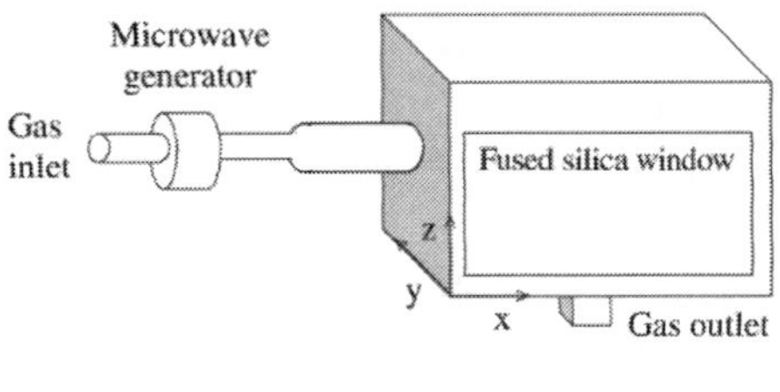

Figure 1. Post-discharge system.

3. Model

A post-discharge system has three main parts: (i) discharge region, (ii) early-afterglow region, which develops downstream from the discharge in the connecting tube between the discharge and the reactor and (iii) late afterglow present in the large volume reactor. The species created in the discharge are carried by the gas flow through the early-afterglow region into the post-discharge reactor. In order to find out the species densities distribution in the post-discharge reactor, a self-consistent model of the whole system needs to be built up, which allows an accurate description of each section.

3.1. Discharge and early-afterglow model

The self-consistent model for the flowing microwave discharge is based on the solutions for the electron Boltzmann equation for the microwave field, coupled to a system of rate-balance equations for the neutral and charged heavy species [42]. Thus the electron energy distribution function (EEDF) is determined, and further the vibrational distribution functions (VDFs) of $N_2(X^1\Sigma_g^+, v)$ and $O_2(X^3\Sigma_g^-, v')$ molecules (with $0 \leq v \leq 45$ and $0 \leq v' \leq 15$), the concentrations of ground state $N(^4S)$ and $O(^3P)$; of the electronic states of molecular and atomic nitrogen and oxygen $N_2(A^3\Sigma_u^+$, $B^3\Pi_g$, $B'^3\Sigma_u^-$, $C^3\Pi_u$, $a'^1\Sigma_u^-$, $a^1\Pi_g$, $w^1\Delta_u)$ $O_2(a^1\Delta_g, b^1\Sigma_g^+)$ and $N(^2D$, $^2P)$; the formed species $NO(X^2\Pi)$, $NO(A^2\Sigma^+)$, $NO(B^2\Pi)$, $NO_2(X,A)$ and O_3; the main positive ions $N_2^+(X^2\Sigma_g^+, B^2\Sigma_u^+)$, N_4^+, O_2^+, O^+ and NO^+; and negative ions O^-.

The reduced sustaining electric field in the plasma is self-consistently calculated by solving the continuity and transport equations for all these ionic species, including O^-, in order to describe the charged particle motion to the wall under the presence of a radial space-charge field. The reduced sustaining electric field in the plasma is determined using the requirement that under steady state conditions the total rate of ionization must compensate exactly for the rate of electron loss by electron-ion recombination plus ambipolar diffusion to the wall. It is worth to note here that this requirement ensures that the quasi-neutrality condition is obtained.

The model calculations have been performed for three typical laboratory conditions as follows: (*i*) $\omega/(2\pi) = 433$ MHz; (*ii*) $\omega/(2\pi) = 915$ MHz; (*iii*) $\omega/(2\pi) = 2450$ MHz. In a surface-wave discharge the electron density decreases from the excitation gap of the launcher to the plasma end, where the electron density reaches its cut-off value. In the case of surface wave mode propagation in a quartz tube, the critical electron densities associated with these frequencies are, respectively, $n_{ec} = 1.17\times10^{10}$ cm^{-3}, 5.20×10^{10} cm^{-3}, and 3.74×10^{11} cm^{-3}.

Since we analyse the early-afterglow downstream from the discharge, we only need to calculate the heavy-species concentrations in the plasma that correspond to a specific value of critical electron density. Within this approach, we obtain a set of initial solutions that is subsequently taken into account in a time-dependent system of master equations, where electronic processes are neglected. This formulation gives the temporal evolution of the populations in the early-afterglow.

The near afterglow that occurs downstream from the discharge, in the same discharge tube, has been considered by cutting-off the excitation by electron impact in the system of master equations. The time-dependent solutions obtained with this model constitute the initial conditions for the present 3D simulation.

3.2. Hydrodynamic model of the post-discharge reactor

The evolution of species densities in the reactor is described with a three dimensional hydrodynamic model. The model is composed of (i) the total mass conservation, (ii) the continuity equations for the different species (2), (iii) the total momentum conservation equation (3), and (iv) the total energy conservation equation (4). The gas is assumed to be a Newtonian fluid. The continuity equations can be written in the following form when the Soret and pressure diffusions are neglected, as well as the Dufour effect:

$$\int_S \rho \mathbf{v} \cdot \mathbf{n} \, dS = 0 \tag{1}$$

$$\int_S \rho y_k \mathbf{v}\cdot\mathbf{n}\, dS \;-\; \int_S \nabla(D_k\rho\, y_k)\cdot\mathbf{n}\, dS \;=\; \int_V m_k\, S_k^{\mathrm{V}}\, dV \;+\; \int_S m_k\, S_k^{\mathrm{S}}\, dS\,, \tag{2}$$

$$\int_S \rho u_i \mathbf{v}\cdot\mathbf{n}\, dS = \int_S \mu\; grad\; u_i\cdot\mathbf{n}\, dS \;-\; \int_S p\; \mathbf{i}_i\cdot\mathbf{n}\, dS\,. \tag{3}$$

$$\int_S \rho T \mathbf{v}\cdot\mathbf{n}\, dS = \int_S \frac{\lambda}{C_p}\; grad\; T\cdot\mathbf{n}\, dS \tag{4}$$

Here ρ denotes the total gas density, $\mathbf{v}$ the gas velocity and $\mathbf{n}$ the unit vector orthogonal to the S surface and directed outwards. Further, y_k denotes the relative mass density ($y_k{=}\rho_k/\rho$), D_k and m_k are the diffusion coefficient and the mass of the species k, and S_k^{V} and S_k^{S} represent the source terms associated with volume and surface reactions, respectively. Since S_k^{S} represents a term taking into account surface losses, this term is considered in (2) only on the last grid point at the proximity of the surface. u_i is the velocity in the i direction, p the static pressure, μ the dynamic viscosity, T is the gas temperature, C_p the specific heat at constant pressure and λ the thermal conductivity.

The term S_k^{V} is the sum of the source terms associated with the various gas phase reactions, which e.g. for a two body reaction has the form

$$S_{k,l}^{\mathrm{V}} = k_{kl}\; n_k\; n_l\;, \tag{5}$$

with k_{kl} denoting the corresponding reaction rate coefficient, n_k and n_l the densities of species k and l involved in this particular reaction, with $n_k{=}y_k\rho/m_k$. The term for surface loss of the atomic species is calculated using the standard procedure

$$S_k^S = -\,\gamma_k\,\frac{v_k}{4}\,n_k\;, \tag{6}$$

where $v_k{=}\sqrt{8k_BT/\pi m_k}$ is the average velocity of k atoms and γ_k is the corresponding atomic surface loss probability. The γ surface loss probability includes all the possible surface reactions, thus it makes possible to describe the loss and creation of species on the surface without a detailed surface kinetic model.

In the afterglow the depopulation of the vibrational levels of the N_2(X, v) ground state molecules and the charged species N_2^+ starts to take place. After a slight increase of the N_2^+(X) and N_2^+(B) ions population in the early-afterglow within the first 10 ms due to the V-V pumping-up mechanism [43], the population of the ions and of the high vibrational levels decrease fast, reaching very low densities in the late-afterglow, with no effect on the densities of most abundant species. Therefore, in the hydrodynamic model of the late afterglow we neglect several excited species and all the ions, which recombine in the early afterglow region, such as: the vibrational molecular distributions $N_2(X^1\Sigma_g^+, v)$ and $O_2(X^3\Sigma_g^-, v')$; the electronic states of molecular and atomic nitrogen $N_2(B^3\Pi_g, B'^3\Sigma_u^-, C^3\Pi_u, a'^1\Sigma_u^-, a^1\Pi_g, w^1\Delta_u)$ and $N(^2D, ^2P)$; the NO_2(A) state; the main positive ions $N_2^+(X^2\Sigma_g^+, B^2\Sigma_u^+)$, N_4^+, O_2^+, O^+ and NO^+; and negative ions O^-. The set of gas phase reactions describing the kinetics of species as considered in the post-discharge chamber has been presented in [44] and [45]. The kinetic scheme for the creation of the NO(A) and NO(B) molecules has been recently revised and presented in [32].

The losses of atomic species on the wall are attributed to the NO formation on the wall and to the atomic re-association resulting in N_2 and O_2, respectively, as presented in [46]. The surface losses of atomic species depend on many parameters, e.g. type of material, cleanliness, morphology, temperature, and surface coverage. In the literature it can be found numerous works dealing with the determination of the surface recombination probabilities of atoms (γ probabilities) for different materials, however, there is a quite large discrepancy between the results obtained by different authors, which could come from a particular surface treatment, the purity of material, surface morphology, and surface properties like crystallinity, structure orientation, oxide or nitride type. The differences of several orders of magnitude found in certain cases may be also associated with the employment of different measuring techniques with different measurement errors. The atomic surface loss probability data are summarized in [46] for three different materials: Pyrex, aluminium and stainless steel.

The continuity equations are discretized using the finite volume method. The linear algebraic equation system so obtained is then solved with Stone's method iteratively using the multigrid method [47]. In our solution three grid levels are used, the finest grid has 80×40×40 control volumes.

3.3. Validity of the models

The validity of the models have been shown in several publications by comparing the modeling results with the experimental one. In the discharge region, as presented in [48], the calculated O and O_2(a) densities were compared at different N_2 percentages in the N_2-O_2 mixture in the case of a discharge generated with 2450 MHz in a 16 mm diameter tube at 1 Torr pressure. The results illustrated in Figures 8 and 9 of [48] have shown a very good agreement between the calculated and experimental data, thus proved the correctness of the model and of the chosen kinetic scheme. The applicability of the early-afterglow model was shown in [44], where the calculated and measured densities of O atoms were compared at the reactor entrance in the afterglow of a discharge generated with 2450 MHz in a 26 mm diameter tube at 2 Torr pressure in N_2-χ% O_2, with χ=1,2; as illustrated in Figure 11 of [44].

The validity of the hydrodynamic model in the reactor has been examined by comparing, under different gas flow rates, the measured UV intensities with the calculated densities of the upper level of the related emitting species. Figure 9 in [32] shows a very good agreement between the measured UV intensity at 247.8 nm (band head of the NO($A^2\Sigma^+$, v=0) to the NO($X^2\Pi$, v=2) transition) and the calculated NO(A) density along the reactor at a total pressure of 2 Torr and at three different N_2 gas flow rates: 400, 550 and 1000 sccm, when the O_2 gas flow rate is set at 8 sccm. The very good agreement found between the calculated and measured data indicates the excellent predicting capabilities of the model.

4. Results and discussion

4.1. Discharge and early-afterglow region

With the discharge model presented above it is possible to calculate the density of different species present in a microwave discharge along the discharge column. Since we are mainly interested in the afterglow, in the density of species leaving the discharge region, the calculations have been carried out only at the end of the discharge column, where the values of the electron density becomes equal to the critical density for surface wave mode propagation in a homogeneous, cold, collisionless plasma, surrounded by a dielectric. Here the species relative densities have been calculated for different discharge conditions by varying the (*i*) discharge tube radius, (*ii*) excitation field frequency, (*iii*) total gas pressure and (*iv*) initial gas mixture composition. These calculations make possible to choose those conditions where, from the application point of view, the optimal species densities are obtained. Here as an example we present the effect of the initial mixture composition on the species densities at the end of the discharge column.

Figure 2 shows the density of N(^{4}S), O(^{3}P), N_2(A), NO(X), NO(A) and NO(B) as a function of O_2% percentage in the N_2-O_2 initial gas mixture at 2 Torr pressure at the end of the discharge generated in a 0.6 cm diameter tube with 2450 MHz microwave field.

At low O_2 percentages the dominant atomic species in the N_2-O_2 discharges are the N(^{4}S) atoms. At low pressures as 2 Torr, the N(^{4}S) atoms are essentially created by direct electron impact dissociation on N_2(X $^1\Sigma_g^+$, v) molecules. With increasing O_2 percentage in the mixture the N atoms density decreases, now it is also controlled by the following populating and depopulating processes: $N_2(X,v\geq13) + O \rightarrow N(^4S) + NO(X)$ and $N(^4S) + NO(X) \rightarrow N_2\ (X,v=3) + O$. However, due to a strong depopulation of $N_2(X,v\geq13)$ levels by VT energy exchange reactions in collisions with $N(^4S)$ atoms at 2 Torr the main populating mechanism for NO(X) is associated with the reaction $N_2(A) + O \rightarrow N(^2D) + NO(X)$. Due to the increase of O atoms density with O_2 percentage, the NO(X) density also increases. The O atoms in the mixture are essentially produced via the reactions $N_2(B^3\Pi_g)+O_2(X^3\Sigma_g^-) \rightarrow N_2(X^1\Sigma_g^+,v=0)+O+O$ and $N(^4S)+NO(X) \rightarrow N_2(X^1\Sigma_g^+,v=3)+O$ and they are mainly lost by re-association on the wall of the discharge tube.

In the N_2-O_2 mixture the UV emitting species are the NO(A) and NO(B) molecules, whose densities are order of magnitude lower than the atoms and NO(X) densities. The densities of these two molecules behave similarly with the O_2 percentage, however the NO(B) density is about three orders of magnitude lower than that of NO(A). This difference can be understood from the analyses of the species populating and depopulating mechanisms. As a matter of fact, the NO(A) state is exclusively populated by collisions of $N_2(A)$ with NO(X): $N_2(A)+NO(X) \rightarrow N_2 + NO(A)$, whereas the destruction of this state occurs by radiative decay (99%), responsible for the emission of NOγ system, and by quenching with O_2 and NO(X), which contribution increases with pressure and with O_2 percentage. The production of NO(B) takes place via the three-body mechanism $N(^4S)+O+(M) \rightarrow NO(B)+(M)$, with $M=(N_2,O_2)$, which in fact is not a single-stage process, since the recombination occurs firstly on a precursor state $NO(a^4\Pi)$ [32,49]. In what concerns the NO(B)destruction, the radiative transfer to NO(X)molecules, responsible for the emission of UV NOβ system, constitutes the main loss process.

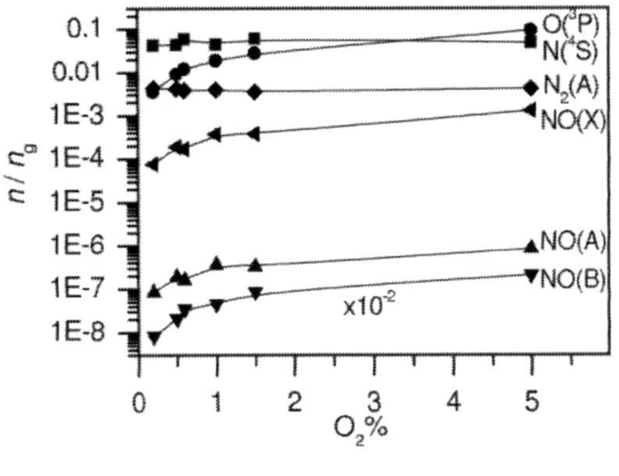

Figure 2. Calculated species densities (relative to gas density) at the end of the discharge column as a function of O_2 percentage added to N_2 at 2 Torr pressure and 2450 MHz excitation field frequency.

The early-afterglow model allows the study of the species densities evolution in time as they flow through the early-afterglow region, which develops downstream from the discharge in the connecting tube between the discharge and post-discharge reactor. In this region, since the excitation field is not present any more, the depopulation of the species starts to take place. The density of species entering the reactor can be controlled by varying the length of the connecting tube, which together with the applied gas flow determines the flight-time of species to the reactor entrance. In this way the density of species, which can cause damage to the treated materials can be minimized. In the following we examine the time evolution of the densities of the discharge effluents (ions and neutrals, metastable state species) as they enter the early-afterglow region and flow through it at 2 Torr with N_2-1%O_2 initial mixture composition.

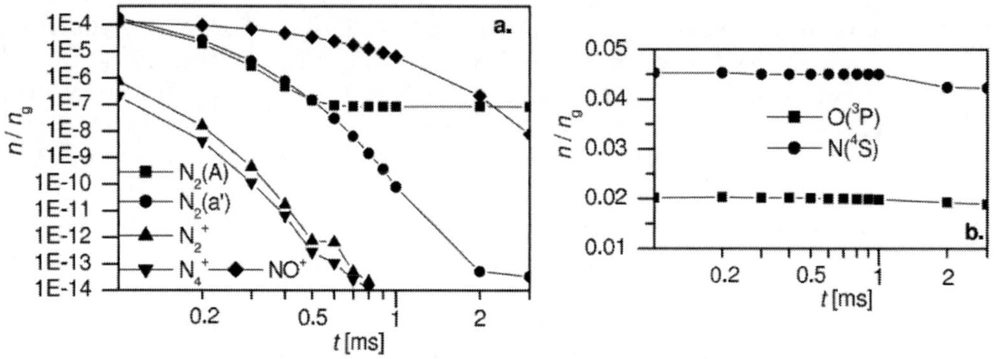

Figure 3. Time evolution of the densities (relative to gas density) of (a) metastables and ions, and (b) atoms at 2 Torr and 2450MHz with 1% O_2 added to N_2.

Figure 3(a) shows the time evolution of the densities of the ions and metastable molecules, which are believed to play a major role in the damage to materials exposed to plasma sterilization. The figure shows that the density of the N_2^+ and N_4^+ ions decreases by more than 8 orders of magnitude within 1 ms, reaching density values that ensure that no damage can result from these species. As far as the NO^+ ion is concerned, its density slightly increases initially, goes through a quasi plateau and then decreases slowly, ending at 2 ms with a value of 2 orders of magnitude lower than its initial one. The NO^+ is a long living ion, populated through the charge exchange of N^+, O^+, N_2^+ and O_2^+ in collision with neutral species. It is the dominating ion in an N_2-O_2 late afterglow [48]. In what concerns the metastable molecules, the density of N_2(a') decreases by 10 orders of magnitude within 2 ms, whereas the density of N_2(A) molecule, after an initial decrease by approximately 4 orders of magnitude over 0.5 ms, reaches a plateau that extends beyond at least 2 ms. The calculations have also shown that varying the O_2% from 0.2 to 2 the density of metastables and ions decrease faster except for the N_2(A) and NO^+ (not shown in figure).

Figure 3(b) shows the time evolution in the early-afterglow of the relative densities of O(^{3}P) and N(^{4}S) atoms, which play important role in the sterilization process. The density of O(^{3}P) does not vary during at least the first 3 ms, while the N(^{4}S) density decreases, but barely, after 1 ms. We further note that, at 1%O_2, there is approximately two times more N than O atoms. The fact that N and O atom densities keep constant in the whole transition zone is strongly supported by the experiments where increasing the distance between the surfatron and the reactor from 3 cm to 10 cm show no significant changes in these densities (Boudam *et al.* [29], Figure 14). As already mentioned these atoms play also an important role in the NO(A) and NO(B) molecules formation. Figure 4 shows the time evolution of the NO(A) and NO(B) densities at 2 Torr with N_2-0.6%O_2 initial mixture composition. In the discharge, also shown by Figure 2, the density of NO(A) molecules dominates over that of NO(B), since here the NO(A) molecules are predominantly created by the two-body reaction of highly populated $N_2(A^3\Sigma_u^+)$ state with the NO($X^2\Pi$) ground state, while the NO(B) molecules are populated through a three-body re-association of N and O atoms. In contrast, in the early afterglow, as shown in Figure 4 the calculated relative density of the NO(A) molecules decreases markedly as a function of the species flight-time t (from t = 0 ms, the end of the discharge, to the beginning of the late afterglow), actually by three orders of magnitude during the first 0.2 ms. This can be explained by the decrease of the N_2(A) state density with time in the early-afterglow, in fact according to the Figure 4 and Figure 3(a) the NO(A) and N_2(A) densities decrease with approximately the same rate during this time period. As far as the density of the NO(B) state is concerned,

Figure 4 shows that it is constant throughout the early-afterglow time period examined. As a consequence of the strong decrease of the density of the NO(A) molecules during that time period, the density of the NO(B) state ends up approximately an order of magnitude higher than that of the NO(A) state after 0.4 ms and at least up to 1 ms, thereby setting the densities of NO(A) and NO(B) states at the very beginning of the late afterglow. Even that the density of NO(A) molecules is considerably lower in the afterglow than that of NO(B), since its radiation frequency is one order of magnitude higher than that of the NO(B), the contribution to the UV emission in the afterglow of both molecules is of the same order. As shown by Figure 4 after 0.4 ms the density of NO(A) and NO(B) molecules behave similarly, this occurs because here both molecules are predominantly created through the three-body re-association of N and O atoms, the only difference is that the two states are populated through different intermedial states, for the detailed kinetic scheme see [32].

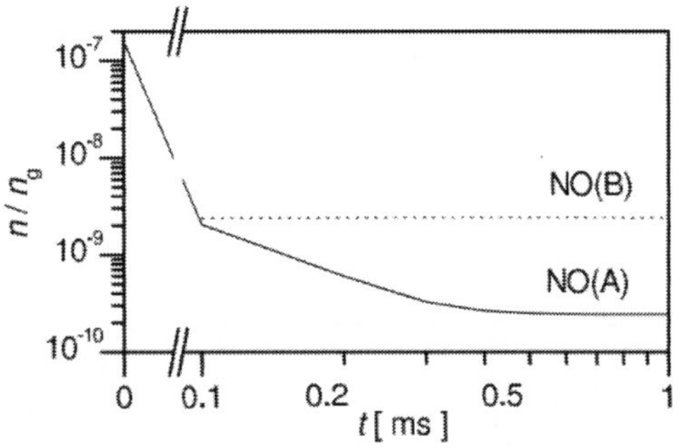

Figure 4. Calculated densities of UV emitting molecules as a function of the species flight-time in the early afterglow region, t.

4.2. Post-discharge region

The hydrodynamic model of the post-discharge reactor allows the calculation of the species densities distribution in the whole reactor. Here we present the effect of the gas flow rate applied and of the total gas pressure on the species density distributions.

4.2.1. Effect of the gas flow Figure 5 shows the calculated density distribution of NO(A), $O(^3P)$ and $N(^4S)$ states in the x-z vertical plane at y=150 mm (symmetry plane) at 2 Torr for N_2 flow rates of 400, 550 and 1000 sccm, with an O_2 flow rate set to 8 sccm. The first effect we can observe is that with gas flow the density distributions become more homogeneous, however due to the species different kinetics the distributions show different behaviours. At 400 sccm the NO(A) density decreases more than three orders of magnitude from the entrance to the opposite wall. This appears related to the corresponding decrease of the N and O atom densities, while the O atoms density decreases about one order of magnitude, the N atoms density is less with two orders of magnitude at the opposite wall comparing to the entrance. With increasing the gas flow rate the density distributions start to become more homogeneous. At 550 sccm the density decrease of the NO(A) molecules is about one order of magnitude in the reactor, similarly to that of the N atoms density, while the O atoms density decreases with about a factor of 2. In the case of a 1000 sccm gas flow, the O atoms density distribution is homogeneous in the whole reactor, while the NO(A) and N atoms densities vary just slightly in a significant part of the reactor. As already shown in the afterglow the NO(A) and NO(B) densities behave similarly, therefore the same distributions are obtained also for the NO(B) as for the NO(A), not shown here in figure. The calculations also show that the density distribution of the UV emitting molecules is mostly determined by that of the N atoms.

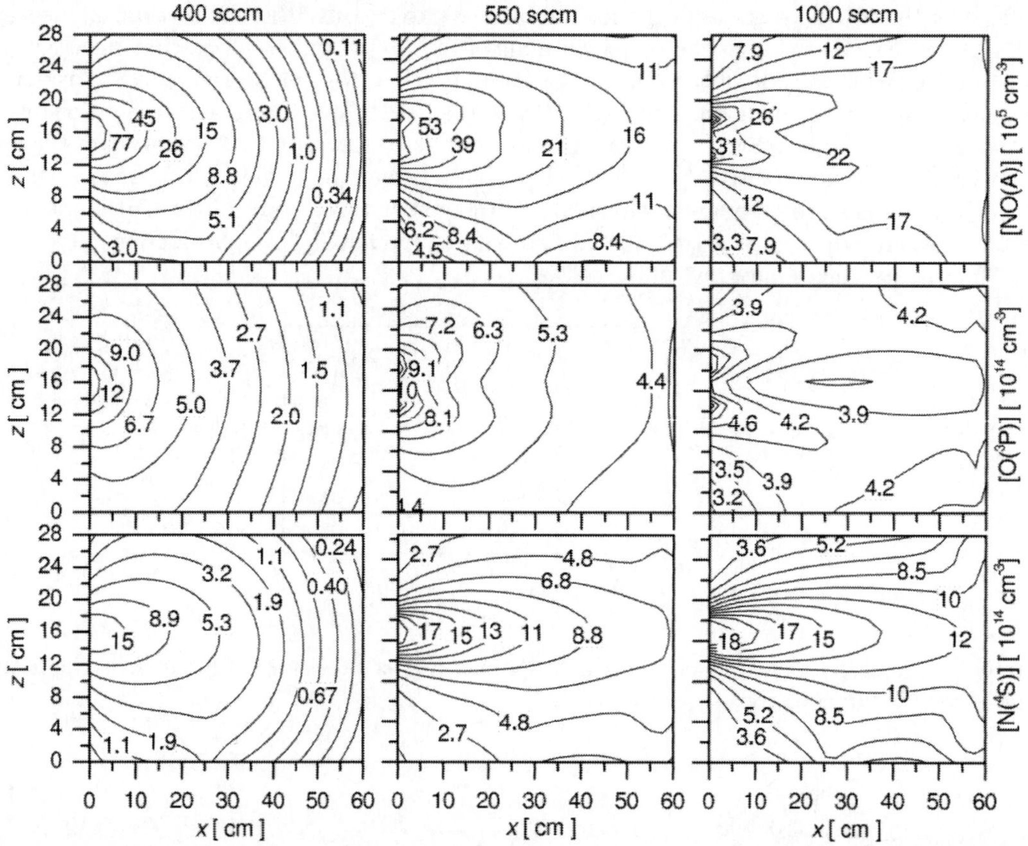

Figure 5. Calculated density distributions in the x-z plane at $y = 150$ mm in the reactor chamber at three N_2 gas flow rates with 8 sccm of O_2 added at a 2 Torr pressure and 2450 MHz field frequency. NO(A) (first row), O(^{3}P) (second row) and N(^{4}S) (third row). The densities are given in 10^5 cm^{-3} units for NO(A) and in 10^{14} cm^{-3} units for O(^{3}P) and N(^{4}S).

4.2.2. Effect of the total gas pressure Figure 6 provides spatially resolved density distribution of NO(A), N(^{4}S) and O(^{3}P) in the $x-z$ symmetry plane at different pressure values at 1000 sccm N_2 and 8 sccm O_2 gas flow rates. It shows that uniformity of the species in the whole reactor volume is obtained only at the lowest pressure investigated, namely 2 Torr.

At 5 and 10 Torr, the density NO(A) state at the discharge (and flow) axis decreases only slightly, due to the high gas flow applied, but it decreases strongly when departing from the entrance axis. A closer look at the calculated NO(A) density at 10 Torr shows that at z positions different than that of the discharge axis, the lowest density can be found near the west wall, where the flow has no effect. The calculations for O and N atoms provide spatially resolved density distribution data, that are not easily accessible experimentally. These will help better understanding the observed spatial behaviour of the NO(A) density that depends on both the N and O atom density distribution. Figure 6 shows that the O-atom density increases with gas pressure but remains spatially almost uniform throughout the reactor chamber with

Journal of Physics: Conference Series **162** (2009) 012008 doi:10.1088/1742-6596/162/1/012008

a slightly higher value along the flow. In contrast, the N-atom density for a given pressure value, is considerably higher at the axis and strongly decreases transversally to it the higher the pressure. The calculated spatial variations of the NO(A) density reflect the N-atom spatial density variations, and not those of O-atoms. In that respect, the stronger decrease of the N atom density with gas pressure is most probably due to reaction: $N(^4S) + N(^4S) + N_2 \rightarrow N_2(B) + N_2$, process that becomes more efficient with increasing pressure. The conclusion drawn from the calculations as concerns the respective role of N and O atoms on the emission intensity stemming from the NO(A) level is in agreement with the experimental data obtained for NO(B), see Figure 15 of Boudam *et al.* [29], which shows that the emission intensity follows the behaviour of N atoms and not that of the O atoms.

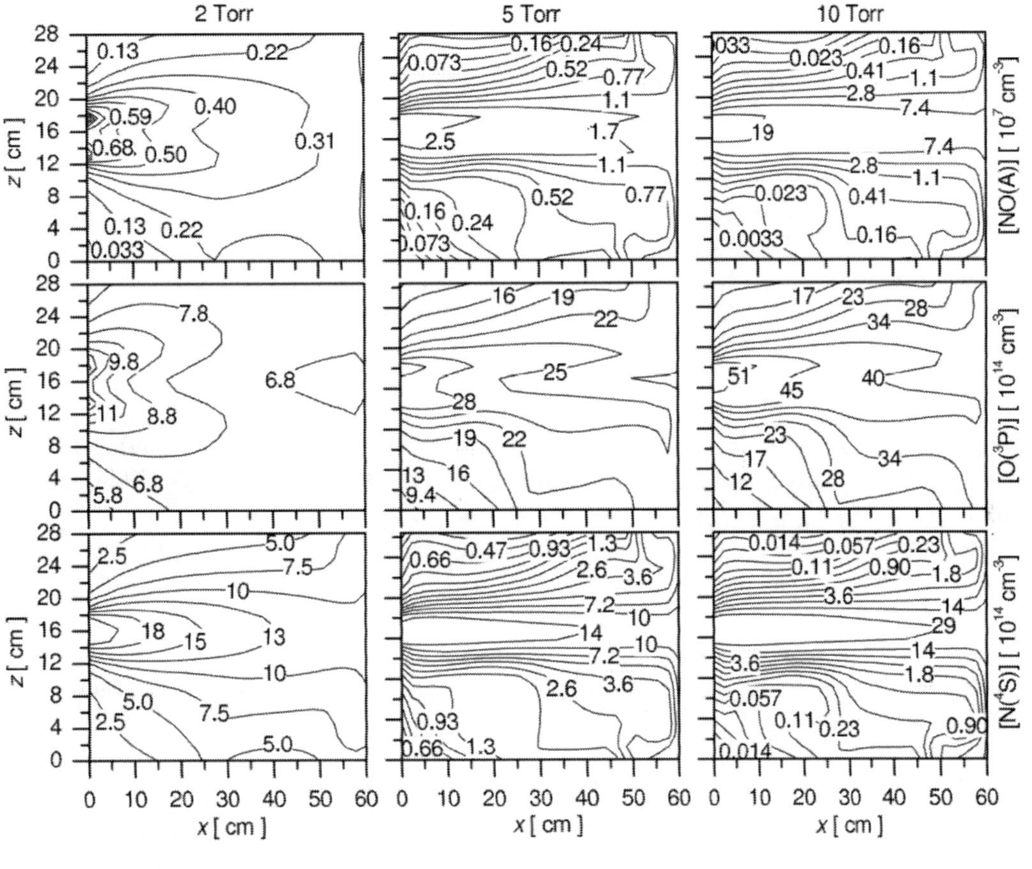

Figure 6. Calculated density distributions in the x-z plane at y = 150 mm in the reactor chamber at three gas pressure at a 1000 sccm N_2 flow rate and with 15 sccm of O_2 added (1.5%) at 2450 MHz. NO(A) (first row), $O(^3P)$ (second row) and $N(^4S)$ (third row). The densities are given in 10^7 cm^{-3} units for NO(A) and in 10^{14} cm^{-3} units for $O(^3P)$ and $N(^4S)$.

5. Conclusions

A low pressure post-discharge system used for plasma sterilization has been studied by means of modeling. The system investigated is composed of three parts: (*i*) a flowing microwave discharge, (*ii*) an early-afterglow region, which develops downstream after the end of the discharge in the same tube, and (*iii*) the late-afterglow region as a result of the entrance of the exit gas in a large volume reactor. We have presented the models valid for these different regions: a kinetic model for the discharge and early-afterglow, and a 3-D hydrodynamic model for the post-discharge reactor. The application of the models have been illustrated with the N_2-O_2 system used for plasma sterilization.

In general trends, we have shown that the kinetic model allows for the possibility of calculating the species densities in the discharge and hence of choosing the best discharge conditions to obtain the required species densities at the end of the microwave discharge. With the kinetic model the time-evolution of the species densities in the early-afterglow can be obtained as well. By modeling it makes possible to determine the optimal length of the connecting tube between the discharge tube and the post-discharge reactor for every discharge condition. This study allows to avoid that the species able to damage the material to be sterilized may enter the reactor, whereas the species playing a role in the sterilization process remain with high enough concentrations. Furthermore, we have shown that with the help of the 3-D hydrodynamic model some important discharge parameters, such as the total gas pressure and gas flow rate can be adjusted in a way that the density distributions of the relevant species for sterilization to be homogeneous in the reactor. We have demonstrated that the modeling is a useful tool for optimizing the post-discharge systems used for plasma sterilization.

Our calculations have shown that in order to achieve a homogeneous density distribution of N and O atoms, and of UV emitting NO(A) and NO(B) molecules in the whole reactor pressures as low as 2 Torr should be used and gas flows as high as 1000 sccm. If the entry of the ions into reactor is aimed to be avoided the connecting tube between the discharge and reactor should be designed in the way to assure that the species flight-time in this region is at least 1 ms.

Acknowledgments

The work has been supported by the Hungarian Science Foundation OTKA through F-67556 and by Portuguese Science Foundation FCT. The authors gratefully acknowledge Professor Michel Moisan and Dr. Bachir Saoudi for several fruitful discussions.

References

[1] Kelly-Wintenberg K, Montie T C, Brickman C, Roth J R, Carr A K, Sorge K, Wadsworth L C, Tsai P P Y 1998 *J. Ind. Microbiology. and Biotechnology* **20** 69
[2] Herrmann H W, Henins I, Park J, Selwyn G S 1999 it Physics of Plasmas **6** 2284
[3] Roth J R, Sherman D M, Gadri R B, Karakaya F, Chen Z, Montie T C, Kelly-Witenberg K 2000 *IEEE Trans. Plasma. Sci.* **28** 56
[4] Laroussi M 2002 *IEEE Trans. Plasma Sci.* **30** 1409
[5] Heise M, Neff W, Franken O, Muranyi P, Wunderlich J 2004 *Plasmas and Polymers* **9** 23
[6] Lee K-Y, Park B J, Lee D H, Lee I-S, Hyun S O, Chung K-H, Park J-C 2005 *Surf. Coat. Technol.* **193** 35
[7] Pointu A M, Ricard A, Dodet B, Odic E, Larbre J and Ganciu M 2005 *J. Phys. D: Appl. Phys.* **38** 1905
[8] Ohkawa H, Akitsu T, Tsuji M, Kimura H, Fukushima K 2005 *Plasma Processes and Polymers* **2** 120
[9] Boudam M K, Moisan M, Saoudi B, Popovici C, Gherardi N and Massines F 2006 *J. Phys. D: Appl. Phys.* **39** 3494
[10] Sato T, Fujioka K, Ramasamy R, Urayama T, Fujii S 2006 *IEEE Trans. On Industry Applications* **42** 399 (2006)
[11] Yu Q S, Huang C, Hsieh F H, Huff H, Duan Y 2006 *Appl. Phys. Lett.* **88** 013903
[12] Fridman G, Peddinghaus M, Ayan H, Fridman A, Balasubramanian M, Gutsol A, Brooks A, Fridman G 2006 *Plasma Chem. Plasma Process.* **26** 425
[13] Lee K et al. 2006 *J. Microbiology* **44** 269
[14] Park B J, Lee D H, Park J-C, Lee I-S, Lee K-Y, Hyun S O, Chun M S, Chung K H 2007 *Surf. Coat. Technol.* **201** 5738
[15] Puać N, Petrović Z Lj, Malović G, Djordjević A, Zivković S, Giba Z, Grubisic D 2006 *J. Phys. D: Appl. Phys.* **39** 3514
[16] Feichtinger J, Schulz A, Walker M, Schumacher U 2003 *Surf. Coat. Technol.* **174-175** 564
[17] Lerouge S, Wertheimer M R, Marchand R, Tabrizian M, Yahia L'H 2000 *J Biomed Mater Res.* **51** 128
[18] Nagatsu M , Terashita F, Nonaka H, Xu L, Nagata T, Koide Y 2005 *Appl. Phys. Lett.* **86** 211502

[19] Rossi F, Kylian O, Hasiwa M 2006 *Plasma Process. Polym.* **3** 431
[20] Hayashi N, Guan W, Tsutsui S, Tomari T, Hanada Y 2006 *Jpn. J. Appl. Phys.* **45** 8358
[21] Xu L, Nonaka H, Zhou H Y, Ogino A, Nagata T, Koide Y, Nanko S, Kurakawi I, Nagatsu M 2007 *J. Phys. D: Appl. Phys.* **40** 803
[22] Moreau S, Moisan M, Tabrizian M, Barbeau J, Pelletier J, Ricard A, Yahia L'H 2000 *J. Appl. Phys.* **88** 1166
[23] Lerouge S, Fozza A C, Wertheimer M R, Marchand R, Yahia L'H 2000 *Plasmas and Polymers* **5** 31
[24] Ricard A, Moisan M and Moreau S 2001 *J. Phys. D: Appl. Phys.* **34** 1203
[25] Philip N, B. Saoudi, M-Ch. Crevier, M. Moisan, J. Barbeau, J. Pelletier 2002 *IEEE Trans. Plasma Sci.* **30** 1429
[26] Monna V, Nguyen C, Kahil M, Ricard A, Sixou M 2000 *IEEE Trans. Plasma Sci.* **30** 1437
[27] Villeger S, Cousty S, Ricard A and Sixou M 2003 *J. Phys. D: Appl. Phys.* **36** L60
[28] Laroussi M 2005 *Plasma Processes and Polymers* **2** 391
[29] Boudam M K, Saoudi B, Moisan M, Ricard A 2007 *J. Phys. D: Appl. Phys.* **40** 1694
[30] Kylian O, Sasaki T, Rossi F 2006 *Eur. Phys. J. Appl. Phys.* **34** 139
[31] Moisan M, Barbeau J, Crevier M-Ch, Pelletier J, Philip N, Saoudi B 2002 *Pure Appl. Chem.* **74** 349
[32] Kutasi K , Saoudi B, Pintassilgo C D, Loureiro J, Moisan M 2008 *Plasma Process. Polym.* **5** 840
[33] Crevier M C, Moisan M, Yahia L, Saoudi B 2002 Proc. Int. Symp. on Advanced Materials for Biomedical Applications (SAMBA) (Montreal, Quebec) 365
[34] Cousty S, Villeger S, Sarette J P, Ricard A, Sixou M 2006 *Eur. Phys. J. Appl. Phys.* **34** 143.
[35] Villeger S, Cousty S, Ricard A 2005 *Plasma Process. Polym.* **2** 709
[36] Ricard A and Monna V 2002 *Plasma Sources Sci. Technol.* **11** 1
[37] Ricard A, Monna V and Mozetic M 2003 *Surf. Coat. Technol.* **174-175** 905
[38] Bockel S, Diamy A M, Ricard A 1995 *Surf. Coat. Technol.* **74-5** 474
[39] Belmonte T, Czerwiec T, Michel H 2001 *Surf. Coat. Technol.* **142-144** 306
[40] Belmonte T, Jaoul C, Borges J N 2004 *Surface Coat. Technol.* **188-189** 201
[41] Belmonte T, Czerwiec T, Garvillet J, Michel H 1997 *Surface Coat. Technol.* **97** 642
[42] Pintassilgo C D, Loureiro J, Guerra V 2005 *J. Phys. D: Appl. Phys.* **38** 417
[43] Sá P A, Guerra V, Loureiro J, Sadeghi N 2004 *J. Phys. D: Appl. Phys.* **37** 221
[44] Kutasi K, Pintassilgo C D, Coelho P J, Loureiro J 2006 *J. Phys. D: Appl. Phys.* **39** 3978
[45] Kutasi K, Pintassilgo C D, Loureiro J, Coelho P J 2007 *J. Phys. D: Appl. Phys.* **40** 1990
[46] Kutasi K, Loureiro J 2007 *J. Phys. D: Appl. Phys.* **40** 5612
[47] Ferziger J H and Peric M *Computational Methods for Fluid Dynamics* 2002, 3rd rev. ed., Springer
[48] Guerra V, Loureiro J 1999 *Plasma Sources Sci. Technol.* **8** 110
[49] Gross R W and Cohen N 1968 *J. Chem. Phys.* **48** 2582

Journal of Physics: Conference Series **162** (2009) 012009 doi:10.1088/1742-6596/162/1/012009

Electric propulsion using ion-ion plasmas

Ane Aanesland, Albert Meige and Pascal Chabert

Laboratoire de Physique et Technologie des Plasmas, Ecole Polytechnique, 91128 Palaiseau, France

E-mail: `ane.aanesland@polytechnique.edu`

Abstract. Recently, we have proposed to use both positive and negative ions for thrust in an electromagnetic space propulsion system. This concept is called PEGASES for Plasma Propulsion with Electronegative GASES and has been patented by the Ecole Polytechnique in France in 2007. The basic idea is to create a stratified plasma with an electron free (ion-ion plasma) region at the periphery of a highly ionized plasma core such that both positive and negative ions can be extracted and accelerated to provide thrust. As the extracted beam is globally neutral there is no need for a downstream neutralizer. The recombination of positive and negative ions is very efficient and will result in a fast recombination downstream of the thruster and hence there is no creation of a plasma plume downstream. The first PEGASES prototype, designed in 2007, has recently been installed in a small vacuum chamber for preliminary tests in our laboratory and the first results have been presented in several conferences. This paper reviews important work that has been used in the process of designing the first PEGASES prototype.

1. Introduction

The scientific history of space propulsion and space exploration can we traced back to the early 20th century, and was mainly driven by three pioneers Robert H. Goddard, Konstantin E. Tsiolkovskiy and Hermann J. Oberth. Their theoretical works also included suggestions on actual space propulsion concepts, and some of them have laid out many of the principles of modern space flights [1, 2, 3]. For example, any method to propel a spacecraft, space vehicle or rocket works on the same basic principle, namely to create thrust by accelerating and expelling mass. The rocket equation based on the conservation of momentum, was derived by Tsiolkovskiy and first published in 1903. This equation is still one of the most important equation for space scientist today,

$$\Delta v = v_e \ln \left(\frac{m_i}{m_f} \right), \tag{1}$$

and it shows that the velocity change, Δv, is determined by the exhaust velocity of the propellant, v_e, and the logarithm of the mass ratio between the initial and final mass of the space craft, m_i and m_f, respectively. The thrust, T, is given by

$$T = v_e \frac{dm}{dt}, \tag{2}$$

where m is the mass of the ejected propellant. Chemical rockets obtain a large thrust by expelling a lot of mass very quickly, and achieve thrust values in the order of $10^3 - 10^7$ N. This high thrust

Journal of Physics: Conference Series **162** (2009) 012009 doi:10.1088/1742-6596/162/1/012009

is needed to overcome earth's gravitation and escape into space. However, the exhaust velocity of chemical rockets is low, only in the order of a few km/s, so this method is very fuel or propellant consuming. The challenge for any kind of space missions (this being everything from satellites in orbit to long inter-stellar explorations) is in general to make the mission as cost effective as possible. One "efficient" way is to reduce the propellant burden and hence the mass of the spacecraft. Equation 1 shows that to use less propellant requires higher exhaust velocities, which is the key limitation of chemical rockets. In this respect, electric propulsion is promising and has become more and more popular in recent years. The exhaust velocity of electric propulsion systems is high and can reach up to 100 km/s for heavy propellant such as Xenon atoms. This means a significant reduction in the propellant burden: Electric propulsion systems use only 1/10th of the fuel used in chemical rockets. The thrust given by electric propulsion is relatively low (so they can not be used to leave earth), however, they provide the large velocity change (Δv) needed for long interplanetary missions.

The two main existing thrusters, already used on space missions, are gridded thrusters (also called ion engines) and Hall effect thrusters (also called closed drift thrusters)[3, 4, 5, and references therein]. In these classical electrostatic and electromagnetic thrusters (illustrated in Figure 1(a), the thrust is provided solely by extracting and accelerating positive ions. To avoid a built up of negative charge on the space vessel that will counteract the accelerating field, a flow of electrons is added to the positive ion beam downstream of the acceleration stage. These electrons are fed from a hollow cathode. To a large extent, the properties of the cathode material, it's physical configuration and the structure of the cathode plasma determine the performance and life of both ion and Hall thrusters [3]. The neutralizer is therefore very important in the present technologies and huge efforts are made to increase their performance and lifetime, but also efforts are put in to develop new concepts where the neutralizer is not needed.

The recombination between ions and electrons is a rather slow process and the presence of electrons downstream adds to the ionization in this region. Hence, even if charge neutrality is insured a downstream plasma with charged particles exists outside of the thruster body, this plasma is known as the plasma plume. The possible damage the plume may cause on the host spacecraft is the main reason why it has been and is still a great concern in using electric propulsion systems to propel a spacecraft rather than using conventional chemical propulsion [6]. One of the major problems with the plasma plume is that the accelerated ions might undergo charge exchange collisions with the slow neutrals, which then produces slow ions that can backscatter and deposit on the thruster body, solar panels, scientific instruments, etc. In the early 1970s a "Porcupine" rocket project was launched to investigate the behavior of an artificial ion beam in the Ionosphere. The results showed that the Xenon ion beam induced high-frequency turbulence as far away as 100 m from the thruster [7, 8]. High and low frequency oscillations in both current and space potential can occur due to the cathode plume interacting with the thruster plume. These oscillations increase the erosion of grids and walls and again limits the lifetime and performance of the thrusters [3, and references therein].

New electric propulsion systems are now intensively studied in order to prolong the lifetime and performances of electrostatic and electromagnetic thrusters. Some of them does not need a neutralizer [9].

1.1. Ion-ion plasma for electric propulsion

Using ion-ion plasma in order to accelerate both positive and negative ions for thrust is a new concept for electric propulsion. By creating a stratified high density plasma with an ion-ion plasma at the periphery allows to extract and accelerate both positive and negative ion beams for thrust. In this way there is no need for a downstream neutralizer and as no electrons are present in the extractors nor downstream of the acceleration stage, there is no downstream plasma plume. The concept was patented by the Ecole Polytechnique early 2007 [10] and is

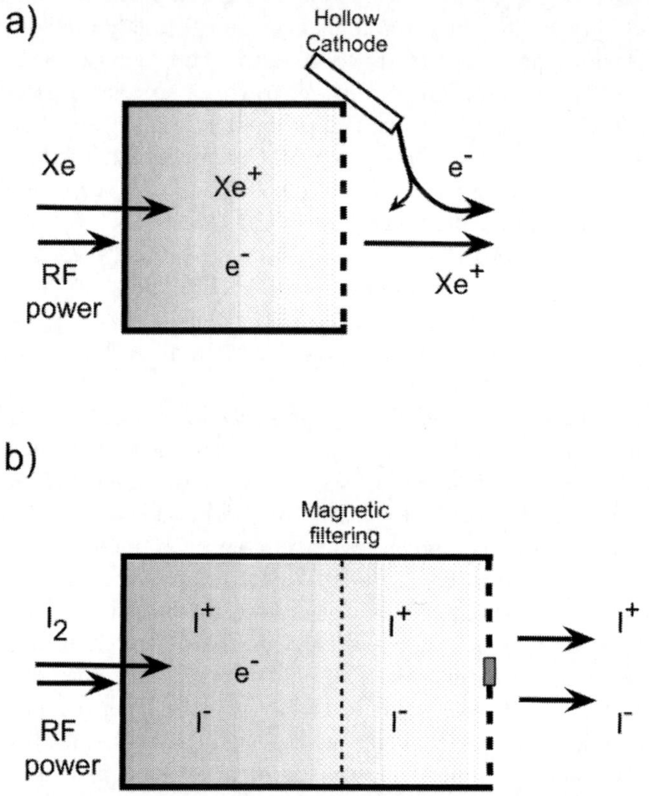

Figure 1. a) Illustration of electropositive plasma thrusters such as the Hall thruster and the gridded ion engine, where a positive xenon beam is neutralized by electrons downstream of the acceleration stage. b) Illustration of the electronegative plasma thruster concept, where both positive and negative ions are accelerated from an electron free region (ion-ion plasma).

called PEGASES for plasma propulsion with electronegative gases. The first prototype was installed in a small vacuum chamber for preliminary tests in our laboratory late 2007 and the first results have been presented at several conferences [11]. A simplified illustration of the system is shown in Figure 1(b). There are many scientific and technological issues that have to be addressed before this concept can be developed into an operating plasma propulsion system in space, and for clarity they can be divided into three stages:

Stage 1) The ionization stage with an electronegative propellant,

Stage 2) Ion-ion plasma formation,

Stage 3) Positive and negative ion acceleration.

Following this introduction on some of the space propulsion challenges this paper will review important experimental, simulation and theoretical works that have come to the design of the first ion-ion plasma thruster prototype. These works are mainly focused on strong electronegative plasmas and in particular on ion-ion plasmas. Section II discusses the ionization stage and the use of electronegative gases, section III the ion-ion formation and section IV the challenges on accelerating both positive and negative ions, and finally a conclusion with the description of the first PEGASES design is given in section V.

2. Stage 1, Ionization stage

2.1. The electronegative propellant

The propellant or fuel has to be an electronegative gas in order to produce both positive and negative ions. The best propellant is presumably iodine, I_2, which is heavy (good for high thrust) and very electronegative and therefore will efficiently produce negative ions. The attachment cross section is maximum and very high at zero (electron) energy. In addition, the ionization thresholds are low, about 9.41 eV for I_2 and 10.5 eV for I, fairly easily leading to fully ionized plasma core at moderate rf power. Iodine is in solid state at room temperature, but with a large vapor pressure, which allows small and light conditioning tanks. Finally, it is inexpensive.

Already in the 1930s and 40s iodine plasmas were known to produce strong electronegative plasmas, and they were therefore used to study the plasma dynamics influenced by negative ions. The works performed during this period has been reviewed in the book by Emeleus and Woolsey [12]. In the 1970's halogen ion sources were investigated as negative ion sources in fusion reactors [13]. Bacal and Doucet [14, 15] obtained what they called *electronegative ion-rich plasmas* with electronegative ion densities of 90 % of the positive ion densities in iodine plasmas. The negative ion densities in iodine was much higher than what was obtained in Oxygen under similar conditions. They found that the upper limit of the electronegative density in their system was limited by the required presence of electrons to ensure an electrostatic confinement of the ion-ion plasma by the ambipolar diffusion of positive ions and electrons.

Recently, iodine has been proposed as a lower cost propellant alternative for ion and hall effect thrusters [16]. The reason for suggesting iodine is based on the same arguments as given above (the low ionization potential of both I_2 and I, the high atomic mass of I, as well as the weight savings associated with the storing a solid fuel with low vapor pressure), but its ability to create negative ions was not considered here. As pointed out by both Dressler *et al* and Tverdokhelbov *et al* [16, 17] an important disadvantage of using Iodine is its corrosiveness. There are therefore currently no active development efforts on iodine propellant for electropositive thrusters [18]. However, given the substantially more corrosive chemical thruster fuels such as hydrazine and ammonia, it is not evident that the corrosiveness of iodine will prevent the use if iodine in space. Not to forget, in space there is no need for sophisticated and fragile vacuum pumps, as in the laboratory, so the corrosiveness and contamination of iodine is therefore not of a great concern when used in space. But for preliminary tests in the laboratory it would be some extra expenses to have high quality vacuum for iodine experiments, and tests should therefore first be made in Cl_2 as suggested by Grisham [19, 13]. Nevertheless, any electronegative gas can be used as propellant.

2.2. Efficiency of ion-ion plasmas

Electronegative gases are molecular gases where energy goes into dissociation and change of vibrational states in addition to ionization and attachment. Hence, creating both positive and negative ions costs more in electrical power than creating only positive ions, although the electrons undergoing attachment collisions have already been created by the electron-positive-ion pair. The recombination between positive and negative ions is very efficient, resulting typically in a lower ion densities in ion-ion plasmas than in electropositive plasmas. One of the big challenges for plasma propulsion with ion-ion plasmas lays therefore in creating high density ion-ion plasmas.

3. Stage 2, Ion-Ion plasma formation

3.1. Ion-ion plasmas

Ion-ion plasmas are electron-free plasmas that consist of positive and negative ions, only (in addition to the neutral particles). In other words, the plasmas dynamics of the negatively charged particles is dominated by heavy "stationary" negative ions rather than by light and

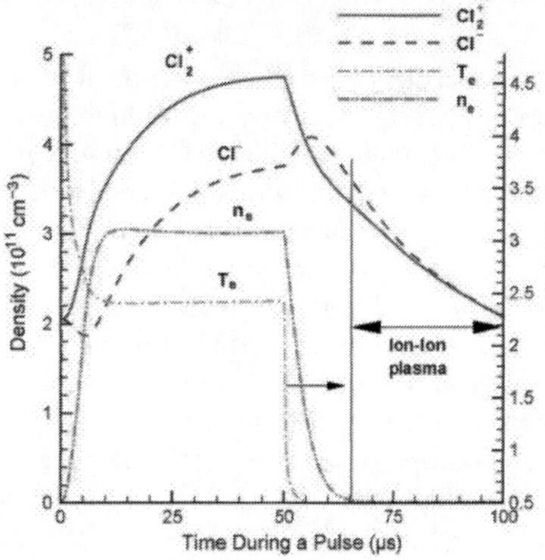

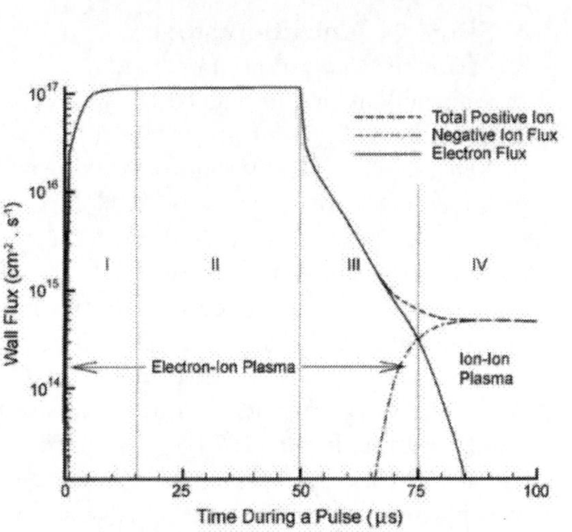

Figure 2. Temporal evolution of (a) ion and electron densities and electron temperature, and (b) ion and electron fluxes striking an unbiased electrode, in a high density pulsed chlorine plasma. Power density 1.0 W/cm3, pressure 20 mTorr, pulse period 100 ms, and duty cycle 50%. Figures taken from [24, 23].

mobile electrons. Such plasmas, with *only* heavy stationary charged particles, are significantly different from the more common electropositive and weakly electronegative plasmas. Ion-ion plasmas have been investigated both experimentally and theoretically (see the following references below), but not to a great extent and there are many remaining physical properties to be explored in this field of plasma physics. Ion-ion plasmas may be formed in the afterglow of pulsed electronegative plasmas, in electron beam generated plasmas and in the periphery of magnetically confined electronegative plasmas (we will come back to this later).

Electronegative plasmas operated in a continuous mode often consist of a core, where negative ions accumulate, surrounded by an electropositive region, with only positive ions and electrons, in front of the positive ion sheath close to the wall. This stratification of electronegative plasmas is due to the trapping of negative ions by the ambipolar electric field set up by the mobile electrons. Thus in this situation the negative ion flux to the wall is essentially zero. Operating the plasma in a pulsed mode destroys this stratified structure in the afterglow and allows the formation of an ion-ion plasma near the walls [20]. Ahn *et al* and Malyshev *et al* [21, 22] recognized that pulsed electronegative discharges can ameliorate anomalous etch profiles, such as notching and other forms of charging damage that occurs in conventional continuous wave discharges, due to the access of the negative ions at the wall. Figure 2 shows a 1D fluid simulation of the temporal evolution of densities, temperatures and ion fluxes to a non biased substrate, in a pulsed chlorine discharge by Midha and Economou [23, 24]. When the power is switched off the electrons respond to the fastest time scale (0.1 μs), electrons are thermalized and their temperature drops very quickly. This temperature drop is then followed by a rapid drop in the electron density and an increase in the negative ion density. The ion-ion plasma is formed on a longer timescale in the order of 10 μs where the ions can react.

For a plasma propulsion system, the ion-ion plasma formed in the afterglow of a pulsed plasma would not be very efficient as the densities drop significantly, as can be seen from figure 2. Pulsing the plasma also add one more element that can fail and makes the system heavier and bigger. Hence, for an electronegative ion thruster a continuous operation is better, but access to the negative ions in the acceleration stage is needed (in contrast to the above continuous

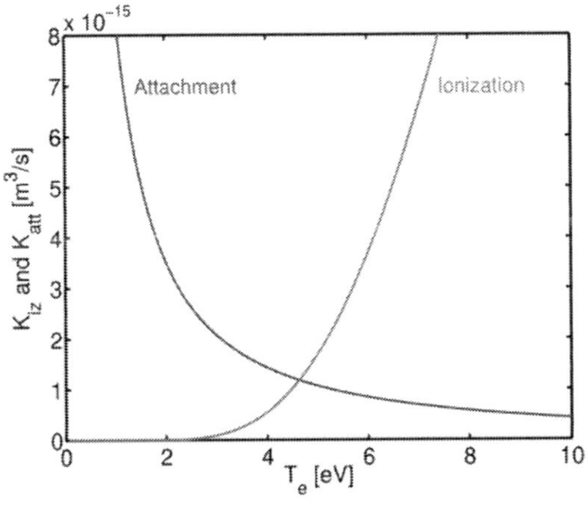

Figure 3. Ionization and attachment rate coefficients as a function of electron temperature in SF_6.

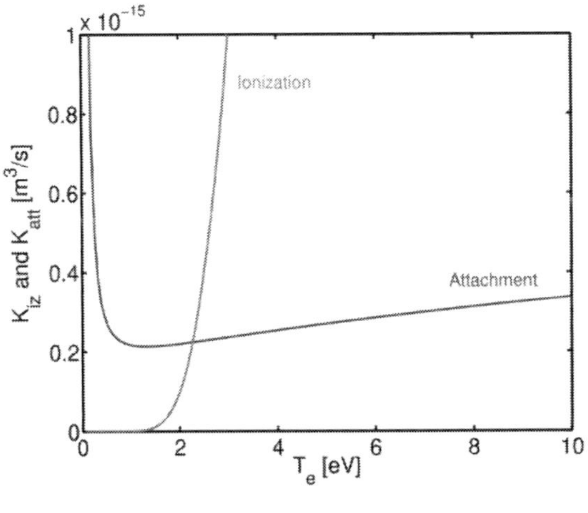

Figure 4. Ionization and attachment rate coefficients as a function of electron temperature in Cl_2.

situation).

Walton *et al* [25, 26] produces large negative ion densities in continuously operated electron-beam-generated plasmas. In these plasmas the electron temperature is very low ($T_e < 1.0$ eV), hence the electron attachment rate is high and becomes comparable to the ionization rate. Figures 3 and 4 show, respectively, the attachment and ionization rates for SF_6 and Cl_2 as a function of electron temperature (cross sections integrated over a maxwellian distribution of electrons). It can be seen that below 1-2 eV the attachment rate becomes comparable and even dominant over the ionization rate. In this case most electrons created are converted to negative ions, hence forming an ion-ion plasma. We will come back to the work by Walton *et al* in section IV, where we discuss the extraction and acceleration of positive and negative ion beams. Producing an ion-ion plasma by electron beams is very efficient, however, the electron beam is fed from a hollow cathode. As we saw in the introduction the hollow cathode has a limited lifetime and we try to get rid of them in space propulsion systems.

3.2. Electron filtering by magnetic fields

Another way to produce an ion-ion plasma region, where the ions are accessible for acceleration, is by using moderate static magnetic fields to confine the electrons along the field lines but allow the ions to move perpendicular to the field, i.e. electron filtering. For example, Walton *et al* [25, 26] use a magnetic fields of the order of 100 - 200 Gauss to collimate the electron beam, while Kawai *et al* [27] use a magnetic fields of the order of 2 kG which confines also the ions in a beam generated plasma. In both of these configurations ion-ion plasmas are achieved in the active phase of operation in both Ar/SF_6 mixtures and CF_4 with ion densities (in the ion-ion region) in the range of 10^9 cm^{-3}. Amemiya *et al* [28] produces an almost electron-free region in a target chamber separated from the source chamber by a magnetic filter consisting of an array of permanent magnets. The source plasma was in this case a capacitively coupled oxygen plasma with a source plasma density of $10^9 - 10^{10}$ cm^{-3} and a target ion-ion plasma density of $10^8 - 10^9$ cm^{-3}. Grisham *et al* [29, 19, 13] generate negative ion beams from a magnetic cusp-confined rf-driven volume production source. The driver plasma is separated from the extractor plane by a pair of permanent magnets with a filter field of around 300 Gcm and a peak value of about 180 G. An ion-ion plasma forms the extractor region between the magnetic filter and

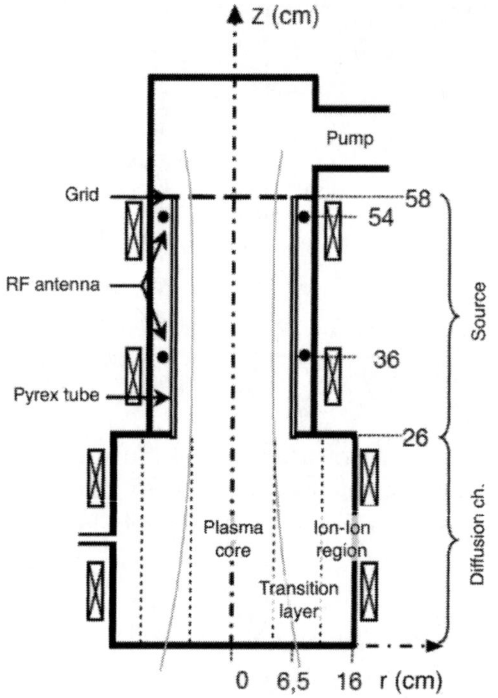

Figure 5. A schematic of the Helicon source at LPTP (Ecole Polytechnique), indicating the plasma core (weakly electronegative plasma), the transition layer and the ion-ion plasma region.

the grids, where the extracted negative ion current (Cl^-) is 85-90 % of the positive ion current (Cl^+ and $Cl_2{}^+$). Considering the position of their diagnostic probe it is likely that the negative ions undergo more loss due to stripping reactions than the positive ions suffer due to charge exchange. The negative ion current density extracted from this source is 12 mA/cm^2 at 30 kV extraction and 15 kW rf.

At the Ecole Polytechnique, we have been studying the stratification of electronegative plasmas due to the magnetic filtering in a helicon source operating in SF_6 and mixtures of SF_6 and argon [30, 31, 32, 11]. These works have led to the design of the first PEGASES prototype, and we will now briefly review these results.

A schematic of this Helicon experiment is shown in Figure 5. The negative ion fraction was characterized using an electrostatic two-probe method [30] and by photodetachment [32]. Figure 6 shows a stratified plasma with: 1) A plasma core where the source electrons are confined by the axial magnetic field and where the plasma potential is greatest. 2) A transition layer with a potential drop of several volts and 3) an ion-ion region near the walls, where the electrons can be neglected. Figure 7 shows Langmuir probe characteristics obtained in the plasma edge/ion-ion region for different radial positions. At $r = 13$ cm the current collected by the probe biased positively is very large compared to the positive ion current collected when the probe is biased negatively. This is due to the presence of hot electrons in addition to negative ions. As the probe is moved towards the edge/radial wall the characteristic becomes symmetrical with almost equal positive and the negative currents (the difference being mainly a result of the mass difference between the positive and negative ions). Here the negative current is dominated by

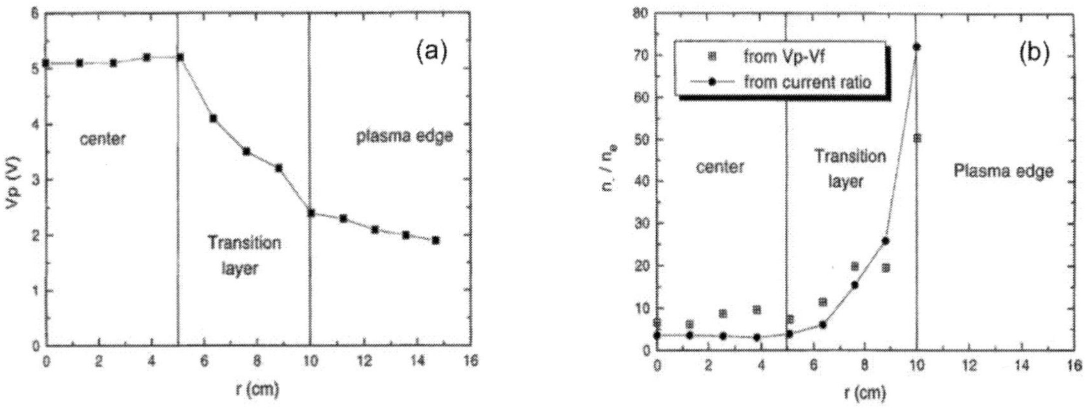

Figure 6. Radial profile of plasma potential (a) and negative ion fraction (b) in the diffusion chamber of the helicon reactor for pure SF6 plasma operating at 400 W rf power. Figures taken from Chabert *et al* [30].

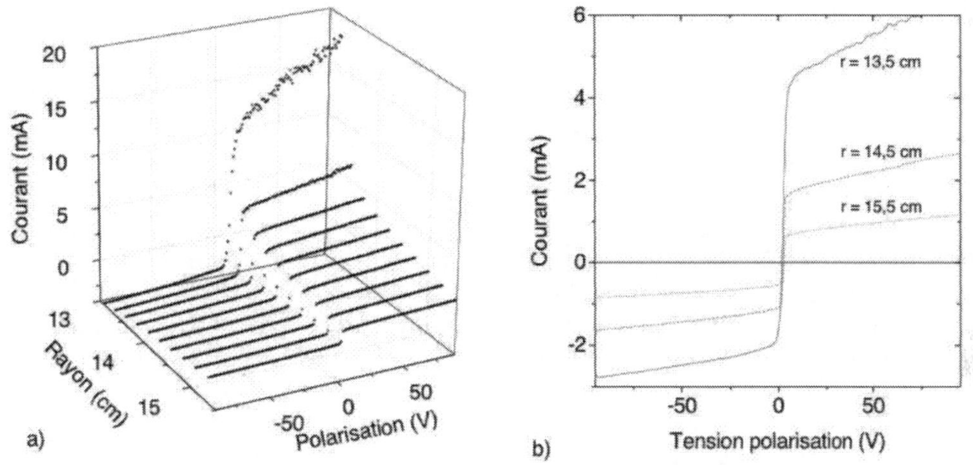

Figure 7. Langmuir probe characteristics for different radial positions in the edge region of the Helicon diffusion chamber operating with a mixture of Ar and SF6 (1:1) at 0.13 Pa (1 mTorr), 600 W rf power and a magnetic field of about 100 G. Figure taken from N. Plihon [32, PhD thesis at Ecole Polytechnique], and a similar figure is published in Corr *et al* [31].

cold negative ions, and shows that electrons have been filtered by the magnetic field. The edge plasma is electron-free and is only composed of positive and negative ions. It was sown that operating the helicon source in inductive mode without a magnetic field creates a highly unstable electronegative plasma in the whole region of the diffusion chamber Corr *et al* [31]. When the magnetic field increases a transition occurs from the unstable electrostatic confinement to a magnetic confinement where the plasma becomes stable and the stratification showed above occurs. Figure 8 shows the ratio of the negative ion fraction in the centre and at the edge of the diffusion chamber as a function of the magnetic field, indicating a strong stratification at higher magnetic fields. The ion densities in the ion-ion region is, also here, about 10^9 cm^{-3}. Figure 7(b) shows that the ion densities drop rapidly by more than 50 % within 1-2 cm. The challenge

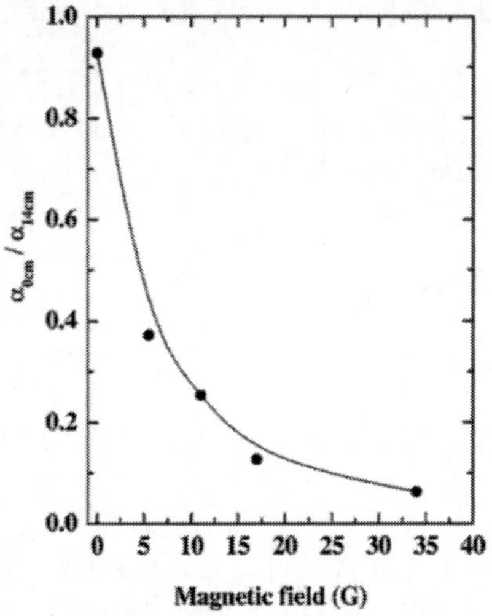

Figure 8. The ratio between the negative ion fraction in the centre (0 cm) and the edge of the diffusion chamber (14 cm) as a function of the magnetic field. Figure taken from Corr *et al* [31].

for efficient utilization of ion-ion plasmas for space propulsion as well as for the etching industry and fusion applications is to be able to increase the ion densities in the ion-ion plasma region.

3.3. Magnetic electron filtering: modeling and simulation
The formation of ion-ion plasmas or highly electronegative plasmas has been investigated by analytical and numerical modeling. The focus has mainly been on the spatio-temporal evolution of pulsed electronegative discharges [33, 34, 23, 20, and references therein]. The formation of ion-ion plasmas by electron magnetic filtering has been very little investigated. Franklin and Snell [35] modeled the positive column with negative ions in a magnetic field. They examined a low pressure plasma (in the mTorr range) where electrons are essentially confined along the field lines in the center of the discharge while ions are not and can move perpendicular to the field. They assumed that the negative ion flux was zero at the boundary wall and that the system was infinitely long along the magnetic field lines. Under these conditions, they obtained solutions with an ion-ion plasma at the periphery. These assumptions do not fit the experimental conditions for the PEGASES electronegative plasma thruster where the system is bounded in the parallel direction and the negative and positive ion fluxes out of the system is finite.

We have recently developed a two-dimensional magnetized plasma fluid simulation to investigate the electron magnetic filtering in electronegative plasmas [36]. The model uses the three first moments of the Boltzmann equation, namely the continuity equation, the conservation of momentum approximated by the drift-diffusion equation and an energy equation for the electrons. The various reaction rates, mobility and diffusion constants (accounting for the presence of an axial magnetic field) are calculated from experimental cross-sections by using a Boltzmann solver. The model is simulating an oxygen plasma following the various species O_2, O, O_2^+, O^- and 14 reactions between them are taken into account (momentum transfer, dissociative attachment, dissociation, ionization, electron impact detachment, dissociative recombination, mutual neutralization etc.). The magnetic field is uniform and parallel to the system axis. Figure 9 shows results obtained with a 30 cm long and 20 cm diameter grounded cylinder

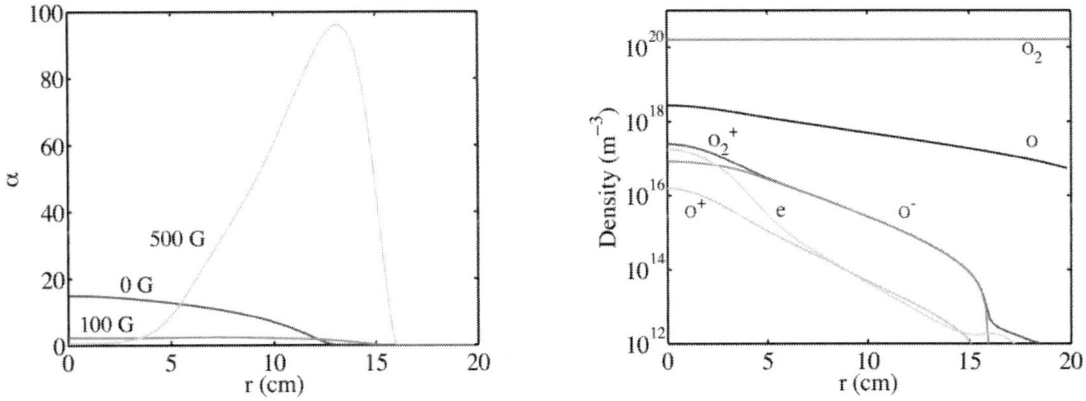

Figure 9. (a) The negative ion fraction $\alpha = n_-/n_e$ as a function of radius. The system is a 30 cm long and 20 cm diameter grounded cylinder with a neutral oxygen pressure of 5 mTorr and inductive power of 50 W. (b) The corresponding particle densities as a function of r in the 500 G case (red=O_2, black=O, blue=O_2^+, pink=O^-, cyan=e and green=O^+.)

where the neutral pressure was 5 mTorr and the inductive power was 50 W with a heating profile maximum on the axis of the cylinder and exponentially decaying from the center of the discharge.

Figure 9(a) shows the radial negative ion fraction $\alpha = n_-/n_e$ for three different magnetic fields and confirms that for low or null magnetic field, the plasma presents the usual stratified structure with an electronegative core (α large) and an electropositive periphery (α almost null), while for a sufficiently large magnetic field strength, the plasma appears to be electronegative along the whole radius. In addition, for $r > 12$ cm the negative ion density becomes several orders of magnitude larger than that of the electrons, hence forming an ion-ion plasma at the periphery of the discharge. Figure 9(b) shows the neutral, ion and electron densities as a function of radius for the 500 G case. The electron density is high in the inductive heating area decreasing rapidly by several orders of magnitude around 5 cm. At this position, the negative ion fraction increases and the negative ion density becomes of the order of the positive ion density (O_2^+). At a radius > 10 cm the ion-ion plasma is formed with an ion density of the order of 10^9 cm^{-3}.

The simulation has been validated for known parameters and will be used to investigate the large parameter space in order to make the ion-ion source as efficient as possible.

4. Extraction and acceleration stage

The sheath, the pre-sheath and their formation are fundamental features in the design of extraction grids for ion beam generation. Due to the lower temperature and greater mass of negative ions compared to electrons, the sheath structure in ion-ion plasmas differs significantly from that of conventional electron-ion plasmas. For the case of the PEGASES prototype we envisage two methods for ion extraction and acceleration via grids. The first is to have two separate, but neighboring grids, one biased negatively to extract a positive beam and one biased positively to extract a negative beam. The second method is to use one grid alternately biased positively and negatively.

Kanakasabapathy *et al* [37, 38] and Walton *et al* [25, 26] have shown that comparable fluxes of positive and negative ions can be extracted from an ion-ion plasma using a low-frequency sinusoidal bias. Kanakasabapathy *et al* extracted ion beams from an ion-ion plasma formed in the afterglow of pulsed plasmas, while Walton *et al* extracted both ion species during the active phase of a pulsed plasma. Midha, Economou and coworkers [39, 40] developed a time-dependent

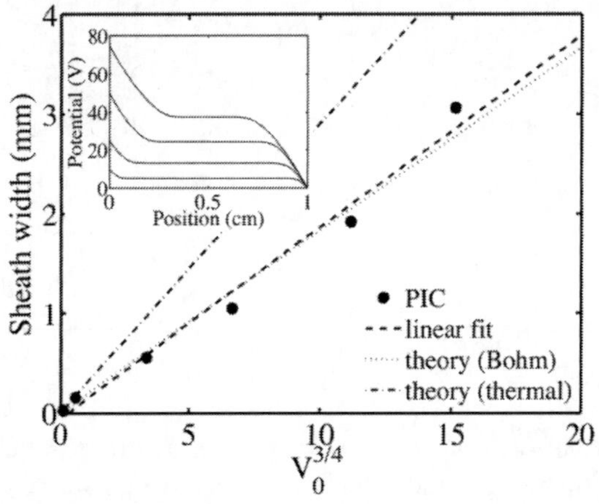

Figure 10. Sheath width as a function of DC bias. The black circles (•) are sheath width from the simulation and the dashed line is the corresponding linear fit, while the dotted and the dotted-dashed lines are the sheath width predicted by the theory assuming a Bohm velocity and a thermal velocity, respectively. The inset shows the electric potential across the plasma for various DC biases. Figure taken from Meige *et al* [41].

fluid model to investigate the dynamics of an ion-ion plasma under the influence of a rf and a dc voltage and found that the sheath structure differed significantly from conventional electron-ion plasmas. When the biased frequency is varied the sheath region shows profound structure changes, where the response of the ion-ion plasma to an rf bias depends on the characteristic time scale for collisions and the ion plasma frequency. To effectively extract positive and negative beams with very high velocities using one biased grid therefore require an optimization of both the frequency and the waveform. We have investigated the size of the sheath and the existence of a pre-sheath via a particle-in-cell simulation [41]. We found that Child-law type sheaths forms in ion-ion plasmas under the influence of a dc bias, but the size is controlled by the reflected ion species rather than by that of the electrons. The sheath formation is within a few microseconds which corresponds to the ion transit time. In addition, there exists the equivalent of a Bohm criterion in ion-ion plasmas (as for classical ion-electron plasmas) where the electron temperature is replaced by the temperature of the ion reflected by the sheath. Figure 10 shows the sheath width s as a function of $V_0^{3/4}$ (black dots). This width is well fitted by a straight line (dashed line) which slope is very close (within 5%) to the theoretical high voltage sheath using the corresponding ion temperature. The dotted-dashed line shows the theoretical sheath width if ions were entering the sheath with a thermal velocity rather than with the sound speed (i.e. if there were no presheath). The consequence of this finding is twofold: the ion velocity at the entrance of the sheath of an ion-ion plasma is larger than the average velocity along one direction of a thermal distribution and, consequently, the sheath size is smaller, which, for the design of acceleration grids, is crucial.

We also observed that even a slight difference in the positive and negative masses (as small as 5%) yields a strong asymmetric potential, where the bulk potential sits much closer to one of the electrodes, hence, getting closer to a conventional electron-ion plasma. This issue needs to be fully understood if molecular gases such as O_2 and SF_6, which forms ions of different masses, are used as propellant in the electronegative thruster.

Figure 11. (a) Photo of the PEGASES plasma thruster prototype operating in Argon (b) The drawing of prototype I.

5. Conclusion

The PEGASES thruster (Plasma propulsion with Electronegative GASES) is a new concept of plasma propulsion where both positive and negative ions are used for thrust. In this way there is no need for a downstream neutralization by electrons. A downstream plasma plume is presumably avoided due to the absence of downstream electrons and the efficient recombination of positive and negative ions (compared to ion electron recombination).

In this paper we have reviewed the works that have lead to the design of the first PEGASES prototype, where the use of electronegative gases, the formation of an ion-ion plasma and extraction and acceleration of ion beams have been discussed.

To conclude, the first PEGASES prototype is shown on figure 11(a) and the schematic of the prototype is shown in Figure 11(b). Note that in this photo PEGASES is operating in Argon (which is not an electronegative gas) for mass efficiency tests and calibration of diagnostics.

The main plasma is produced in a long cylinder, 20 cm long and 4 cm in diameter. Four solenoids create a longitudinal magnetic field of moderate magnitude. Typically 100-200 Gauss is obtained with dc currents of 1-3 Amperes. This field confines the electrons but not the ions. The plasma is generated by a three loop radio frequency (13.56 MHz) antenna coupled to the plasma either in inductive or helicon mode. 50 W to 2 kW rf power is used. As the thruster is immersed completely into the vacuum chamber, the plasma is pulsed above 400 W to avoid overheating. The ion-ion plasma is formed in the two extractors, and will be accelerated via biased grids. These grids are not yet installed. Preliminary results using SF_6 and O_2 as a propellant indicates a strong electronegative plasma in the extractors, with a negative ion fraction ($\alpha = n_-/n_e$) of 30 in SF_6 and 20 in O_2 under similar conditions. Results from the prototype will be published in the near future.

Acknowledgments

This research was supported by a Marie Curie Intra-European Fellowship within the 6th European Community Framework Program, grant nr: 039757 and by ANR (Agence National de la Recherche) under Contract No. ANR-06-JCJC-0039.

References

[1] Stuhlinger E 1964 *Ion Propulsion for Space Flight* (New York: McGraw-Hill)

[2] Jahn R G 1968 *Physics of Electric Propulsion* (New York: McGraw-Hill)
[3] Goebel D M and Katz I 2008 *Fundamentals of Electric Propulsion: Ion and Hall Thrusters* (John Wiley & Sons) URL `http://descanso.jpl.nasa.gov/SciTechBook/SciTechBook.cfm`
[4] Kaufman H R 1985 *AIAA Journal* **23** 78–87
[5] Zhurin V V, Kaufman H R and Robinson R S 1999 *Plasma Sources Science Technology* **8** 1–+
[6] Boyd I D 2001 *Journal of Spacecraft and Rockets* **38** 381–387
[7] Pottelette R, Illiano J M, Bauer O H and Treumann R 1984 *Journal of Geophysical Research* **89** 2324–2334
[8] Pottelette R, Treumann R A, Bauer O H and Lebreton J P 1998 *Journal of Geophysical Research* **93** 14,688–14,698,
[9] Charles C, Boswell R, Alexander P, Costa C, Sutherland O, Pfitzner L, Franzen R, Kingwell J, Parfitt A, Frigot P E, Del Amo J and Saccoccia G 2008 *IEEE Transactions on Plasma Science* **36** 1196–1197
[10] Chabert P 2007 *WO 2007/065915 A1*
[11] Aanesland A, Leray G and Chabert P 2008 *44th AIAA/ASME/SAE/ASEE Joint Propulsion Conference and Exhibit, Hartford, CT, July 21-23* AIAA–2008–5198
[12] Emeleus K G and Woolsey G A 1970 *Discharges in Electronegative Gases* (Taylor and Francis LTD (London))
[13] Grisham L 2008 *Plasma Science, IEEE Transactions on* **36** 1512–1518 ISSN 0093-3813
[14] Bacal M and Doucet H J 1973 *Plasma Science, IEEE Transactions on* **1** 91–99 ISSN 0093-3813
[15] Doucet H J 1970 *Physics Letters A* **33** 283–284
[16] Dressler R A, Chiu Y H and Levandier D J 2000 *AIAA, Aerospace Sciences Meeting and Exhibit*
[17] Tverdokhlebov O S and Semenkin A V 2001 *AIAA/ASME/SAE/ASEE Joint Propulsion Conference and Exhibit*
[18] Kieckhafer A and King L 2007 *41st AIAA/ASME/SAE/ASEE Joint Propulsion Conference and Exhibit,*
[19] Grisham L R, Kwan J and Westenskow G 2007 *Nuclear Instruments and Methods in Physics Research Section A: Accelerators, Spectrometers, Detectors and Associated Equipment* **577** 267
[20] Ashida S and Lieberman M A 1997 *Japanese Journal of Applied Physics* **36** 854–861
[21] Ahn T H, Nakamura K and Sugai H 1996 *Plasma Sources Science and Technology* **5** 139–144
[22] Malyshev M V, Donnelly V M, Colonell J I and Samukawa S 1999 *Journal of Applied Physics* **86** 4813–4820
[23] Midha V and Economou D J 2000 *Plasma Sources Science Technology* **9** 256–269
[24] Economou D J 2007 *Applied Surface Science* **253** 6672 – 6680
[25] Walton S G, Leonhardt D, Fernsler R F and Meger R A 2002 *Applied Physics Letters* **81** 987–989
[26] Walton S G, Leonhardt D, Fernsler R F and Meger R A 2003 *Applied Physics Letters* **83** 626–628
[27] Kawai R and Mieno T 1997 *Japanese Journal of Applied Physics* **36** L1123–L1125
[28] Amemiya H 1991 *Japanese Journal of Applied Physics* **30** 2601 – 2605
[29] Grisham L R, Kwan J W, Hahto S K, Hahto S T, Leung K N and Westenskow G 2006 *Review of Scientific Instruments* **77** 03A501
[30] Chabert P, Sheridan T E, Boswell R W and Perrin J 1999 *Plasma Sources Science Technology* **8** 561–566
[31] Corr C S, Plihon N and Chabert P 2006 *Journal of Applied Physics* **99** 103302 (pages 6)
[32] Plihon N 2006 *Stabilité et structure électrique d'une décharge inductive en gaz électronégatif* Ph.D. thesis Ecole Polytechnique
[33] Lichtenberg A J, Kouznetsov I G, Lee Y T, Lieberman M A, Kaganovich I D and Tsendin L D 1997 *Plasma Sources Science and Technology* **6** 437–449
[34] Kaganovich I D, Economou D J, Ramamurthi B N and Midha V 2000 *Physical Review Letters* **84** 1918–1921
[35] Franklin R N and Snell J 1999 *Journal of Physics D Applied Physics* **32** 1031–1038
[36] Meige A, Hagelaar G and Chabert P 2008 *24th summer school and International Symposium on the Physics of Ionized gases* (Novi Sad, Serbia)
[37] Kanakasabapathy S K and Overzet L J 1998 *Plasma Sources Science Technology* **7** 289–297
[38] Kanakasabapathy S K, Overzet L J, Midha V and Economou D 2001 *Applied Physics Letters* **78** 22–24
[39] Midha V and Economou D J 2001 *Journal of Applied Physics* **90** 1102–1114
[40] Midha V, Ramamurthi B and Economou D J 2002 *Journal of Applied Physics* **91** 6282–6287
[41] Meige A, Leray G, Raimbault J L and Chabert P 2008 *Applied Physics Letters* **92** 061501 (pages 3)

Journal of Physics: Conference Series **162** (2009) 012010 doi:10.1088/1742-6596/162/1/012010

The Electrical Asymmetry Effect - A novel and simple method for separate control of ion energy and flux in capacitively coupled RF discharges

U. Czarnetzki[1], B. G. Heil[1,4], J. Schulze[1], Z. Donkó[2], T. Mussenbrock[3], and R. P. Brinkmann[3]

[1]Institute for Plasma and Atomic Physics, Ruhr-University Bochum, Germany
[2]Research Institute for Solid State Physics and Optics of the Hungarian Academy of Science, Budapest, Hungary
[3]Institute for Theoretical Electrical Engineering, Ruhr-University Bochum, Germany
[4]Current address: Richardt Patents & Trademarks, Leergasse 11, D-65343 Eltville am Rhein, Germany

E-mail: fjschulze@hotmail.com

Abstract.
If a temporally symmetric voltage waveform is applied to a capacitively coupled radio frequency (CCRF) discharge, that contains one or more even harmonics of the fundamental frequency, the sheaths in front of the two electrodes will necessarily be asymmetric even in a geometrically symmetric discharge. Optimally this is achieved with a dual-frequency discharge driven at a phase locked fundamental frequency and its second harmonic, e.g. 13.56 MHz and 27.12 MHz. An analytical model, a hybrid fluid/Monte-Carlo kinetic model as well as a Particle in Cell (PIC) simulation show that this Electrical Asymmetry Effect (EAE) leads to the generation of a DC self bias as a function of the phase between the applied voltage harmonics in geometrically symmetric as well as asymmetric discharges. The DC self bias depends almost linearly on the phase angle and the role of the electrodes (powered and grounded) can be reversed. At low pressures the EAE is self-amplifying due to the conservation of ion flux in the sheaths. By tuning the phase, precise and convenient control of the ion energy at the electrodes can be achieved, while the ion flux remains constant. The maximum ion energy can typically be changed by a factor of about three at both electrodes. At the same time the ion flux is constant within $\pm 5\%$.

Journal of Physics: Conference Series **162** (2009) 012010 doi:10.1088/1742-6596/162/1/012010

1. Introduction

Capacitively coupled radio frequency (CCRF) discharges play an important role for industrial applications such as etching and deposition processes for semiconductor fabrication (chip production, solar cells, etc.). In this context separate control of the ion flux and the ion energy at the substrate surface is essential. The ion energy controls the individual processes taking place at the wafer surface and the ion flux determines the throughput of a given process [1].

Separate control of ion flux and ion energy cannot be achieved in conventional single frequency discharges, since both parameters are controlled by the applied voltage amplitude [2]-[33]. Besides hybrid discharges [34] the usual method to solve this problem is the use of dual-frequency CCRF discharges operated at two substantially different frequencies, typically 2 MHz and 27 MHz, applied to one or more electrodes [20],[35]-[46]. Usually the low frequency voltage is much higher than the high frequency voltage. The ion energy is mainly controlled by the low frequency component. The ion flux is mainly controlled by the high frequency component, since electron heating is more efficient at higher frequencies. However, recent investigations have shown, that there can be a strong coupling between the two frequencies, that might limit separate control of ion flux and energy [20], [40]-[45].

Here we propose a completely novel method to achieve separate control of ion energy and flux at the electrode surfaces in a simple and almost ideal way in dual-frequency CCRF discharges operated at similar frequencies, namely 13.56 MHz and 27.12 MHz [47, 48, 49]. If one electrode is driven at a fundamental frequency and its second harmonic with variable phase between the two voltage waveforms, a DC self (electrical asymmetry) bias will develop as a function of the phase angle even in a geometrically symmetric CCRF discharge. This variable DC self bias controls the ion energy at the electrodes. Therefore, by adjusting the phase the ion energy can be controlled. As only the relative phase between the voltage harmonics and not their amplitude is changed, the ion flux at the electrodes remains constant.

This Electrical Asymmetry Effect (EAE) is investigated here in three different ways: An analytical proof and model, a hybrid hydrodynamic and Monte-Carlo kinetic model [47, 48, 50], and a self consistent particle-in-cell simulation (PIC) [49]. All three models yield very similar results with the PIC simulation revealing further insight into kinetic details. The paper is structured in the following way: In section 2 it will be demonstrated that the sheaths in a CCRF discharge operated at a fundamental frequency and one or more even harmonic are necessarily asymmetric. In section 3 the analytical model to describe the EAE is presented and the results are compared to a hybrid hydrodynamic and Monte-Carlo kinetic model. In section 4 the results of a self-consistent PIC simulation are presented and separate control of ion energy and flux via the EAE is demonstrated. In section 5 conclusions are drawn.

Journal of Physics: Conference Series **162** (2009) 012010 doi:10.1088/1742-6596/162/1/012010

2. Electrical Asymmetry in CCRF discharges driven at a fundamental frequency and one or more even harmonics

The two sheaths in a CCRF discharge are symmetric only if the following condition holds [47, 48]:

$$V_{sp}(\omega t) = -V_{sg}(\omega t + \pi) \tag{1}$$

where V_{sp} and V_{sg} are the voltage drops across the sheath at the powered and grounded electrode, respectively. The voltage drop across the discharge, V_{tot}, is assumed to be the sum of both sheath voltages. This assumption is verified by the PIC simulation [49].

$$V_{tot} = V_{sp} + V_{sg} \tag{2}$$

V_{tot}, V_{sp}, and V_{sg} can be decomposed into their Fourier components:

$$V_{tot} = \sum_{n=-N}^{N} t_n e^{in\omega t} \tag{3}$$

$$V_{sp} = \sum_{n=-\infty}^{\infty} p_n e^{in\omega t} \tag{4}$$

$$V_{sg} = \sum_{n=-\infty}^{\infty} g_n e^{in\omega t} \tag{5}$$

Equations 2 through 5 imply that:

$$t_n = p_n + g_n \quad \forall \quad |n| \leq N \tag{6}$$

$$0 = p_n + g_n \quad \forall \quad |n| \geq N \tag{7}$$

Substituting equations 4 and 5 into the criterion for symmetric sheaths (equation 1) yields:

$$\sum_{n=-\infty}^{\infty} p_n e^{in\omega t} = -\sum_{n=-\infty}^{\infty} g_n e^{in(\omega t+\pi)} \tag{8}$$

This implies that the sheaths are only electrically symmetric if:

$$p_n = -g_n e^{in\pi} \tag{9}$$

$e^{in\pi} = 1$ whenever n is even and $e^{in\pi} = -1$ whenever n is odd. Using equations 9 and 6 yields:

$$t_n = \begin{cases} 0 \quad \forall \quad n = 0, 2, 4, 6, ... \\ 2p_n \quad \forall \quad n = 1, 3, 5, 7, ... \end{cases} \tag{10}$$

Based on equation 10 it is clear that the amplitudes of the fundamental and any odd harmonic can be freely chosen in order to obtain electrically symmetric sheath. However, there must not be any even voltage harmonic. If there is an even harmonic, the sheaths will necessarily be electrically asymmetric.

3. Modeling of the EAE

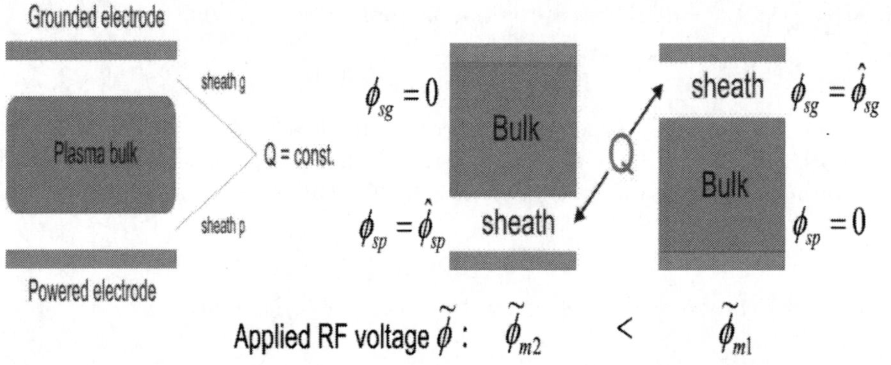

Figure 1. Sketch of a CCRF discharge (left) and location of the net positive charge in the discharge Q at the phase of complete sheath collapse at the grounded (middle) and powered electrode (right). ϕ_{sp} and ϕ_{sg} are the sheath voltages at the powered and grounded electrode, respectively. $\hat{\phi}_{sp}$, $\hat{\phi}_{sg}$ are the maxima of the sheath voltages. $\tilde{\phi}$ is the applied RF voltage and $\tilde{\phi}_{m1}$, $\tilde{\phi}_{m2}$ are the maximum and minimum of the applied voltage waveform, respectively.

Figure 1 shows the typical setup of a CCRF discharge of two plane parallel electrodes, one powered and one grounded, of either the same (geometrically symmetric) or different (geometrically asymmetric) surface areas. The discharge consists of a quasineutral plasma bulk and two sheaths adjacent to each electrode. The net positive charge in the entire discharge Q is located in the sheaths. Here Q is assumed to be constant within one RF period, which has been demonstrated to be a good approximation [16]. The voltage balance for such a CCRF discharge is:

$$\tilde{\phi} + \eta = \phi_{sp} + \phi_{sg} \tag{11}$$

where $\tilde{\phi}$ is the applied RF voltage, η the DC self-bias and ϕ_{sp}, ϕ_{sg} are the sheath voltages at the powered and grounded electrode, respectively. Here the floating potential and the voltage drop across the bulk are neglected and the voltage drop across the discharge is assumed to be determined only by the sum of both sheath voltages. A PIC simulation has demonstrated that this assumption is justified [49]. At two distinct phases within one RF period one sheath is completely collapsed at one of the electrodes (see figure 1). If the sheath is collapsed at the grounded electrode, the applied RF voltage $\tilde{\phi}$ will be minimum ($\tilde{\phi} = \tilde{\phi}_{m2}$), no voltage will drop across the sheath at ground ($\phi_{sg} = 0$) and maximum voltage will drop across the sheath at the powered electrode ($\phi_{sp} = \hat{\phi}_{sp}$). Vice versa if the sheath is collapsed at the powered electrode, the applied RF voltage will be maximum ($\tilde{\phi} = \tilde{\phi}_{m1}$), no voltage will drop across the sheath at the powered electrode ($\phi_{sp} = 0$) and maximum voltage will drop across the sheath at ground ($\phi_{sg} = \hat{\phi}_{sg}$). The voltage balance of equation 11 at these two distinct phases is given by:

$$\tilde{\phi}_{m2} + \eta = \hat{\phi}_{sp} \tag{12}$$

$$\tilde{\phi}_{m1} + \eta = \hat{\phi}_{sg} \tag{13}$$

The maximum sheath voltages $\hat{\phi}_{sp}$ and $\hat{\phi}_{sg}$ can be calculated by integrating Poisson's equation at the corresponding phase:

$$\hat{\phi}_{sp} = -\frac{1}{2e\varepsilon_0}\left(\frac{Q_{mp}}{A_p}\right)^2 \frac{I_{sp}}{\bar{n}_{sp}} \tag{14}$$

$$\hat{\phi}_{sg} = -\frac{1}{2e\varepsilon_0}\left(\frac{Q_{mg}}{A_g}\right)^2 \frac{I_{sg}}{\bar{n}_{sg}} \tag{15}$$

Here e is the elementary charge, ε_0 the dielectric constant, $Q_{mp,mg}$ the maximum charge in the sheath at the powered and grounded electrode, respectively, $A_{p,g}$ the surface area of the respective electrode, $\bar{n}_{sp,sg}$ the mean ion density in the respective sheath and $I_{sp,sg}$ the respective sheath integral.

$$I_s = 2\int_0^1 p_s(\xi)\xi d\xi \tag{16}$$

with $\xi = x/s_m$ and $p_s(\xi) = n_i(x)/\bar{n}_i$. Here s_m is the maximum sheath width and n_i is the ion density.
Under the assumption of constant total charge Q the ratio of equations 14 and 15 yields:

$$\varepsilon = \left|\frac{\hat{\phi}_{sg}}{\hat{\phi}_{sp}}\right| = \left(\frac{A_p}{A_g}\right)^2 \frac{\bar{n}_{sp}}{\bar{n}_{sg}} \frac{I_{sg}}{I_{sp}} \tag{17}$$

ε is called the symmetry parameter [48]. Using equations 17, 12, and 13 the following expression for the DC self bias is derived, that depends only on the extremes of the applied RF voltage waveform and the symmetry parameter:

$$\eta = -\frac{\tilde{\phi}_{m1} + \varepsilon\tilde{\phi}_{m2}}{1+\varepsilon} \tag{18}$$

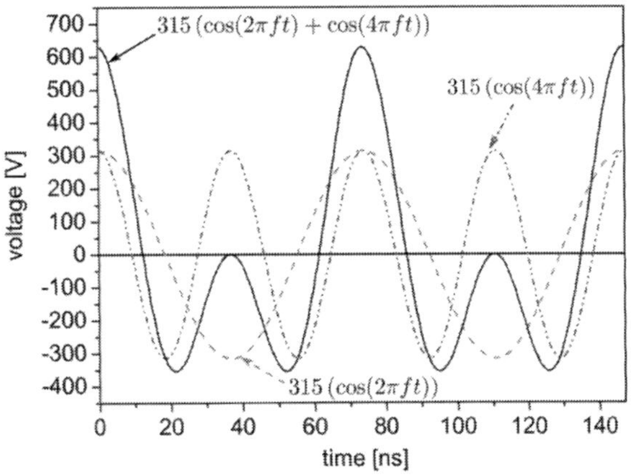

Figure 2. Voltage waveform $\tilde{\phi}(t) = 315\,(\cos(2\pi ft) + \cos(4\pi ft))$, where $f = 13.56\,\mathrm{MHz}$, applied to the discharge for two RF periods (solid black line). The absolute values of the positive and negative extremes and the voltages across the two sheaths are different. Therefore, a DC self bias develops under these conditions [49].

From now on we consider a geometrically symmetric CCRF discharge ($A_p = A_g$) in argon operated at 2.7 Pa with an electrode gap of $d = 6.7\,\mathrm{cm}$, similar to conditions of an experimental investigation by Godyak and Piejak in a single frequency CCRF discharge [18, 19, 21, 47, 50, 51]. The following RF voltage waveform is applied to one of the electrodes:

$$\tilde{\phi}(t) = 315\,\mathrm{V}\,(\cos(2\pi \mathrm{f t} + \theta) + \cos(4\pi \mathrm{f t})) \tag{19}$$

with $f = 13.56\,\mathrm{MHz}$. θ is the phase angle between the harmonics. Figure 2 shows this voltage waveform as well as the fundamental cosine function and its second harmonic for $\theta = 0°$. Each of the two cosine functions is harmonically symmetric ($\tilde{\phi}(\varphi+\pi) = -\tilde{\phi}(\varphi)$, $\varphi = \omega t$, $\tilde{\phi}$ corresponds to the applied voltage), but the sum of the two is not. The sum is symmetric with respect to $\varphi = \pi$ ($\tilde{\phi}(\varphi) = \tilde{\phi}(-\varphi)$). The absolute values of the positive and negative extremes are different. Therefore, according to equation 18 at $\theta = 0°$ a DC self bias will be generated ($\varepsilon \leq 1$).

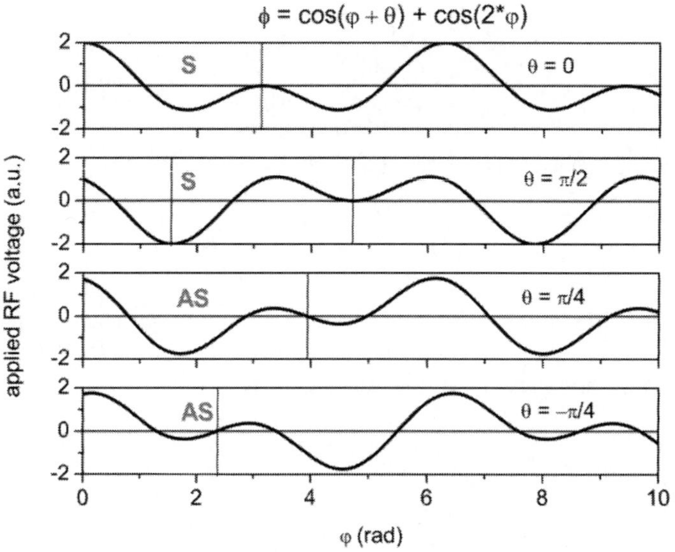

Figure 3. RF voltage waveform according to equation 19 at different phase angles θ [48].

By tuning the phase angle θ from 0° to 90° the symmetry of the applied RF voltage waveform is changed from symmetric (S, $\theta = 0°$) to anti-symmetric (AS, $-\tilde{\phi}(\varphi) = \tilde{\phi}(-\varphi)$, $\theta = 45°$) and back to symmetric (S, $\theta = 90°$). This is shown in figure 3. In case of an anti-symmetric waveform the absolute values of the positive and negative extremes are the same and no DC self bias will develop. At $\theta = 90°$, $\eta_{(90°)} = -\eta_{(0°)}$. Already based on this simple graphical analysis it is obvious that by adjusting the phase between the applied voltage harmonics the DC self bias can be changed (assuming $\varepsilon = 1$).

In order to calculate η explicitly ε needs to be known. ε is calculated by using the charged particle densities resulting from the Brinkmann sheath model [52, 53]. Then equation 19 is solved for its extrema $\tilde{\phi}_{m1}$ and $\tilde{\phi}_{m2}$. Using equation 18 this procedure yields the DC self bias at different phase angles θ.

Figure 4 shows the symmetry parameter resulting from the Brinkmann sheath model and a PIC simulation, that will be discussed in detail later. Obviously, the symmetry parameter changes with the phase from $\theta = 0°$ to $\theta = 90°$. At low pressures the sheath is collisionless and a finite DC self bias leads to different mean sheath voltages. Due to flux continuity different mean sheath voltages lead to different mean ion densities in both sheaths. This causes ε to deviate from unity. According to equation 18 such a dependence of ε on the phase leads to a self-amplifications of the EAE [48].

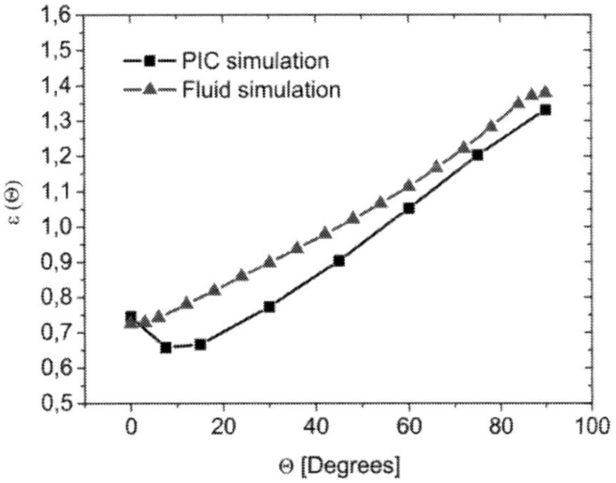

Figure 4. Symmetry parameter defined by equation 17 as a function of the phase angle θ (blue line (triangles) - fluid simulation [48], black line (squares) - PIC simulation) [49].

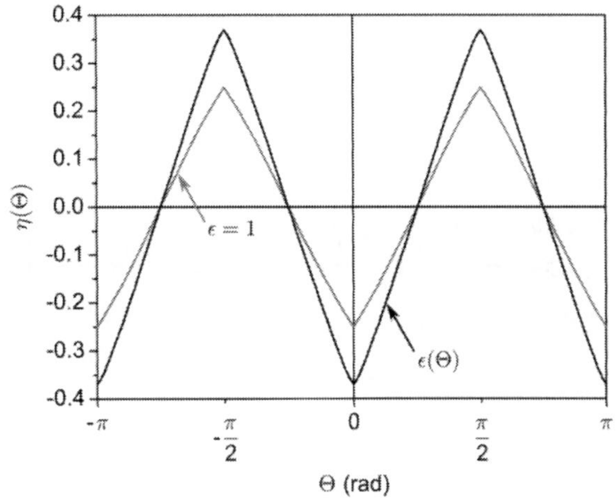

Figure 5. Normalized DC self bias as a function of the phase between the applied voltage harmonics resulting from equation 18 using $\varepsilon = 1$ and ε resulting from the Brinkmann sheath model (figure 4) [48].

Figure 5 shows the normalized DC self bias as a function of the phase between the applied voltage harmonics resulting from equation 18 using $\varepsilon = 1$ and $\varepsilon(\theta)$ resulting from the Brinkmann sheath model (figure 4). The self-amplification is obvious.

The Brinkmann sheath model can also be used to directly calculate the DC self bias. Figure 6 shows a comparison between η resulting from the analytical model, the Brinkmann sheath model, and a PIC simulation. All approaches yield similar results. The DC self bias is an almost linear function of θ.

Similar to the way the maximum sheath voltages were calculated (equations 14,15) the sheath voltages can also be calculated as a function of time within one RF period by using the temporary sheath edge instead of the maximum for the integration of Poisson's equation. Figure 7 shows the resulting sheath voltages as a function of the phase within one low frequency RF period.

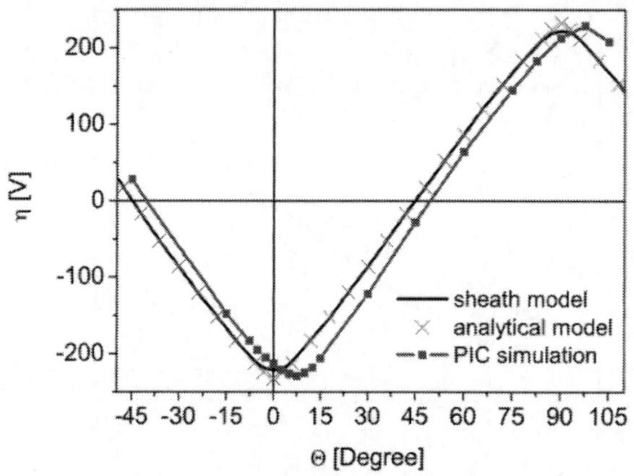

Figure 6. DC self bias η calculated by the analytical model using ε from the Brinkmann sheath model (red markers, [48]), directly by the Brinkmann sheath model (solid black line, [48]), and by a PIC simulation (blue markers and solid line) [49].

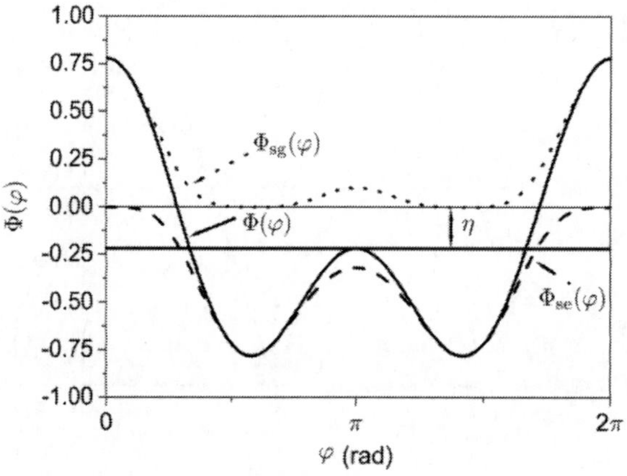

Figure 7. Sheath voltages as a function of the phase within one low frequency RF period resulting from the analytical model. The total voltage drop across the discharge and the DC self bias are also shown [48].

Similar results are also obtained from the Brinkmann sheath model.

The partial derivative of the sheath voltage ϕ_s with respect to the charge in the sheath q is given by [48]:

$$\frac{\partial \phi_s}{\partial q^2} = -\frac{1}{2e\varepsilon_0 \bar{n}(s) A^2} \tag{20}$$

Here $\bar{n}(s)$ is the average ion density in the sheath depending on the position of the sheath edge s, A is the electrode surface area. Equation 20 corresponds to a quadratic charge-voltage relation. The left plot in figure 8 shows the charge voltage relation resulting from the Brinkmann sheath model (solid line) at $\theta = 0°$ for the geometrically symmetric discharge discussed here. To a good approximation it agrees with a quadratic relation (dashed line). Such a quadratic charge voltage relation has been assumed for capacitive discharges before [13] and has been

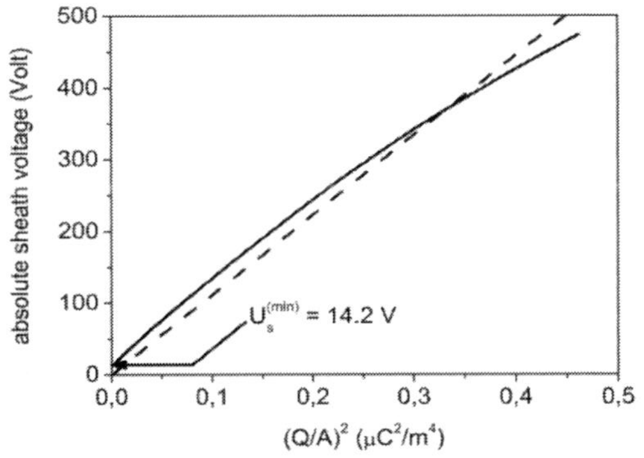

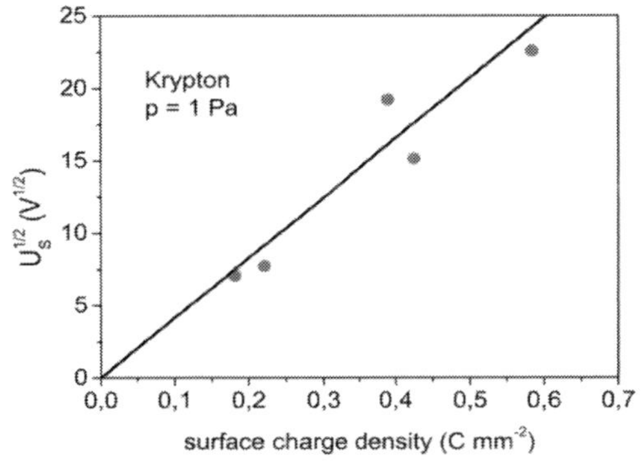

Figure 8. Left: Charge voltage relation resulting from the Brinkmann sheath model (solid line) at $\theta = 0°$ for the geometrically symmetric discharge discussed here. The simple quadratic relation is also shown (dashed line) [48]. Right: Measured charge voltage relation in a geometrically strongly asymmetric single frequency CCRF discharges operated at 13.56 MHz, 1 Pa in krypton [19, 21].

measured by Fluorescence Dip Spectroscopy in krypton in a single frequency geometrically strongly asymmetric CCRF discharge before [19, 21]. Slight deviations from an exactly quadratic relation are caused by the actual density profile in the sheath. At large sheath width $\bar{n}(s)$ in equation 20 is larger than at small sheath width.

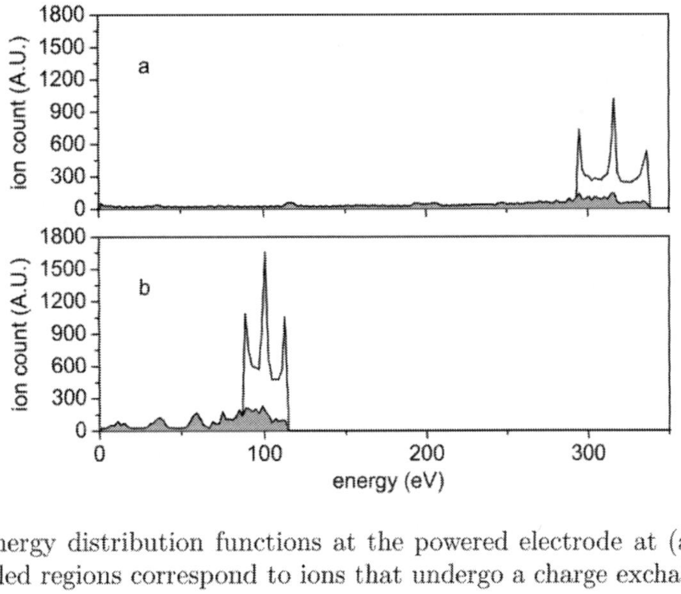

Figure 9. Ion energy distribution functions at the powered electrode at (a) $\theta = 0°$ and (b) $\theta = 90°$. The shaded regions correspond to ions that undergo a charge exchange collision [48].

The variable DC self bias (see figure 6) allows convenient and simple control of the ion energy distribution function (IEDF) at both electrodes. Here an ion Monte Carlo simulation coupled to the Brinkmann sheath model is used to calculate the IEDFs at the electrodes as a function of the phase between the applied voltage harmonics. Figure 9 shows the IEDF at the powered electrode for different phase angle θ. At $\theta = 0°$ (a) the bias is strongly negative and the mean sheath voltage at the powered electrode is high. Therefore, the IEDF is shifted to higher energies. At $\theta = 90°$ (b) the bias is strongly positive and the mean sheath voltage at the powered electrode is low. Therefore, the IEDF is shifted to lower energies.

4. PIC simulation of the EAE

Here we describe a geometrically symmetric dual-frequency CCRF discharge in argon using a one-dimensional (1d3v) bounded plasma particle-in-cell simulation code, complemented with a Monte Carlo treatment of collision processes (PIC/MCC, [20, 54, 55]). The electrodes are assumed to be infinite, planar and parallel. In our implementation of the PIC simulation, one of the electrodes is driven by a voltage specified by equation 19, while the other electrode is grounded. The electron impact cross sections for the collision processes are taken from [56]. For the positive ions, elastic collisions with the gas atoms are divided into an isotropic and a backward part [57]. The cross sections for these collisions are taken from [57, 58]. Metastable atoms are not taken into account. The bias voltage η is determined in the simulation in an iterative way to ensure that the charged particle fluxes to the two electrodes averaged over one low frequency RF period become equal. Electrons are reflected from the electrode surfaces with a probability of 0.2 and the secondary electron emission coefficient is taken to be $\gamma = 0.1$ for most of our simulations. The number of superparticles in the simulations is of the order of 10^5. From the trajectories of the particles followed in the PIC simulation as well as from the collision events we derive the spatio-temporal distributions of several plasma parameters.

The DC self bias and the symmetry parameter resulting from the PIC simulation at 2.7 Pa and an electrode gap of 6.7 cm have already been shown in figures 6 and 4, respectively. Both show good agreement with previously performed model results.

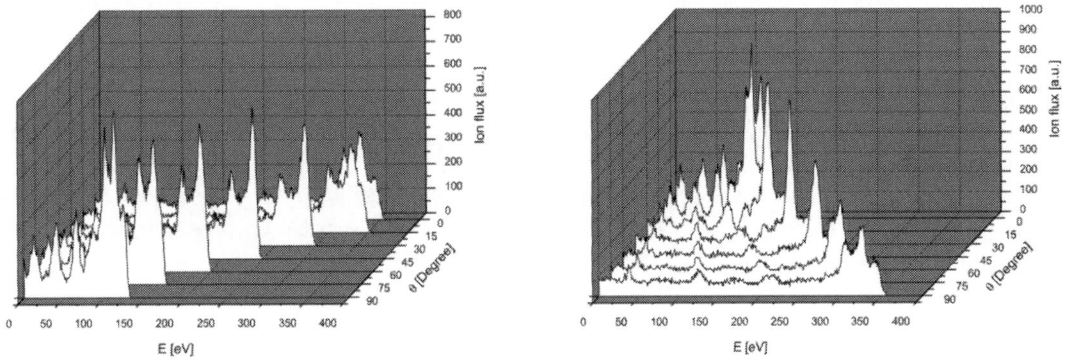

Figure 10. Ion flux-energy distributions at the powered (left) and grounded (right) electrode as a function of the phase angle θ of equation 19 calculated by the PIC simulation under the conditions mentioned in the text [49].

Figure 10 shows the ion flux-energy distribution at each electrode as a function of θ resulting from the PIC simulation. By changing θ from 0° to 90° the maximum ion energy at each electrode can be changed by a factor of about three. Furthermore, the role of each electrode can be reversed electrically. Figure 10 agrees well with the distribution functions calculated in the frame of the fluid/Monte Carlo model (see figure 9). The local maxima of the distribution functions at low energies are again caused by ions, that undergo charge exchange collisions in the sheath.

The analytical and the fluid/Monte Carlo model presented in the previous section are both not self consistent. In both models the mean plasma density is an input parameter and is not calculated self-consistently. Therefore, separate control of ion energy and flux cannot be investigated using these models. However, the PIC simulation calculates the plasma density and the ion flux self-consistently and can, therefore, be used to investigate this separate control.

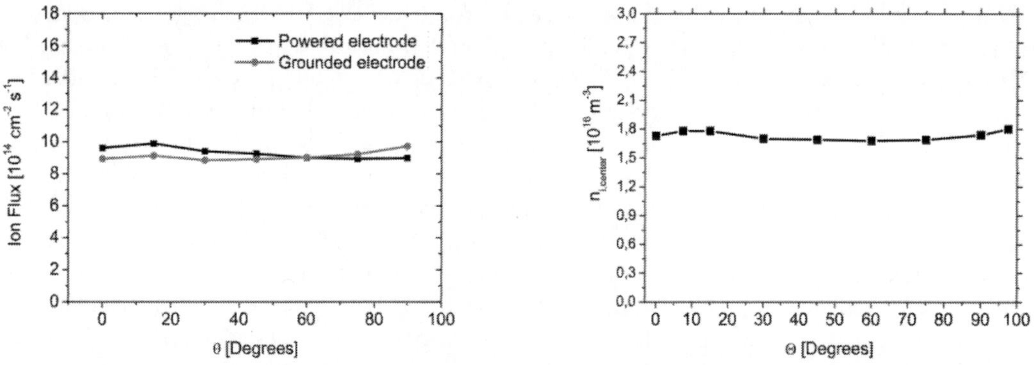

Figure 11. Left: Ion fluxes at the powered and grounded electrode as a function of the phase angle θ in equation 19 calculated by the PIC simulation [49]. Right: Ion density in the discharge center as a function of the phase angle θ (PIC) [49].

The left plot of figure 11 shows the ion flux resulting from the PIC simulation at both electrodes as the phase angle θ is varied from $0°$ to $90°$. The ion flux is constant within $\pm 5\%$, while the maximum ion energy changes by a factor of three as θ changes (see figure 10). The observed stability of the ion flux is within the range of tolerance for most industrial applications [59, 60]. The right plot of figure 11 shows the ion density in the discharge center calculated by the PIC simulation. The ion density is basically constant as the phase angle changes. Based on the above results the EAE easily allows separate control of ion energy and ion flux by keeping the applied voltage constant and changing the phase angle θ.

In the analytical model the total voltage across the discharge is assumed to be the sum of the voltages across both sheaths (equation 2). Generally, the total voltage across the discharge is the sum of the applied voltage, V_{AC}, and the DC self bias, η:

$$V_{total} = \tilde{\phi} + \eta \tag{21}$$

The PIC simulation shows that the assumption corresponding to equation 2 is correct under the conditions investigated here: At the low pressure of 20 mTorr, the voltage drop across the plasma bulk is negligible and the total voltage across the discharge is indeed the sum of both sheath voltages.

For the PIC simulation the voltage is an input parameter and the voltage amplitude is kept constant. However, in experiments the applied power and not the voltage is usually set externally. Therefore, it is important to examine how the absorbed power changes as a function of θ while the voltage amplitude is kept constant. The mean power density absorbed by electrons and ions, $\bar{p}_{\mathrm{e,i}}$, results from an integration of the space and time resolved dissipated power density $p_{\mathrm{e,i}}$ directly resulting from the PIC simulation:

$$\bar{p}_{\mathrm{e,i}} = \frac{1}{d \cdot T_{\mathrm{lf}}} \int_0^{T_{\mathrm{lf}}} \int_0^d p_{\mathrm{e,i}} \, dx \, dt \tag{22}$$

Here T_{lf} is the duration of one $1\,\omega$ RF period and d is the electrode gap. The result for the total dissipated power as well as the electron and ion components are shown in figure 12. The absorbed power density is essentially constant and does not differ from its mean value by more than about 6%. This means, that keeping the applied voltage amplitude constant in the simulation corresponds to a good approximation to keeping the power constant. The small

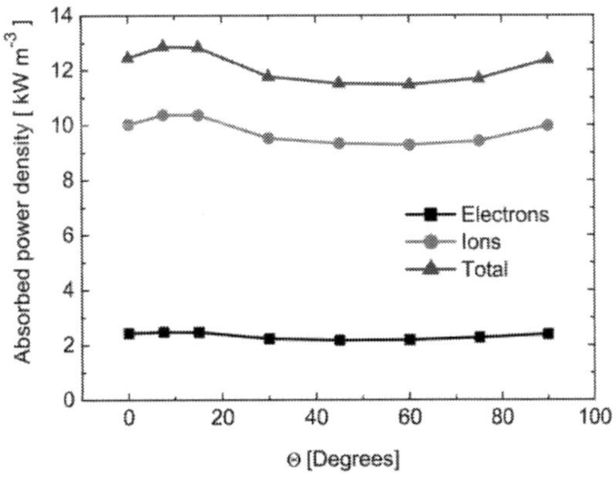

Figure 12. Power absorbed by electrons and ions and total power absorbed as a function of the phase angle θ (PIC) [49].

modulations of the absorbed power reflect the small modulations of the ion flux (figure 11). Therefore, the ion flux might change even less, if the power is kept constant.

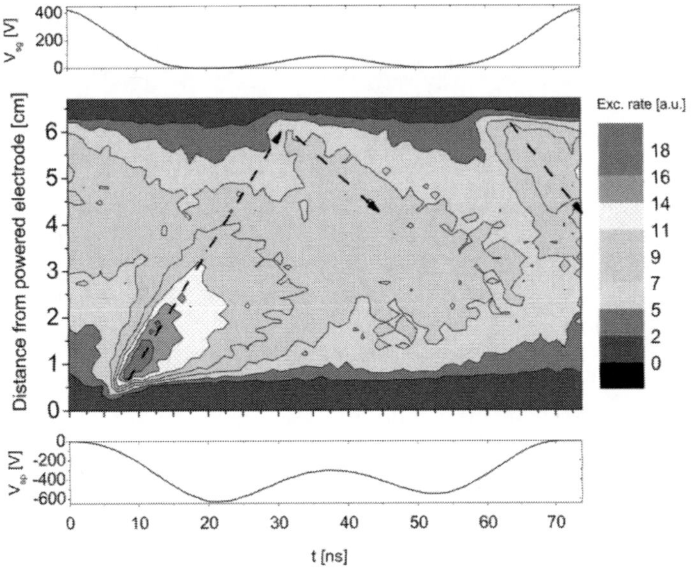

Figure 13. Spatio-temporal plot of the total excitation rate of argon atoms at $\Theta = 7.5°$ calculated by the PIC simulation. The voltages across the sheath at the powered electrode, V_{sp}, and across the sheath at the grounded electrode, V_{sg}, are also shown [49].

The spatio-temporal total excitation rate of argon atoms at $\Theta = 7.5°$ (phase of strongest DC self bias) is shown in figure 13. The generation of beams of highly energetic electrons by the expanding sheath at both electrodes is observed. Such electron beams have been investigated experimentally and theoretically before [16, 18, 19, 20, 21, 22, 24, 25, 42, 44, 47, 49]. As the discharge is asymmetric at this phase angle (see figure 4), the sheath width and sheath expansion velocities are different at each electrode. Therefore [2, 3], the observed electron beams at each electrode are different. The strongest excitation is caused by the initial sheath expansion of the sheath at the powered electrode.

5. Conclusions

The Electrical Asymmetry Effect - a novel and simple method to achieve efficient separate control of ion energy and ion flux at the electrodes in capacitively coupled radio frequency discharges - has been discovered. If a temporally symmetric voltage waveform is applied to a capacitively coupled radio frequency (CCRF) discharge, that contains one or more even harmonics of the fundamental frequency, the sheaths in front of the two electrodes will necessarily be asymmetric even in a geometrically symmetric discharge. The EAE was analyzed by an analytical model, a fluid/Monte Carlo model and a PIC simulation.

The analytical model showed that a DC self bias is generated even in a geometrically symmetric discharge if the extremes of the applied RF voltage waveform are different. If a fundamental frequency and its second harmonic are applied to one electrode, the symmetry of the applied voltage waveform can be changed by adjusting the phase angle between the applied voltage harmonics. Consequently, the DC self bias can be controlled by simply changing the phase and the role of both electrodes (powered and grounded) can be reversed.

The fluid/Monte Carlo model demonstrated that this variable DC self bias provides a unique opportunity to control the IEDF at both electrodes via the phase angle. It also showed that the mean ion densities in both sheaths are different at low pressures and at phase angles of non-vanishing DC self bias due to flux continuity. Therefore, the symmetry parameter deviates from unity and the EAE is self-amplifying.

A self consistent PIC simulation finally showed that the EAE allows efficient separate control of ion energy and flux at the electrode surfaces. The ion flux is found to be constant within $\pm 5\,\%$ when changing the phase angle, while the maximum ion energy is changed by a factor of about three. This result is particularly interesting for industrial applications, since it allows efficient separate control of ion flux and energy using an easily applied technique. In many cases existing process chambers could be easily modified to make use of the EAE by simply replacing the matching and power supply. Using this technique reactor sizes and, consequently, costs can be reduced, since high area ratios are no longer needed. Furthermore, limitations of the separate control of ion flux and energy by the frequency coupling in conventional dual-frequency CCRF discharges operated at substantially different frequencies can be avoided.

A method for controlling the ion energy based upon this effect is patent pending (PCT application No PCT/EP2008/059133).

Acknowledgements

This work has been funded by the DFG through GRK 1051, the Ruhr University Research School and the Hungarian Scientific Research Fund through grants OTKA-T-48389 and OTKA-IN-69892.

References

[1] Lieberman M A and Lichtenberg A J 2005, Principles of Plasma Discharges and Materials Processing, 2nd. ed., Wiley Interscience, NJ: Wiley
[2] Lieberman M A 1988 *IEEE Trans. on Plasma Sci.* **16** No. 6 638
[3] Lieberman M A and Godyak V A 1998 *IEEE Trans. on Plasma Sci.* **26** 955
[4] Surendra M and Graves D B 1991 *Phys. Rev. Lett.* **66** 1469
[5] Turner M M 1995 *Phys. Rev. Lett.* **75** 1312
[6] Gozadinos G, Turner M M and Vender D 1995 *Phys. Rev. Lett.* **87** 135004
[7] Kaganovich I D 2002 *Phys. Rev. Lett.* **89** 265006
[8] Kaganovich I D, Polomarov O V and Theodosioue C E 2006 *IEEE Trans. on Plasma Sci.* **34** No. 3 696
[9] Klick M 1996 *J. Appl. Phys.* **79** 3445
[10] Klick M, Kammeyer M, Rehak W, Kasper W, Awakowicz P and Franz G 1998 *Surf. Coat. Technol.* **98** 1395
[11] Mussenbrock T and Brinkmann R P 2006 *Appl. Phys. Lett.* **88** 151503
[12] Mussenbrock T and Brinkmann R P 2007 *Plasma Sources Sci. Technol.* **16** 377-385

[13] Czarnetzki U, Mussenbrock T and Brinkmann R P 2006 *Phys. Plasmas* **13** 123503
[14] Lieberman M A, Lichtenberg A J, Kawamura E, Mussenbrock T and Brinkmann R P 2008 *Phys. Plasmas* **15** 063505
[15] Mussenbrock T, Brinkmann R P, Lieberman M A, Lichtenberg A J and Kawamura E, 2008 *Phys. Rev. Letters* **101** 085004
[16] Donkó Z, Schulze J, Luggenhölscher D, and Czarnetzki U 2008 submitted to Phys. Rev. Letters
[17] Ziegler D, Mussenbrock T and Brinkmann R P 2008 *Plasma Sources Sci. Technol.* **17** 045011
[18] Schulze J, Heil B G, Luggenhölscher D, Mussenbrock T, Brinkmann R P and Czarnetzki U 2008 *J. Phys. D FTC* **41** 042003
[19] Schulze J, Heil B G, Luggenhölscher D, Brinkmann R P and Czarnetzki U 2008 *J. Phys. D* **41** 195212
[20] Schulze J, Donkó Z, Heil B G, Luggenhölscher D, Mussenbrock T, Brinkmann R P and Czarnetzki U 2008 *J. Phys. D* **41** 105214
[21] Schulze J, Kampschulte T, Luggenhölscher D and Czarnetzki U 2007 *IOP Conf. Series* **86** 012010
[22] Schulze J, Heil B G, Luggenhölscher D and Czarnetzki U 2008 *IEEE Trans. on Plasma Sci.* **36** 1400
[23] Salabas A, Marques L, Jolly J and Alves L L 2004 *J. Appl. Phys.* **95** 9 4605-4620
[24] Vender D and Boswell R W 1990 *IEEE Trans. Plasma Sc.* **18** 725
[25] Wood B P 1991 *PhD thesis* University of California at Berkely
[26] Tochikubo F, Suzuki A, Kakuta S, Terazono Y and Makabe T 1990 *J. Appl. Phys.* **68** 11 5532
[27] Petrović Z Lj, Tochikubo F, Kakuta S, and Makabe T 1992 *J. Appl. Phys.* **71** 2143
[28] Czarnetzki U, Luggenhölscher D and Döbele H F 1999 *Plasma Sources Sci. Technol.* **8** 230
[29] Gans T, Schulz-von der Gathen V and Döbele H F 2004 *Europhysics Letters* **66** 232
[30] Belenguer Ph, and Boeuf J P 1990 *Phys. Rev. A* **41** 4447
[31] Sato A H and Lieberman M A 1990 *J. Appl. Phys.* **68** 12 6117
[32] Vender D and Boswell R W 1992 *J. Vac. Sci. Technol. A* **10** 4 1331
[33] Turner M M and Hopkins M B 1992 *Phys. Rev. Letters* **69** 24 3511
[34] Rauf S and Kushner M J 1999 *IEEE Trans. on Plasma Science* **27** 1329
[35] Boyle P C , Ellingboe A R and Turner M M 2004 *Plasma Sources Sci. Technol.* **13** 493-503
[36] Boyle P C, Ellingboe A R and Turner M M 2004 *J. Phys. D.* **37** 697
[37] Kitajima T, Takeo Y, Petrovic Z L and Makabe T 2000 *Appl. Phys. Lett.* **77** 489
[38] Denda T, Miyoshi Y, Komukai Y, Goto T, Petrovic Z L and Makabe T 2004 *J. Appl. Phys.* **95** 870
[39] Lee J K, Manuilenko O V, Babaeva N Yu, Kim H C and Shon J W 2005 *Plasma Sources Sci. Technol.* **14** 89
[40] Kawamura E, Lieberman M A and Lichtenberg A J 2006 *Phys. Plasmas* **13** 053506
[41] Turner M M and Chabert P 2006 *Phys. Rev. Lett.* **96** 205001
[42] Gans T, Schulze J, O'Connell D, Czarnetzki U, Faulkner R, Ellingboe A R and Turner M M 2006 *Appl. Phys. Lett.* **89** 261502
[43] Schulze J, Gans T, O'Connell D, Czarnetzki U, Ellingboe A R and Turner M M 2007 *J. Phys. D* **40** 7008-7018
[44] Schulze J, Donkó Z, Luggenhölscher D and Czarnetzki U 2009 *Plasma Sources Sci. Technol.* in print
[45] Semmler E, Awakowicz P and von Keudell A 2007 *Plasma Sources Sci. Technol.* **16** 839
[46] Salabas A and Brinkmann R P 2005 *Plasma Sources Sci. Technol.* **14** 2 53-59
[47] Heil B G, Schulze J, Mussenbrock T, Brinkmann R P and Czarnetzki U 2008 *IEEE Trans. on Plasma Sci.* **36** 1404
[48] Heil B G, Czarnetzki U, Brinkmann R P and Mussenbrock T, 2008 *J. Phys. D* **41** 165202
[49] Donkó Z, Schulze J, Heil B G, Czarnetzki U 2009 *J. Phys. D* **42** 025205
[50] Heil B G, Brinkmann R P and Czarnetzki 2008 *J. Phys. D* **41** 225208
[51] Godyak V A, Piejak R B and Alexandrovich B M 1992 *Plasma Sourc. Sci. Technol.* **1** 36
[52] Kratzer M, Brinkmann R P, Sabisch W and Schmidt H 2001 *J. Appl. Phys.* **90** 2169
[53] Brinkmann R P 2007 *J. Appl. Phys.* **102** 093302
[54] Donkó Z and Petrović Z 2006 *Japanese Journal of Applied Physics* **45** 8151
[55] Donkó Z and Petrović Z 2007 *J. Phys.: Conf. Series* **86** 012011
[56] Phelps A V, Petrović Z Lj 1999 *Plasma Sources Sci. Technol.* **8** R21
[57] Phelps A V 1994 *J. Appl. Phys.* **76** 747
[58] Phelps A V, *http://jilawww.colorado.edu/~avp/collision_data/* unpublished
[59] Tsuboi H, Itoh M, Tanabe M, Hayashi T and Uchida T 1995 *Jpn. J. Appl. Phys.* **34** 2476
[60] Chinzei Y, Ogata M, Sunada T, Itoh M, Hayashi T, Shindo H, Itatani R, Ichiki T and Horiike Y 1999 *Jpn. J. Appl. Phys.* **37** 4572

Space and phase resolved optical emission in mode transitions of radio-frequency inductively coupled plasmas

D O'Connell, K Niemi, M Zaka-ul-Islam, T Gans
Centre for Plasma Physics, Queen's University Belfast, Northern Ireland, UK
E-mail: d.oconnell@qub.ac.uk

Abstract. Inductively coupled radio-frequency plasmas can be operated in two distinct modes. At low power and comparatively low plasma densities the plasma is sustained in capacitive mode (E-mode). As the plasma density increases a transition to inductive mode (H-mode) is observed. This transition region is of particular interest and governed by non-linear dynamics, which under certain conditions results in structure formation with strong spatial gradients in light emission. These modes show pronounced differences is various measureable quantities e.g. electron densities, electron energy distribution functions, ion energy distribution functions, dynamics of optical light emission. Here the transition from E- to H- mode in an oxygen containing inductively coupled plasma (ICP) is investigated using space and phase resolved optical emission spectroscopy (PROES). The emission, measured phase resolved, allows investigation of the electron dynamics within the rf cycle, important for understanding the power coupling and ionization mechanisms in the discharge. The temporal variation of the emission reflects the dynamics of relatively high-energy electrons. It is possible to distinguish between E- and H-mode from the intensity and temporal behaviour of the emission.

1. Introduction

Inductively coupled radio frequency (rf) discharges have proven technological significance, for surface modification, fabrication of modern integrated circuit devices, and thin film deposition [1], [2]. Rf plasmas in oxygen or oxygen containing plasmas have wide applications in material processing such as photo resist etching, chemical vapour deposition, and oxidation. Other technologies include for example thin oxide growth at low temperature for the fabrication of gate oxide of thin film transistors [3].

It is well known that inductively coupled plasmas (ICP) operate in two distinct regimes. At lower powers and plasma densities the discharge operates capacitively in so-called E-mode. When the plasma density increases beyond a critical value the discharge switches to inductive H-mode. This switch is accompanied by sudden changes in several plasma and circuit parameters such as the electron density and electron energy distribution function (EEDF), the coil current and the optical emission signal. [1], [2], [4]. This transition regime between the two modes is a popular parameter space for technological processes, due to improved control over the ion energy and flux onto the substrate. However, it is important to note that this transition is complex and the dominating mechanisms are not fully understood or rather ambiguous. Power dependence of the optical emission intensity, electron density and temperature, ion density, metastable density, gas temperature, electrical characteristics and magnetic field, has been reported in the literature through the E to H transition region [5] - [16]. But detailed knowledge, in particular, of the spatial contribution of various plasma formation mechanisms is crucial, to control essential plasma parameters. Different mechanisms responsible for the transition from E-mode to H-mode are discussed in a number of theoretical and experimental studies [17] - [20].

2. Experimental Setup

The optical emission from an inductively coupled plasma operated in oxygen is investigated using both space and phase resolution to identify the transition between E- and H-modes and investigate the important relevant sustaining mechanisms.

2.1. Plasma System - GEC reactor

The plasma system is an ICP GEC reactor driven at a radio-frequency of 13.56 MHz, with a planar antenna coil configuration. The rf power via a matching network is coupled into the plasma through a dielectric quartz window so that the plasma is not in direct contact with the antenna. The oscillating currents in the antenna generate a time dependent magnetic field inducing an electric field in the plasma. However, the antenna can also act as an electrode. Therefore, both inductive and capacitive power coupling mechanisms can coexist in the plasma.

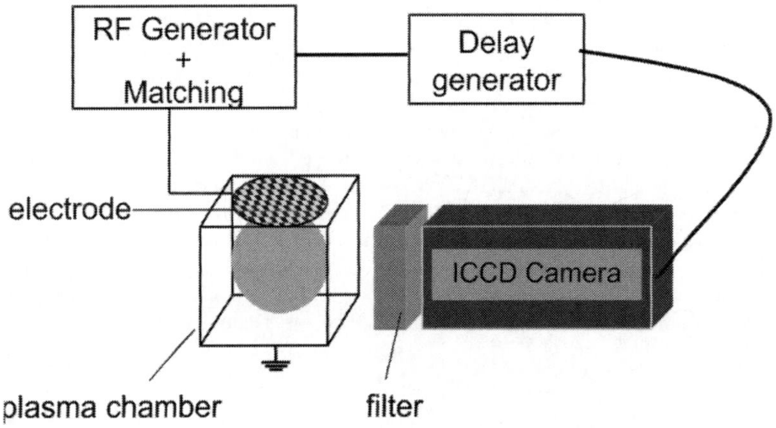

Figure 1. Experimental setup for phase and space resolved optical emission from the plasma

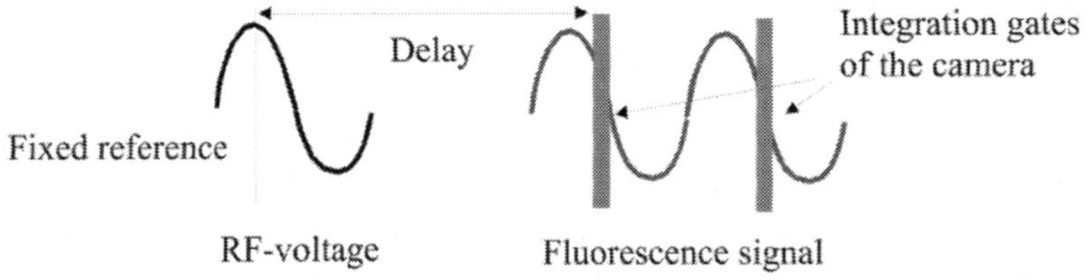

Figure 2. Delay generator scheme for phase resolved measurements synchronized with the RF power supply of the plasma

2.2. Diagnostics - Optical Emission Spectroscopy (OES) and Phase Resolved OES

The transition from the capacitive to inductive mode is investigated using optical emission spectroscopy. Time integrated optical emission over the RF cycle is measured using a calibrated spectrograph (300 - 1050 nm, Ocean Optics).

Phase resolved optical emission spectroscopy (PROES) allows investigation of the electron dynamics within the rf cycle, important for understanding the power coupling mechanisms in the discharge. An experimental setup for PROES and the synchronization to the RF power supply are shown in figure 1 and 2, respectively. PROES provides non-invasive access with a spatial resolution of microns [21], in both axial and radial direction, limited in the present case to 5 mm due to integration of the signal of adjacent pixels.

Two emission lines are investigated in the following. 844 nm ($O(3p^3P \rightarrow 3s^3S)$) with a lifetime of 34.7 ns and 777 nm ($O(3p^5P \rightarrow 3s^5S)$) with lifetime of 27.1 ns. This emission is detected using a special fast, gate-able, intensified CCD camera (PicoStar, LaVision). Rather than using a monochromator to select the wavelength, an interference filter is used since it combines relatively high transmission and 2-dimensional spatial resolution although here the second dimension is used to collect more photons. Light is collected from the region below the antenna with the camera synchronised to the rf generator, and measurements are performed with a 2 ns gate width. Phase resolved measurements over the entire rf cycle (74 ns) were obtained using a variable delay between the camera gate and the rf trigger that could be varied in increments of 2 ns (figure 2). For each phase measurement, the emission is integrated over several million rf cycles to obtain an adequate signal to noise ratio. The extremely high repetition rate of the camera accepts sequential rf-cycles for integration allowing high signal-to-noise ratio.

3. Phase resolved emission and excitation

The temporal variations of the optical emission within the RF cycle are caused by temporal changes in the electron energy distribution function (EEDF). Highly excited states in atoms or molecules are excited by electrons in the tail of the EEDF where the modulation is particularly pronounced. Temporal variations can, therefore, be observed in the optical emission from these high electronically excited atoms or molecules.

Electron impact excitation out of the ground state is described by the excitation function $E_i(t)$. The excitation function $E_i(t)$ can be determined from the measured number of photons per unit volume and unit time $n_{Ph,i}(t)$

$$E_i(t) = \frac{1}{n_0 A_{ik}} \left(\frac{dn_{Ph,i}(t)}{dt} + A_i n_{Ph,i}(t) \right) \tag{1}$$

where $n_{Ph,i}(t) = A_{ik} n_i(t)$ is given by the transition probability A_{ik} of the observed emission and the population density of the investigated state $n_i(t)$, n_0 is the ground state density [22]. The effective decay rate A_i takes into account spontaneous emission, radiation trapping and quenching.

$$A_i = \sum_k A_{ik} g_{ik} + \sum_q k_q n_q \tag{2}$$

where g_{ik} is the so-called escape factor and k_q the quenching coefficient with the species q of density n_q. For the 844 nm line the following quenching coefficients were used with $k_{O_2} = 9.4 \pm 0.5 \times 10^{-10} cm^3 s^{-1}$ and $k_{Ar} = 0.14 \pm 0.007 \times 10^{-10} cm^3 s^{-1}$ [23]. For the 777 nm line quenching with O_2 is $10.8 \pm 1.8 \times 10^{-10} cm^3 s^{-1}$ [24].

The time dependence of the excitation function, reflecting the electron dynamics, is strongly dependent on the power coupling mechanisms.

Journal of Physics: Conference Series **162** (2009) 012011 doi:10.1088/1742-6596/162/1/012011

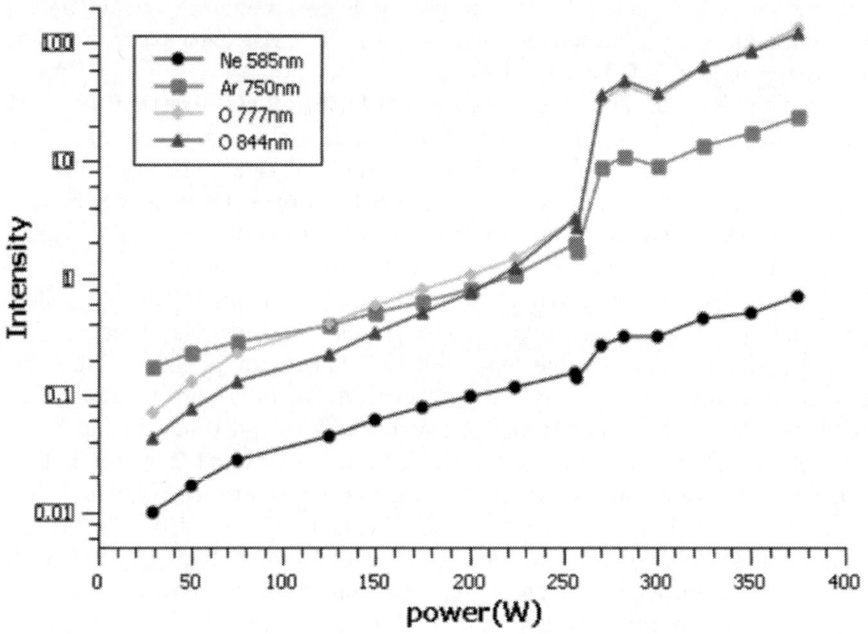

Figure 3. Intensity of various emission lines from the plasma as a function of power, illustrating the different operation modes in the plasma

4. Optical emission from RF ICP

Figure 3 shows the optical emission of various emission lines as a function of RF power from an oxygen containing ICP (with an admixture of neon and argon). With increasing power the intensity of the emission lines increases with a jump around 270 W as the discharge transits from capacitive to inductive mode. Different emission lines exhibit different relative percentage changes in the emission intensity. This indicates a complex change in electron density, electron energy distribution function and degree of dissociation in the different modes.

5. Space resolved emission

Figure 4 shows the time integrated space resolved emission of the 777 nm line at 50 W. There is a clear distinct maximum, axially close to the quartz and radially in the center of the discharge. This indicates capacitive coupling where the plasma is sustained through electron acceleration during the expansion of the plasma sheath. At low powers the antenna acts as an electrode and the large voltage at the center of the antenna gives rise to a high potential sheath, causing excitation and ionization.

The time integrated emission of the 777 nm line before the transition regime at 150 W is plotted in figure 5 as a function of space. The radial profile is similar to that at 50 W, while the axial differs slightly, with more excitation further into the bulk plasma. With increasing power the voltage drop across the antenna increases and secondary electrons produced at the quartz surface can be accelerated through this large sheath potential subsequently inducing excitation and ionization. The increased excitation at longer axial distances from the quartz into the plasma is due to very energetic electrons with large penetration depths. These energetic secondary electrons already become obvious in the time integrated images at 150 W.

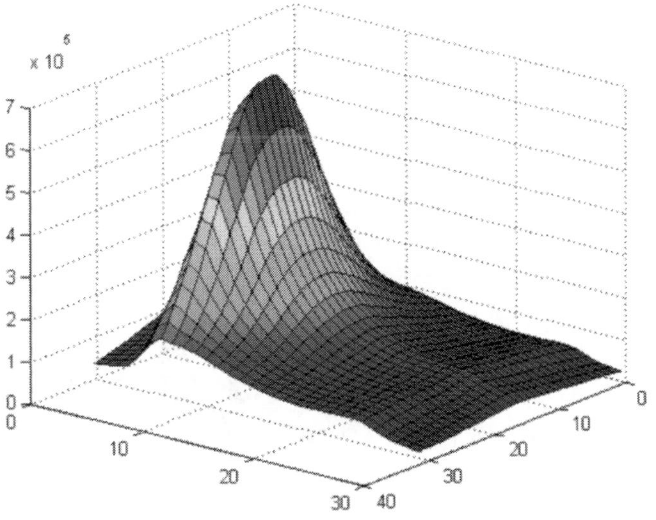

Figure 4. Time integrated space resolved optical emission at 50 W in capacitive mode

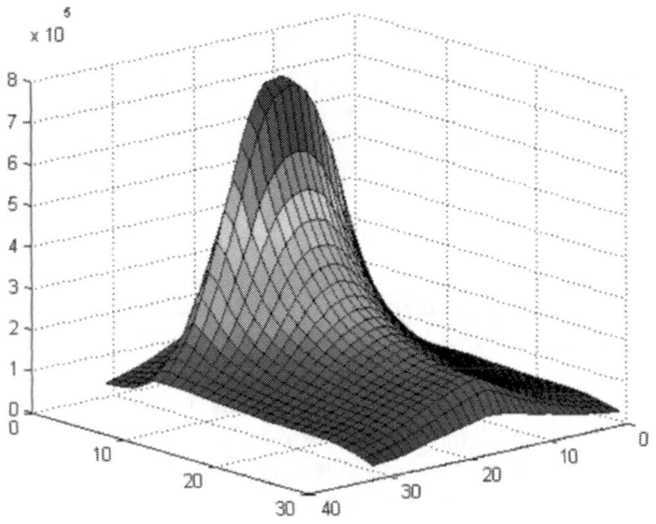

Figure 5. Time integrated space resolved optical emission at 150 W in the mode transition regime

Figure 6 shows the time integrated emission of the 777 nm line at a higher power of 350 W. The radial profile of the plasma emission in inductive mode is quite different to that at lower powers in capacitive mode. In inductive mode the antenna excites a decaying electromagnetic wave in the plasma. The induced electric field in the plasma does not propagate entirely through

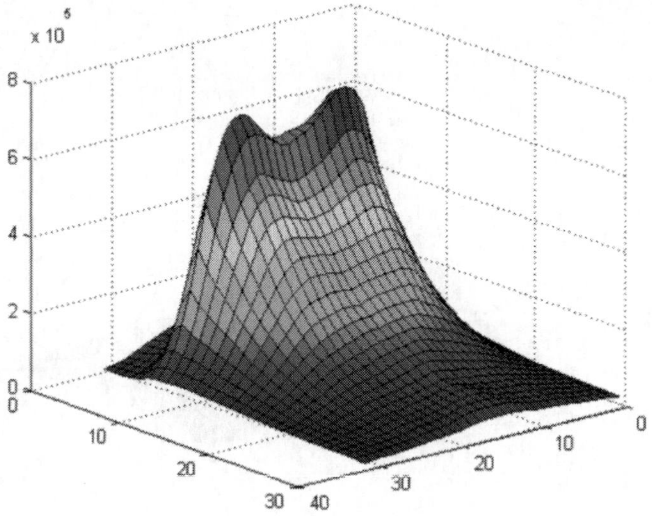

Figure 6. Time integrated space resolved optical emission at 350 W in inductive mode

the plasma volume, but into the plasma electrons within a small region of the skin depth layer just below the dielectric window. The planar configuration of the antenna geometry deposits the power into the plasma in a ring shaped volume. This arises since the induced electric field is zero on the plasma axis rising to a maximum with increasing antenna radius and decreasing again toward the chamber wall. This ring-shaped structure, illustrating the current path parallel to the plane of the antenna, is clearly observable in the optical emission profile in figure 6.

6. Electron dynamics

As explained above the emission within the rf cycle of the discharge is measured. The temporal variation of the emission, reflects the dynamics of electrons with energies greater than the excitation threshold of 10.73 eV for 777 nm and 10.98 eV for 844 nm. It is possible to distinguish between E- and H- mode from the intensity and temporal behaviour of the emission [25].

Figure 7 shows a plot of the phase resolved excitation for different RF powers over one RF cycle. At low powers in capacitive mode there is one pronounced excitation peak at a phase of 52 ns. This is due to the sheath expansion mechanism as discussed above where electrons are accelerated in the main expansion phase of the sheath. In addition to this there is also a second more minor excitation mechanism at a phase of 26 ns. This can be attributed to heavy particle collisional excitation close to the surface, previously observed in hydrogen [26] and also oxygen [27].

With increasing power at 100 W an additional broad excitation maxima arises at around 2 to 10 ns that even dominates the excitation above 250 W. This corresponds to the energetic electrons, which cause the excitation structure, that penetrates far into the plasma bulk in figure 5.

Increasing the power further the discharge transits into inductive mode, where the dynamics is characterised by two excitation maxima in each rf cycle (approx. 36 ns and 72 ns). In comparison to capacitive mode the two excitation structures are symmetric. These are due to the induced electric currents in the plasma. The induced electric field is opposite in sign during each half of the RF cycle, thus resulting in electron acceleration along the ring-shaped trajectory in opposite directions during each half of the RF cycle and correspondingly two excitation maxima.

The dynamics of electrons and corresponding excitation and ionization mechanisms in the different modes are very different and the transition from capacitive to inductive mode is a multi-step process. Energetic electrons, produced during the transition regime, provide additional excitation and ionization and thus the critical increase in electron density required for the induced currents in inductive mode to be sustained in the plasma.

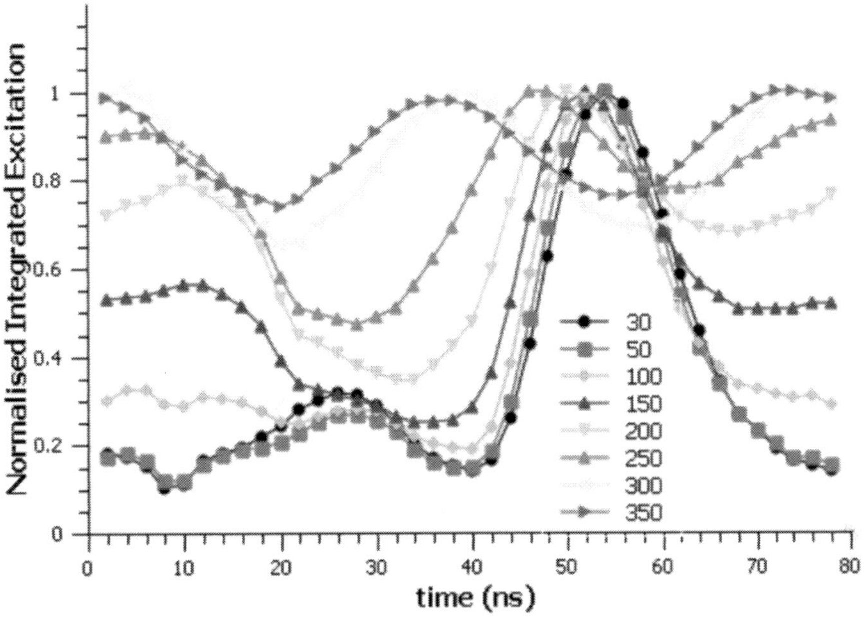

Figure 7. Phase resolved optical emission for various powers

7. Acknowledgements

The authors would like to thank Bill Graham for his support and valuable discussions.

8. References

[1] M A Lieberman and A Lichtenberg, Principles of plasma discharges and materials processing, New York: Wiley-Interscience, 1994
[2] T Makabe, Z Lj Petrovic, Plasma Electronics: Applications in microelectronic device fabrication, New York: Taylor & Francis Group, 2006
[3] Y Kawai, N Konishi, J Watanabe, T Ohmi, 1994, Appl. Phys. Lett. , Vol. 64, p. 22235
[4] B Chapman, Glow discharge processes: Sputtering and plasma etching, US: John Wiley and Sons, 1980
[5] U Kortshagen, N D Gibson, J E Lawler, 1996, J. Phys. D, Vol. 29, p. 1224
[6] I M El-Fayoumi, I R Jones, M M Turner, 1998, J. Phys. D , Vol. 31, p. 3082
[7] M Abdel-Rahman, T Gans, V Schulz-von der Gathen, H F Döbele, 2005, Plasma Sources Sci Technol., Vol. 14, p. 51-60
[8] G Cunge, B Crowley, D Vender, M M Turner, 1999, Plasma Sources Sci.Technol., Vol. 8, p. 576
[9] K N Ostrikov, S Xu, M Y Yu, 2000, J. Appl. Phys., Vol. 88, p. 2268
[10] Y Miyoshi, Z Petrovic, T Makabe, 2002, J. Phys. D, Vol. 35, p. 454
[11] D B Graves, T Czerwiec, 2004, J. Phys. D, Vol. 37, p. 2827

[12] A M Daltrini, S A Moshkalev, M J R Monteiro, E Besseler, A Kostryukov, M Machida, 2007, J. Appl. Phys., Vol. 101, p. 073309
[13] M Osiac, T Schwarz-Selinger, D O'Connell, B Heil, Z Lj Petrovic, M M Turner, T Gans, U Czarnetzki, 2007, Plasma Sources Sci Technol, Vol. 16, p. 355
[14] T Gans, M Osiac, D O'Connell, V A Kadetov, U Czarnetzki, T Schwarz-Selinger, H Halfmann, P Awakowicz, 2005, Plasma Sources Sci Technol., 2005, Vol. 47, p. A353-A360
[15] M. Lee, K H Lee, D Hyun, C Chung, 2007, Appl. Phys. Lett. , Vol. 90, p. 191502
[16] A M Daltrini, S A Moshkalev, T J Morgan, R B Piejak, W G Graham. 2008, Appl Phys Lett, Vol. 92, p. 061504
[17] U Kortshagen, N D Gibson, J E Lawler, 1996, J. Phys. D, Vol. 29, p. 1224
[18] M M Turner and M A Lieberman, 1999, Plasma Sources Sci. Technol., Vol. 8, p. 313
[19] G Cunge, B Crowley, D Vender, M M Turner. 1999, Plasma Sources Sci.Technol., Vol. 8, p. 576
[20] V I Demidov, C A DeJoseph, Jr., A A Kudryavtsev, 2004, Plasma Sources Sci. Technol., Vol. 13, p. 600
[21] J Waskoenig, D O'Connell, V Schulz-von der Gathen, J Winter, S J Park, J G Eden, 2008, Appl. Phys. Lett., Vol. 92 10, p. 101503
[22] D O'Connell, T Gans, E Semmler, P Awakowicz, 2008, Appl. Phys. Lett., Vol. 93 8, p. 081502
[23] K Niemi, V Schulz-von der Gathen, H F Döbele, 2005, Plasma Sources Sci. Technol., Vol. 14 , p. 375
[24] P J Dagdigian, B E Forch, A W Miziolek, 1988, Chem. Phys. Lett., Vol. 148, p. 299
[25] M Abdel-Rahman, V Schulz-von der Gathen, T Gans, 2007, J. Phys. D: Appl. Phys., Vol. 40, p. 1678-1683
[26] T Gans, CC Lin, V Schulz-von der Gathen, H F Döbele, 2001, J Phys D: Appl Phys, Vol 34 8, p. L39-42
[27] K Dittmann, D Drozdov, B Krames, J Meichsner, 2007, J Phys D: Appl Phys, Vol. 40 21, p. 6593-6600
[28] M Schulze, D O'Connell, T Gans, P Awakowicz, A von Keudell, 2007, Plasma Sources Sci. Technol., Vol. 16 4, p. 774-784

Journal of Physics: Conference Series **162** (2009) 012012 doi:10.1088/1742-6596/162/1/012012

Influence of plasma and radio wave treatment on paper structure

I I Filatova[1], V V Azharonok[1], V A Dlugunovich[1], O Tsaruk[1], I V Voschula[1], N A Fomin[2] and N B Bazylev[2]

[1]B I Stepanov Institute of Physics NAS Belarus, Nezavisimosti Ave., 68, 220072 Minsk, Belarus

[2]Heat & Mass Transfer Institute NAS Belarus, P Brovki Str., 15, 220072 Minsk, Belarus

E-mail: filatova@imaph.bas-net.by

Abstract. A paper for offset printing was exposed to radio frequency discharge plasma and radio frequency electromagnetic field. Optical response and structural features of paper before and after the treatment have been investigated by several techniques: digital laser speckle photography, laser Stokes polarimetry and attenuated total reflectance (ATR). The changes of statistical distributions of speckle patterns has been observed for plasma treated paper as a result of increase of correlative dimensions of structural heterogeneity of modified paper. It is shown also that the radio wave treatment leads to anisotropy reduction in the paper structure that is more apparent if the magnetic field lines are oriented transversely to the plane of a paper sheet.

1. Introduction

Last years low-temperature plasma and radio wave treatment has been confirmed to be effective techniques for modification of natural polymers like cellulose based materials because of the possibility of a predictable change in their physical and chemical properties for various practical application [1-9]. In particular, plasma processing of cellulose based paper changes its layer properties [7-9]. It is reported that in a magnetic field changes occur in light transmission and the efficiency of small-angle scattering of polarized radiation by polymers belonging to the class of cellulose derivatives [6]. According to idea about the mechanisms of action of an external magnetic field on diamagnetic materials, the appearance of some orientational effects in extended polymer macromolecules (reorientation or a change in their conformational structure) is connected with anisotropy in the diamagnetic susceptibility of polymer fibers [10-12].

In this paper we study an influence of low-temperature radio frequency (rf) plasma as well as rf electromagnetic field on a paper for offset printing that is a complex material consisted of sulphite bleached wood cellulose with other substances including alum, rosin, glue, kaolin and fillers. Interest to such problem is connected with the purposeful change of a paper structure with the aim of paper samples identification and their operating characteristics improvement for different application.

2. Experimental conditions

2.1. Experimental set-up

Paper sheets (thickness about 70 μm, size (10 x 10) мм^2) were processed with rf helium plasma at pressure 0.5–5 Torr and exposed to rf electromagnetic field at atmospheric pressure under conditions of the prebreakdown mode operation of a capacity coupled rf discharge.

The experimental set-up is schematically depicted in figure 1. The discharge was operated on frequency 5.28 MHz between two parallel rectangular copper electrodes with dimension of 40 x 5 mm^2. The distance between electrodes was 3 mm. The samples were placed on the grounded electrode. Duration of the process was 1 – 5 min.

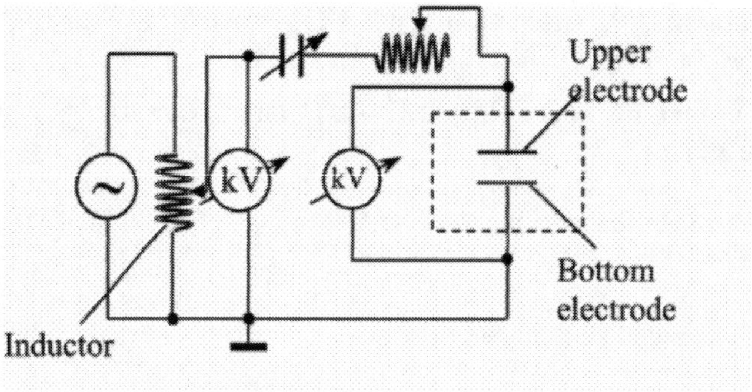

Figure 1. The experimental set-up.

The experimental conditions for magnetic field treatment were as follows: the alternator frequency was 5.28 MHz, the root-mean-square value of magnetic field strength was H= 590 A/m. The paper sheets were oriented transversely or parallel to the force lines of the magnetic component **H** (figure 2). Duration of the exposure to electromagnetic field was 15 min.

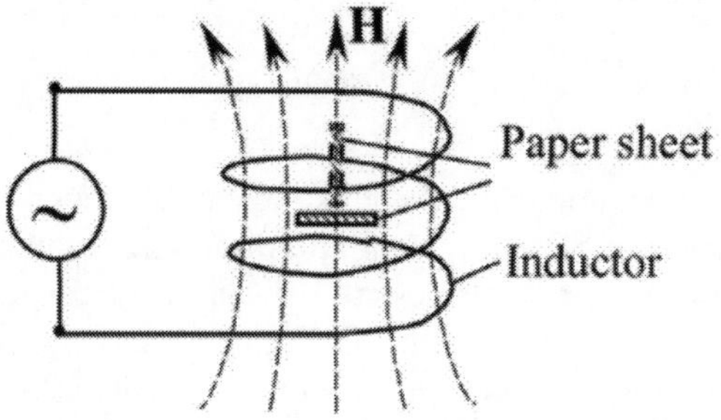

Figure 2. A scheme of the magnetic inductor.

Structural features of paper before and after the treatment were studied by digital laser speckle photography, laser goniophotometric Stokes polarimeter and attenuated total reflectance (ATR).

2.2. Digital laser speckle photography

The microstructure changes in plasma treated paper were tested by the methods of digital laser speckle photography [13]. The speckle-field recorded from the paper was formed as a result of 3D-interference of repeatedly scattered coherent radiation. The speckle images were recorded with a high resolution CCD camera Nikon D70S (the resolution of the matrix is 3008 x 2000 pixels). The statistical functions of the first order have been analyzed for speckle images recorded before and after the treatment of paper. The details of the data processing have described in [14].

2.3. A laser goniophotometric Stokes polarimeter

A block diagram of the measurement set-up using a laser goniophotometric Stokes polarimeter is shown in figure 3. He-Ne laser was the source of probing radiation with wavelength 632.8 nm. The

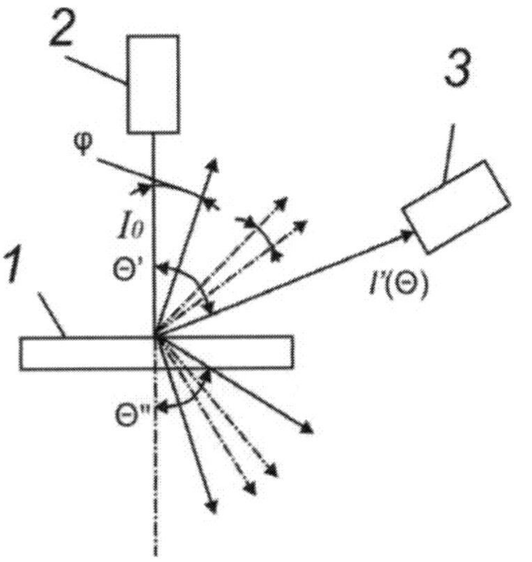

Figure 3. Block diagram of the setup for studying the properties of specimens using the polarization method: 1: paper specimen; 2) He-Ne laser; 3) photo-electron multiplier.

diameter of the laser beam was ~5 mm and the power of the generated radiation was 12 mW. Using polarimetric plates, the electric vector of the linearly polarized laser radiation was oriented parallel (||) or perpendicular (⊥) to the plane of incidence. The intensity I and the Stokes parameters (P_1, P_2, P_3) of the studied radiation were recorded using a one-channel setup with rotating quarter-wave (λ/4) phase plate and a fixed analyzer [15]. The distribution of the intensity of the probing laser radiation, directed at an angle $\varphi = 5°$ relative to the normal to the surface of the material, that was reflected by the specimen (I') and transmitted through it (I'') was measured in the plane of incidence in the angular range Θ'(Θ") 5-85°. The solid angle of the detection system (Δω) within which the scattered radiation was picked up was $7.2 \cdot 10^{-3}$ sr, which corresponded to an angular resolution in the plane of observation $\Delta\Theta \approx 5°$. The field of view of the photodetector was greater than the size of the irradiated spot on the specimen, but was smaller than the area of the specimen itself for all measurement angles. This made the dependence of the radiation power reflected by a Lambert surface on the observation angle Θ take the form of a function proportional to cos Θ [16]. The relative error in measurement on the intensity of the reflected (transmitted) laser radiation was -5%, while the absolute error in determination of the degree of polarization was no greater than 0.06. As the photometric parameters characterizing the studied material, we used distribution functions over the detection angles for the bidirectional reflection coefficients $F'(\Theta')$ and the bidirectional transmission coefficients $F''(\Theta'')$ [17, 18]:

$$F'(\Theta') = \Delta I_r(\Theta')/I_{0\lambda} \cos\Theta' (\Delta\omega),$$

$$F''(\Theta'') = \Delta I_\tau(\Theta'')/I_{0\lambda} \cos\Theta'' (\Delta\omega),$$

where ΔI_r and ΔI_τ are the intensities of laser radiation reflected in the 0' direction and transmitted in the Θ'' direction, detected within solid angle $\Delta\omega$ of the detection system; $I_{0\lambda}$ is the intensity of the laser radiation incident on the specimen, polarized in the plane of incidence or orthogonal to it. We also determined the indicatrices $f(\Theta') = \Delta I_r/\Delta\omega$, of the reflected laser radiation power, normalized to the values $f(\Theta' = 5°)$, and also the degree of polarization of the reflected and transmitted laser radiation [19]:

$$P=(P_1^2+P_2^2+P_3^2)^{1/2}/I.$$

In order to estimate the effect of structural inhomogeneities on the surface and within the volume of the sheet of paper on formation of reflected radiation field the values of $f'(\Theta')$ of the indicatrices of the reflected laser radiation power were multiplied by the degree of its polarization $P(\Theta')$ or by the quantity $(1 - P(\Theta'))$. In the first case, we obtained the indicatrices of the radiation power for the polarized component of the radiation reflected by the surface of the paper when it is probed by radiation polarized in the plane of incidence $f_r^{(S)} = P_\parallel(\Theta') f_\parallel(\Theta')$ or orthogonal to it $f_r^{(S)} = P_\perp(\Theta') f_\perp(\Theta')$; in the second case, we obtained the corresponding indicatrices of the radiation power for the unpolarized component of the reflected radiation, formed by scattering within the volume of the paper: $f_r^{(V)} = (1 - P_\parallel(\Theta'))/ f_\parallel(\Theta')$ and $f_r^{(V)} = (1 - P_\perp(\Theta'))/ f_\perp(\Theta')$.

3. Results and discussion

3.1. Plasma treatment

The histograms of intensity in the speckle-fields of paper before $(G^*(I_{pq}))$, after $(G^{**}(I_{pq}))$ the plasma treatment and the difference between them $(\Delta = G^{**}(I_{pq}) - G^*(I_{pq}))$ are shown in figure 4. The shape of histogram changes after the treatment. It is obvious that the correlative volumetric dimensions of paper micro heterogeneity increase for plasma modified paper. The main reason for the paper microstructure changes is the porosity increase of the paper because of the micro discharges occurred in micro pores between the cellulose fibrils in plasma.

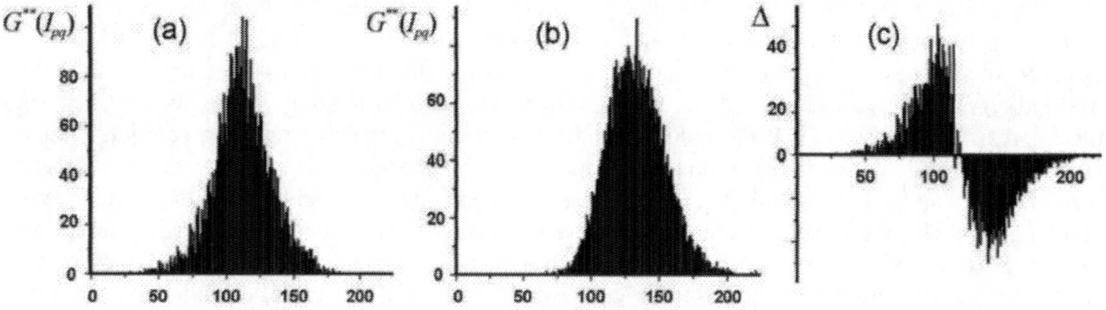

Figure 4. Histograms of intensity in speckle-field of untreated (a), plasma treated (b) paper and the difference between them (c).

3.2. Radiowave treatment

The surface and volumetric components of indicatrices of light power reflected by paper is shown in figure 5.

We should note the differences in the shape of the indicatrices of the reflected radiation power $f_r^{(S)}$ (especially for parallel orientation of the magnetic component of the field to the plane of the specimens and before exposure to the electromagnetic field) and the scattered laser radiation power as a function of the azimuth of its polarization and the orientation of the specimen relative to the force lines of the magnetic component of the field (Figure 5). However, for $f_r^{(V)}$ these differences appear to a lesser degree due to multiple scattering of light within the volume of the paper, which leads to depolarization and broadening of the indicatrices of the radiation power transmitted through the volume of the paper.

As a result of exposure to a high-frequency electromagnetic field with parallel orientation of the magnetic component force lines **H** to the surface of the specimen, the anisotropy of the surface layer of the paper decreases (Figure 5b). Exposure to a field with the perpendicular orientation of magnetic component force lines to the surface of the specimen, completely removes the anisotropy, and the surface layer of the paper becomes practically isotropic (Figure 4c). Similar trends also appear in the shape of the indicatrices $f_r^{(V)}$ (Figure 4e, f). Thus exposure to a high-frequency electromagnetic field, the force lines of which are perpendicular to the surface of the studied paper specimens, removes the anisotropy not only in the surface layer of the paper but also within its volume. So exposure to a high-frequency electromagnetic field leads to disorientation of the cellulose fibers making up the matrix of the paper, and causes a decrease in their degree of crystallization.

It is shown also that the effect of the magnetic component of an rf electromagnetic field on the paper appeared in the frequency shift in the ATR spectra (in the range of 1300 – 1500 cm^{-1}) (figure 6). It may be connected with orientation effects of cellulose macromolecules and change their conformation structure.

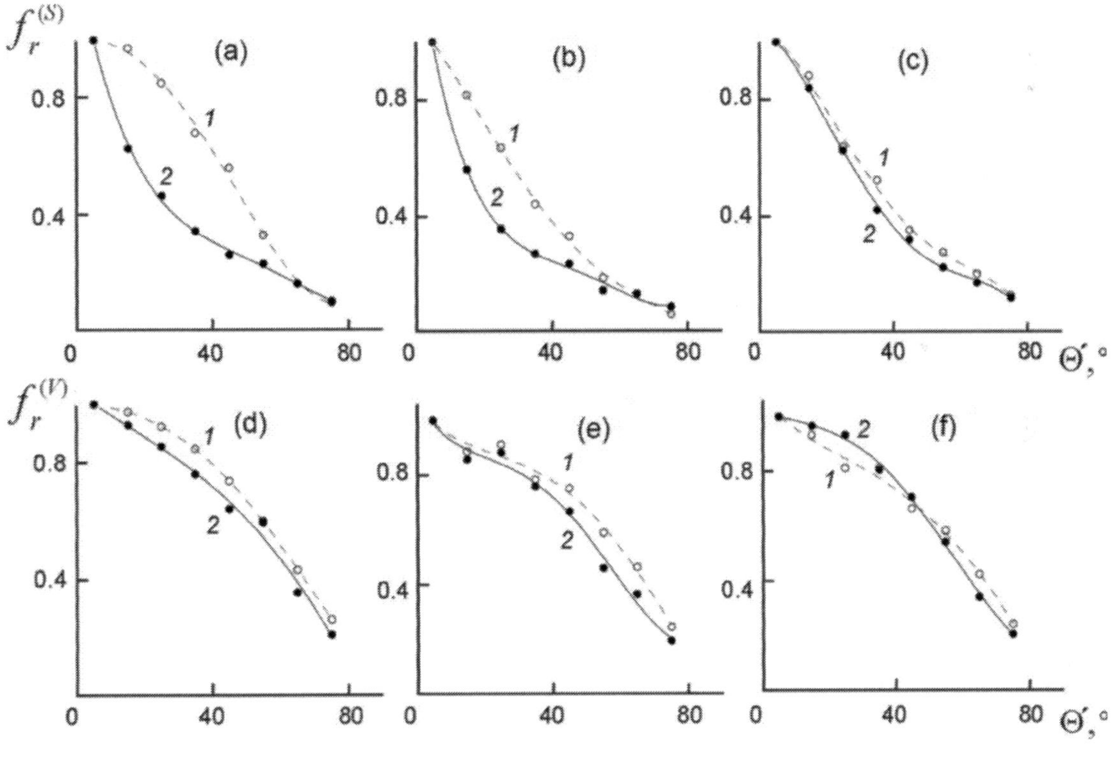

Figure 5. The surface (a-c) and volumetric (d-f) components of indicatrix of the radiation power: reflected by the untreated specimen (a, d) and after its exposure to an electromagnetic field with parallel (b, e) and transversal (c, f) orientation of paper sheet to the magnetic lines; probing laser radiation polarized in the plane of incidence (*1*) and orthogonal to it (*2*).

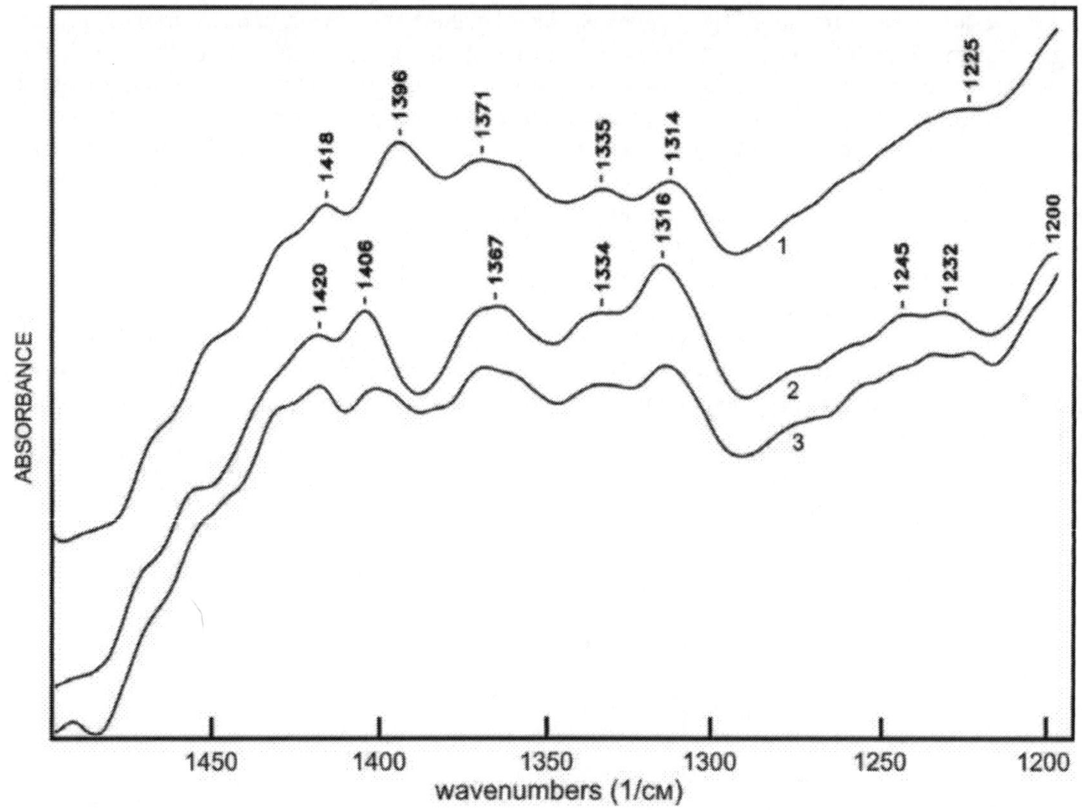

Figure 6. ATR spectra of paper in the range of 1200-1500 cm^{-1}: 1- untreated sample; 2- after exposure to rf electromagnetic field; 3- after treatment and artificial ageing by thermal processing at 120° C.

4. Conclusions

The effect of radio frequency plasma and radio frequency electromagnetic field treatment on some physical and chemical properties of a paper for offset printing is studied. Plasma treatment changes the microstructure of paper. The exposure of paper to the rf magnetic field decreases the structural order in the material, causes a decrease in their degree of crystallization, changes the shape of surface component of indicatrices of the reflected laser radiation power with the wavelength of 632.8 nm, induces a frequency shift in the ATR IR spectra (in the range of 1300 – 1500 cm^{-1}).

The anisotropy of the diamagnetic susceptibility of the cellulose molecule leads to the fact that when the force lines of the magnetic component of the electromagnetic field are oriented perpendicular to the surface of the paper, the decrease in structural ordering of the material is more substantial than for the parallel orientation.

References

[1] Deslandes Y, Pleizier G, Poire E, Sapieha S, Wertheimer M R and Sacher E 1998 *Plasmas and Polymers* **3** 61
[2] De Geyter N, Morent R and Leyes C 2006 *Plasma Sources Sci. Technol.* **15** 78
[3] Laguardia L, Vassallo E, Cappitelli F, Mesto E, Cremona A, Sorlini C and Bonizzoni G 2005 *Appl. Surf. Science* **252** 1159
[4] Assender H E and Windle A H 1996 *Polymer* **37** 371
[5] Kimura T, Yamato M, Koshimizu W and Kawai T 1990 *Chem. Lett.* **10** 1057
[6] Gul' V E, Khanchich O A and Savchenko N A 1995 *Mekhanika Kompozitsionnykh Materialov i*

Konstruktsii **1** No. 2 124
[7] Hua Z Q, Sitaru Q, Denes F and Young R A 1998 *Plasmas and Polymers* **2** No. 3 196
[8] Mukhopadhyay S M, Joshi P, Datta S, Zhao J G and France P 2002 *J. Phys. D: Appl. Phys.* **35** 1927
[9] Deslandes Y, Pleizier G, Poire E, Sapieha S, Wertheimer M R and Sacher E 1998 *Plasmas and Polymers* **3** No. 2 61
[10] Kimura F, Kimura T, Tamura M and Hirai A 2005 *Langmuir* **21** 2034
[11] Kimura T, Yamato M, Koshimizu W et al. 2000 *Langmuir* **16** 858
[12] J. Sugiyama, H. Chanzy, and G. Maret 1992 *Macromolecules* **25** 4232
[13] Fomin N 1998 *Speckle photography for fluid mechanics measurements* (Berlin: Springer Verlag) pp. 248
[14] Doerffer P, Szwaba R, Szumsla J-A, Telega J, Fomin N A, Lavinskaya EI and Bazylev N B 2005 *Physics of Shock Waves, Combustion, Detonation and Non-Equilibrium Processes* (Minsk: HMTI NASB) p 198
[15] Dlugunovich V A, Snopko V N and Tsaryuk O V 1999 *Zh. Prikl. Spektr.* **66** 869
[16] Kavaya M J, Menzies R T and Yaner D A 1983 *Appl. Opt.* **22** 2619
[17] Nicodemus E 1970 *Appl. Opt.* **9** 1474
[18] Kholopov K *Opt. Zh.* **61** 30 [in Russian]
[19] Snopko V N 1992 *Polarization Characteristics of Optical Radiation* [in Russian] (Minsk: Navuka i Tekhnika)

Journal of Physics: Conference Series **162** (2009) 012013 doi:10.1088/1742-6596/162/1/012013

The dynamics of radio-frequency driven atmospheric pressure plasma jets

L Schaper[1], S Reuter[1], J Waskoenig[1], K Niemi[1], V Schulz-von der Gathen[2], T Gans[1]

[1]Centre for Plasma Physics, Queen's University Belfast, Northern Ireland, UK
[2]Center for Plasma Science and Technology, Ruhr-University Bochum, Germany

t.gans@qub.ac.uk

Abstract. The complex dynamics of radio-frequency driven atmospheric pressure plasma jets is investigated using various optical diagnostic techniques and numerical simulations. Absolute number densities of ground state atomic oxygen radicals in the plasma effluent are measured by two-photon absorption laser induced fluorescence spectroscopy (TALIF). Spatial profiles are compared with (vacuum) ultra-violet radiation from excited states of atomic oxygen and molecular oxygen, respectively. The excitation and ionization dynamics in the plasma core are dominated by electron impact and observed by space and phase resolved optical emission spectroscopy (PROES). The electron dynamics is governed through the motion of the plasma boundary sheaths in front of the electrodes as illustrated in numerical simulations using a hybrid code based on fluid equations and kinetic treatment of electrons.

1. Introduction

Non-thermal non-equilibrium atmospheric pressure plasmas are commonly characterized by reduced spatial dimensions of the confining structures, e.g. electrodes, stabilizing the discharge and preventing the transition to a 'thermal' discharge. Typical dimensions vary from a few micrometres up to a few millimetres. The wide range of these discharges is, therefore, often sub-summarized under the topic 'microplasmas' [1–5].

High concentrations of atoms in the order of 10^{15} cm^{-3} and a low gas temperature significantly below 100°C, suitable for many applications especially in localized modifications of sensitive surfaces, can be provided by jet-like devices without the requirement of complicated and expensive vacuum systems [6]. Many of these discharges operate in a mixture of a noble gas, preferably helium, and a small molecular component selected depending on the envisaged application [7]. Despite enormous potential for technological applications, the fundamentals of these stable homogeneous non-equilibrium plasmas at ambient pressure are only poorly understood. This is mainly due to the complexity of these discharges composed of electrodes in the close vicinity of the confining walls and a mixture of neutral and charged atomic and molecular components. Furthermore, radiation can play an important role in the discharge balance. These jets are typically divided into the core plasma where plasma processes are driven and the effluent with reactive components suitable for technological applications.

The discharge concept adopted here is the atmospheric pressure plasma jet (APPJ) introduced by Selwyn and co-workers [8] in 1998. This technically relatively simple capacitively coupled device is typically operated at an excitation frequency of 13.56 MHz and an electrode spacing of about 1 mm. A mixture of a noble gas as the base gas and a small percentage addition of a molecular component serves as the operating gas. The molecular admixture is limited by instabilities in the discharge. At typical gas velocities of some 10 ms^{-1} the atomic oxygen concentration in the effluent of an APPJ is in the order of $n_O \sim 10^{15}$ cm^{-3} [9]. These reactive oxygen species are a prominent candidate for the surface modifying properties of the jet [10, 11].

Figure 1 shows a photograph of a typical coaxial APPJ illustrating the potential for technological applications of the effluent at comparatively low gas temperature. The front view picture shows the geometrical design of the coaxial APPJ with an inner radio-frequency driven rod-electrode and an outer grounded electrode.

Figure 1: Photograph of a coaxial APPJ. The side view illustrates the potential for technological applications of the effluent at low gas temperature. The front view shows the geometrical design with an rf driven inner rod and an outer grounded electrode.

Figure 2 shows a photograph illustrating the capability for localised surface modifications using APPJs. A simple example is the observable significant change in wettability of a treated Petri-dish. Efficient exploitation, further development, and tailoring towards future high benefit technological applications require detailed understanding of the dynamics of APPJs. This includes the complex dynamics of plasma production and energy carrying species in the plasma core. Crucial aspects are energy transport mechanisms from the core plasma to the effluent, in particular in the vicinity of their interface region.

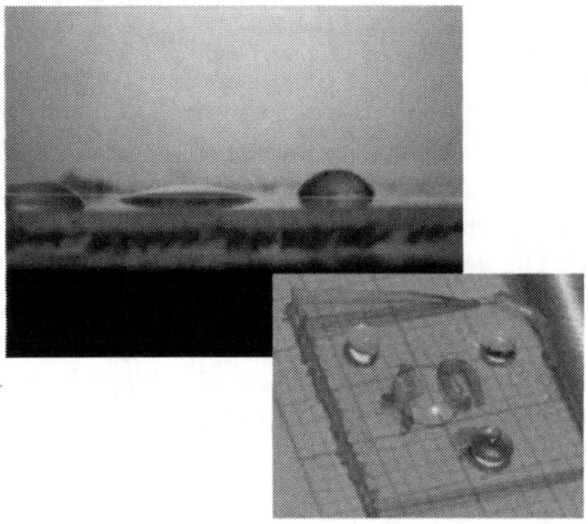

Figure 2: Photograph illustrating the change in wettability of a locally treated Petri-dish.

The sketch of figure 3 shows the constituents determining the dynamics of energy dissipation in the core plasma and energy transport mechanisms to the effluent region. The plasma chemistry inside the discharge volume is strongly influenced by the dynamics of the plasma boundary sheaths in front of the electrodes and corresponding surface processes. The plasma boundary sheaths energize electrons which dominate plasma ionization and the production of electronically excited particles through electron impact. Subsequent Penning ionization and three body collisions, due to the high pressure collisional environment, also play important roles in the particle and power balance of the core plasma. Energy transport mechanisms to the effluent region include transport of radicals, excited particles, and radiation. It has been shown that the effluent region is practically free of charged particles [8].

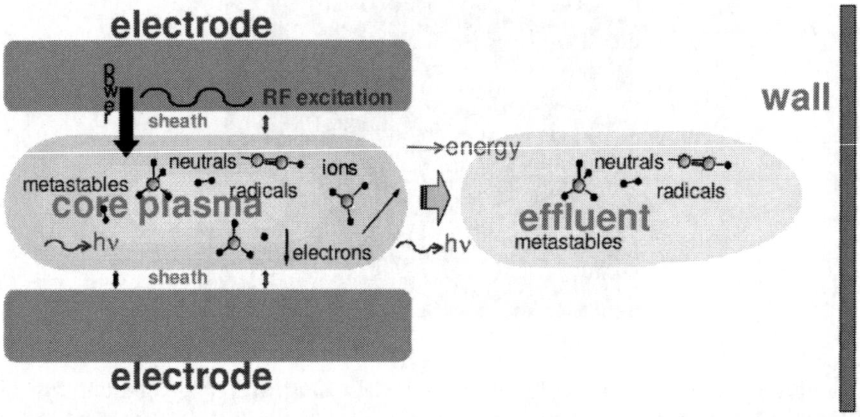

Figure 3: Sketch of the components determining the dynamics in the plasma core and energy transport mechanisms from the core to the effluent region. The wall represents a surface to be treated in technological applications.

2. Measurements in the effluent region

Ground state atomic oxygen radicals and (V)UV radiation play key roles in energy transport mechanisms in the effluent region and are also crucial energy carriers for technological surface modifications. Measurements in the effluent region of a planar APPJ are discussed in the following.

Advantages of the planar configuration over the coaxial design shown in figure 1 are better diagnostics access to the discharge core and the simpler geometry allows better comparison with modeling and simulation results. A detailed description of the planar APPJ can be found in S Reuter et al. [12]. The two parallel stainless steel electrodes are 80 mm in length and 40 mm in width. The electrode gap is 1.1 mm. One electrode is connected to a 13.56 MHz radio-frequency power supply, while the other is grounded. The equal area electrode design results in symmetric plasma operation with corresponding plasma boundary sheaths in front of both electrodes. The APPJ is operated with helium feed gas and small oxygen admixtures.

The atomic oxygen ground state density is measured by TALIF spectroscopy. Tunable UV laser radiation is used to excite oxygen atoms by simultaneous absorption of two UV photons. The fluorescence radiation emitted when the atoms revert to an energetically lower state is measured in order to gain information about the atomic oxygen density. The measurements are calibrated with TALIF measurements on xenon, according to earlier investigations on the concentric APPJ [9, 13]. Collisional quenching, temperature effects and vignetting of the signal are essential aspects to consider in the analysis of the measurements [9, 12].

Figure 4 shows the atomic oxygen ground state density distribution in the effluent of the planar APPJ for a helium flux of 2 $m^3 h^{-1}$, 0.5 vol% O_2 admixture and an RF power of 150 W (reading at the power supply). The effluent stays confined over several centimetres. The initial atomic oxygen ground state density directly at the APPJ's nozzle is as high as $\sim 1 \times 10^{16}$ cm^{-3}. Due to the planar APPJ's simple geometry, these measurement results close to the nozzle are more precise than earlier measurements on the concentric version of the APPJ [9, 13]. Furthermore, the TALIF measurements show that at 10 cm distance from the nozzle still ~1% of the initial atomic oxygen density remains.

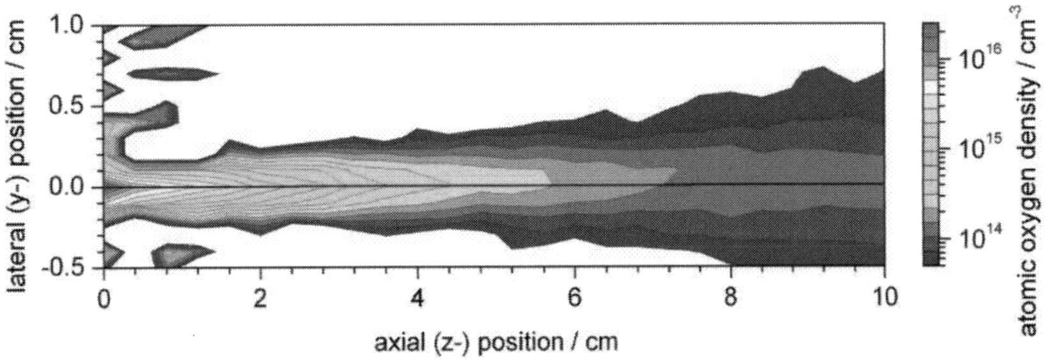

Figure 4: 2-dimensional spatial map of absolute densities of ground state atomic oxygen in the effluent region of a planar APPJ. The measurements are carried out in a controlled helium/oxygen environment using absolutely calibrated two-photon absorption laser induced fluorescence spectroscopy (TALIF).

VUV radiation in the effluent is measured using a 0.2 m monochromator (Minuteman Laboratories Inc. 302VM, grating 1200 mm^{-1}) and a solar blind photomultiplier with a CsTe-photocathode (EMI G-26H315, MgF_2 window, spectral response 110 to 340 nm). The main spectral features in the VUV spectrum are the Schumann–Runge bands of O_2 and the atomic oxygen line at $\lambda = 130$ nm.

Figure 5 shows that considerable VUV and UV radiation - highest at the jet's nozzle - can be detected even at several centimetres distance from the nozzle. An atomic oxygen line at $\lambda = 130$ nm and an O_2 emission at $\lambda = 181$ nm are presented. (V)UV radiation measured off axis is below the detection limit of the measurement setup. It can be concluded that the major fraction of the (V)UV radiation originates from the jet's discharge region and is not produced in the effluent. The radiation is not absorbed by the surrounding air, because the jet emits a plume of helium (and molecular oxygen), in which VUV radiation can be transported. The differing progressions of the two emission peaks' intensities result from the difference in absorption by the molecular oxygen [14].

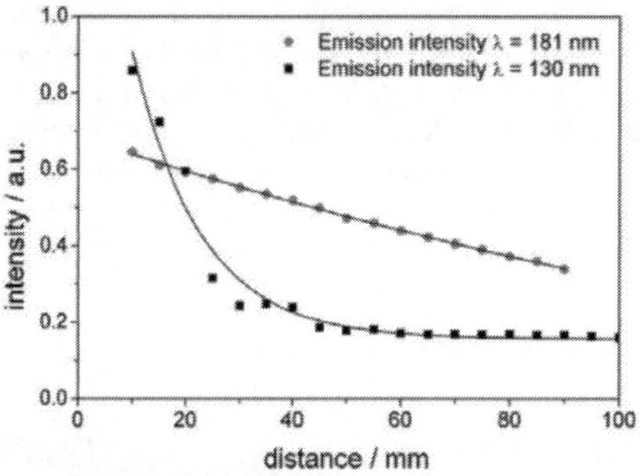

Figure 5: Spatially resolved VUV emission in the effluent region of the planar APPJ. The main spectral features are the Schumann-Runge bands of O_2, exemplarily shown at λ= 181 nm, and the atomic oxygen line at λ = 130 nm.

3. Measurements in the plasma core and the interface region

Measurements of the plasma dynamics in the core region are carried out in the so-called μ-APPJ. The μ-APPJ is an especially designed microscale version of the APPJ providing excellent access for optical diagnostics, in particular to the core plasma. Details of the design can be found in references [13], [15], and [16]. The reduced discharge dimensions ($1x1x30\,mm^3$) significantly reduce the required power input (5 – 20 W) and gas flow (1 slm) keeping the gas velocity in the same range of some $10\,ms^{-1}$. Figure 6 is a photograph of the μ-APPJ in operation. Optical emission is observable from the core plasma, interface region, and effluent region. The red square indicates the region investigated by space and phase resolved optical emission spectroscopy (PROES).

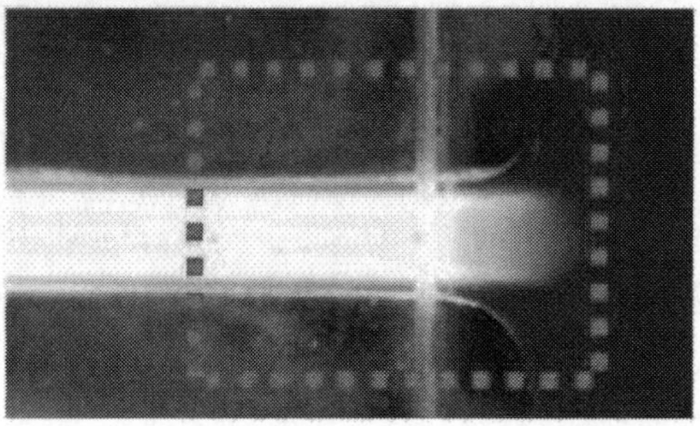

Figure 6: Photograph of the electrode configuration and optical plasma emission of the μ-APPJ. The red square indicates the measurement region for space and phase resolved optical emission spectroscopy (PROES).

PROES provides deep insight into the ionization and excitation dynamics of radio-frequency plasmas with excellent spatial and temporal resolution - down to microns [17] and nanoseconds [18]. A typical PROES experiment is described in reference [19]. Figure 7 displays spectrally integrated (ICCD sensitivity range: 300 – 800 nm) 2-dimensional spatially resolved images at different phases of the radio-frequency cycle. The images exhibit a pronounced dynamics - particularly in the plasma core. A symmetric plasma boundary sheaths dynamics, typical for capacitively coupled plasmas with equal area electrodes, is clearly visible. Interesting structures are observable in the interface region to the effluent. These structures may be explained by enhanced electric fields at the edges of the electrodes.

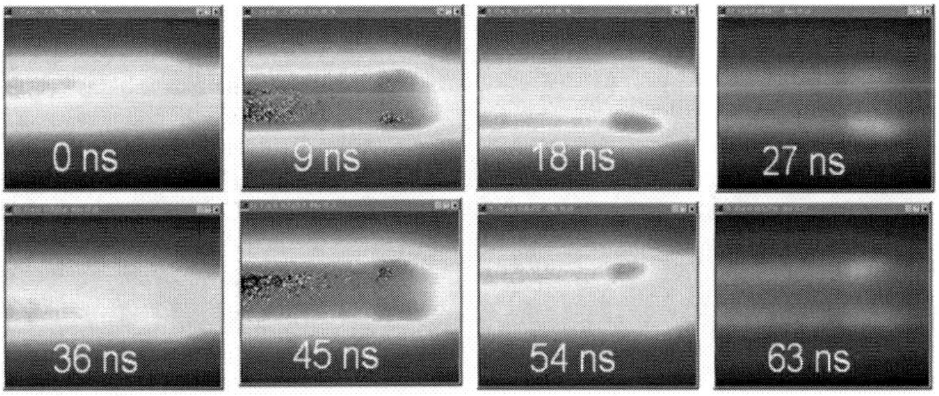

Figure 7: 2d-PROES measurements on the µ-APPJ. The images at different phases exhibit pronounced dynamics during the rf-cycle. The plasma is geometrically symmetric with plasma boundary sheath dynamics in front of both electrodes.

Details of the excitation and ionization dynamics in the plasma core are shown in figure 8. From this spatio-temporal map conclusions can be drawn on the electron kinetics within the discharge between the electrodes. A first observation is the excellent symmetry of the discharge showing three pairs of emission features (indexed as 1–3) of equal intensity and distance to the electrodes. The respective emission at the opposite electrode is displaced by half of the period (~ 37 ns). The structures 1 and 2 in the symmetric half-phases of the cycle can be attributed to the emission induced by electrons accelerated by the expanding and collapsing sheaths. They are typical for a volume ionization-based α-mode discharge. The sheath collapse is also correlated with an electric field reversal facilitated by the reduced mobility of electrons in the collision dominated environment of atmospheric pressure plasmas. Feature 3 can be attributed to electron acceleration and multiplication within the high voltage plasma boundary sheaths. Initial electrons starting this mechanism in the sheath regions can be gamma-electrons from the electrode surfaces or electrons created through Penning ionization.

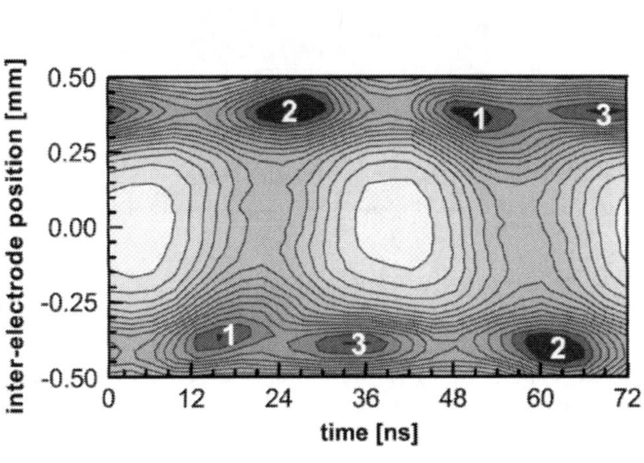

Figure 8: Excitation and ionization dynamics in the plasma core of the µ-APPJ. Energetic electrons are observable during sheath expansion (1), sheath collapse (2), and maximum sheath expansion (3). The intensity rises from bright to dark colours.

Journal of Physics: Conference Series **162** (2009) 012013 doi:10.1088/1742-6596/162/1/012013

4. Numerical simulation of the electron dynamics

Numerical computer simulations can provide more detailed insight in the plasma dynamics, in particular the electron and plasma boundary sheath dynamics. Here we present a one dimensional hybrid model across the discharge gap of the μ-APPJ with rf-excitation using the finite element method. The electrons are described kinetically and all other plasma species (ions, neutrals, excited particles) hydro-dynamically. The transport coefficients for electrons and chemical reaction rates involving electrons are calculated in advance using the Boltzmann solver BOLSIG+ [20] and stored in look-up tables depending on the mean electron energy. The governing fluid equations are Poisson's equation, the continuity equations of the considered species in the drift-diffusion approximation and the electron energy equation. The numerical simulations are done for Helium (He) with an added impurity of 0.1% Nitrogen (N_2). The species e, He, He^*, He^+, He_2^*, He_2^+, N_2, and N_2^+ are taken into account. He and N_2 are treated as constant background gases, all other species densities are determined using a set of chemical reactions equivalent to the one in reference [21].

The current voltage characteristics of atmospheric pressure glow discharges exhibits multiple solutions for a given voltage in the transition region from α-to-γ-mode operation [22]. Therefore, often the external current is given in numerical simulation rather than the voltage as under experimental conditions. By specifying the power rather than the current it is still possible to simulate a discharge with an externally applied voltage. This can be realised using a feedback control loop for the power coupled into the plasma [23].

Figure 9 shows a typical space and phase resolved map of the electron density in the μ-APPJ. The voltage amplitude is 245 V and the maximum electron density is 1.16 10^{17} m^{-3}. The symmetric character of the plasma boundary sheaths in front of both electrodes is clearly visible. The rapid movement of the sheath edges energises electrons during sheath expansion and sheath contraction - as observed in the PROES measurements.

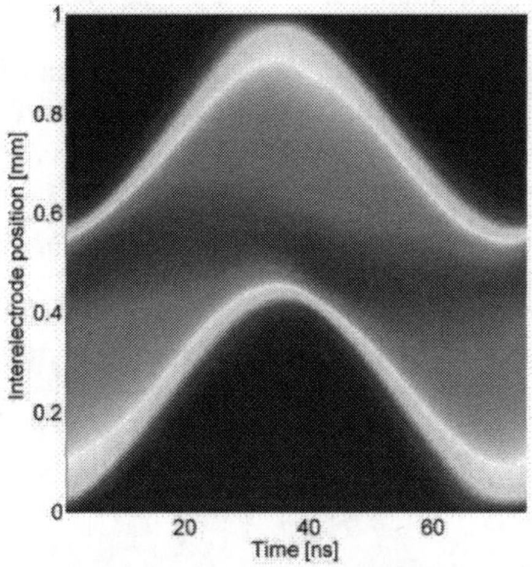

Figure 9: Numerical simulation of the electron dynamics in the discharge gap of the μ-APPJ operated in helium with a nitrogen impurity of 0.1%. The voltage amplitude is 245 V and the maximum electron density is 1.16 10^{17} m^{-3}. Red colour indicates high and blue low electron densities, respectively. The symmetric character of the plasma boundary sheaths in front of both electrodes is clearly visible.

Acknowledgement

Experiments have been carried out at the University Duisburg-Essen and Ruhr-University Bochum. The authors wish to deeply thank HF Döbele for his invaluable support. The numerical simulations have been generously supported by Y Sakiyama and DB Graves from the University California Berkeley. We would also like to greatly acknowledge very helpful discussions with MG Kong and F Iza from Loughborough University.

References

[1] Schoenbach K H, El-Habachi A, Shi W and Ciocca M 1997 Plasma Sources Sci. Technol. 6 468
[2] Eden J G et al 2003 J. Phys. D: Appl. Phys. 36 2869
[3] Baars-Hibbe L, Sichler P, Schrader C, Lucas N, Gericke K H and Büttgenbach S 2005 J. Phys. D: Appl. Phys. 38 510
[4] Becker K H, Eden J G and Schoenbach K H (ed) 2005 J. Phys. D: Appl. Phys. 38 issue 11, cluster issue on microplasmas
[5] Mahony CMO, Gans T, Graham WG, Maguire PD, Petrovic Z Lj 2008 Appl. Phys. Lett. 93(1) 011501
[6] Foest R, Kindel E, Lange H, Ohl A, Stieber M and Weltmann K-D 2007 Contrib. Plasma Phys. 47 119
[7] Léveillé V and Coulombe S 2005 Plasma Sources Sci. Technol. 14 467
[8] Schütze A, Jeong J Y, Babayan S E, Park J, Selwyn G S and Hicks R F 1998 IEEE Trans. Plasma Sci. 26 1685
[9] Niemi K, Schulz-von der Gathen V and Döbele H F 2005 Plasma Sources Sci. Technol. 14 375
[10] Goree J, Lin B and Drake D 2006 J. Phys. D: Appl. Phys. 39 3479
[11] Kuwabara A, Kuroda S and Kubota H 2007 Plasma Sci. Technol. 9 181
[12] Reuter S, Niemi K, Schulz-von der Gathen V, Döbele HF 2009 Plasma Sources Sci. Technol. 18 015006
[13] Niemi K, Reuter S, Schaper L, Knake N, Schulz-von der Gathen V and Gans T 2007 J. Phys.: Conf. Ser. 71 012012
[14] Watanabe K, Inn E C Y and Zelikoff M 1953 J. Chem. Phys. 21 1026–30
[15] Schulz-von der Gathen V, Buck V, Gans T, Knake N, Niemi K, Reuter S, Schaper L and Winter J 2007 Contrib. Plasma Phys. 47 510–9
[16] Schulz-von der Gathen V, Schaper L, Knake N, Reuter S, Niemi K, Gans T and Winter J 2008 J. Phys. D: Appl. Phys. 41 194004
[17] Waskoenig J, O'Connell D, Schulz-von der Gathen, Winter J, Park SJ, Eden JG 2008 Appl. Phys. Lett. 92(10) 101503
[18] Gans T, Lin CC, Schulz-von der Gathen V, Döbele HF 2001 J. Phys. D: Appl. Phys. 34(8) L39
[19] Abdel-Rahman M, Gans T, Schulz-von der Gathen V, Döbele HF 2005 Plasma Sources Sci. Technol. 14(1) 51
[20] Hagelaar GJM, Pitchford LC 2005 Plasma Sources Sci. Technol. 14(4) 722
[21] Sakiyama Y, Graves DB 2007 J. Appl. Phys. 101(7) 073306
[22] Shi JJ, Kong MG 2005 J. Appl. Phys. 97(2) 023306
[23] Sakiyama Y, Graves DB 2006 J. Phys. D: Appl. Phys. 39(16) 3451

Journal of Physics: Conference Series **162** (2009) 012014 doi:10.1088/1742-6596/162/1/012014

Progress report: Direct injection of liquids into low-pressure plasmas

Matthew Goeckner, Daisuke Ogawa, Iqbal Saraf and Lawrence Overzet

Electrical Engineering, University of Texas at Dallas, Richardson, TX USA

goeckner@utdallas.edu

Abstract. While laboratory based plasmas are always in contact with solid surfaces (often vacuum chambers) they have historically been formed in gas environments. In more recent times, the use of plasmas has grown to include plasma contact with liquids including biological items. Inevitably the plasmas in contact with liquids had been at or near atmospheric pressures. This need not be the case. We have developed a novel method for injecting liquids directly into low-pressure discharges. As such, this technique opens new areas of possible industrial use for plasmas. For example, we have injected inorganic nano-particles into argon plasma by suspending them in hexane (or ethanol) as a high vapor pressure liquid carrier. As a result, we believe that metals, dielectrics, superconductors, aromatics, proteins, viruses, etc. could all potentially be injected into low-pressure plasma environments using this technique. The resulting films indicate the ability to synthesize nano-structured composites. Here we examine some of the basic phenomenon that are observed both experimentally and theoretically.

1. Introduction

Liquids have been used for processing in other systems but not in the same manner as is being reported here. For example, Senateur, et al. [1] show direct liquid injection into an MOCVD tool. In that technique the liquid (solvent, solute) is injected onto a heated, porous, moving tape in an area of the tool that is relatively cool. The tape, with liquid, is moved toward a warmer (~300 °C) portion of the tool, causing the solvent to evaporate. A 'vector' gas is introduced in a position along the path of travel such that the evaporating solvent s sweep always from the process area. The tape, now containing only the solute, reaches the process area wherein solvent evaporates and is transported to the substrate. Such techniques have been used to grow $YBa_2Cu_3O_7$ superconductors. Plasma spray methods are also well known for introducing liquids (solvent, solute) into plasmas. However these operate at much higher pressure and much higher neutral temperatures. See for example Ref [2]. Finally M Coppins has reported the production of liquid droplets in 'dusty' plasmas. None of these exhibit our desired situation: Low pressure, Low neutral temperature, Plasma interactions, resulting in thin film deposition from fragile or non-volatile precursors.

2. Experimental efforts

We have build a simple apparatus that combines a automotive fuel injector with traditional low pressure, low temperature plasmas, see figure 1[3]. This allows us to inject a wide range of materials directly into a discharge, including solvents with non-volatile or fragile precursors. Examples might include solids and DNA or viruses. This technique is inexpensive to setup and we find that we can control the liquid insertion to a reasonable level. This new technique also opens new areas of plasma science research. For example we find that the plasma reacts in fascinating ways to the pressure wave induced by the injection process. This is seen in the emission from the discharge as shown Figure 2.

We have used the system to inject various species. Typical of those are hexane and ethanol. In figures 3 and 4 we show the deposition as a function of time. Specifically we have used injector pulse widths of 1.3 and 2 ms. At 1 atmosphere backing pressure and 1 Hz pulses on the injector, these times correspond to flow rates of 70 and 110 μL/min. In addition to the liquid injection we also have 20 SCCM of Ar as a buffer gas, 10 W or rf (13.56 MHz) power and heated the chamber to 60 °C. The pumping system was set to achieve ~ 400 mTorr.

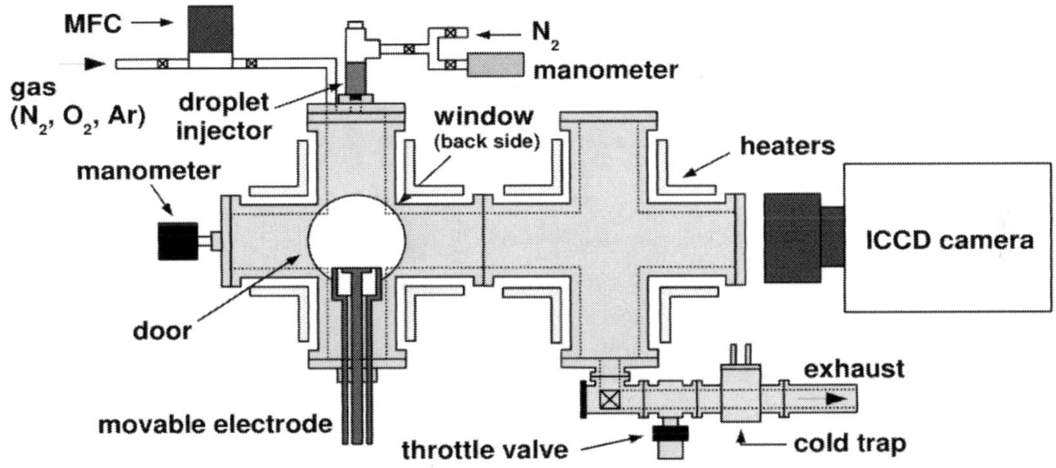

Figure 1. System layout for the liquid injection tool. The heart of the system is an automotive fuel injector, allowing the input of various liquid mixtures.

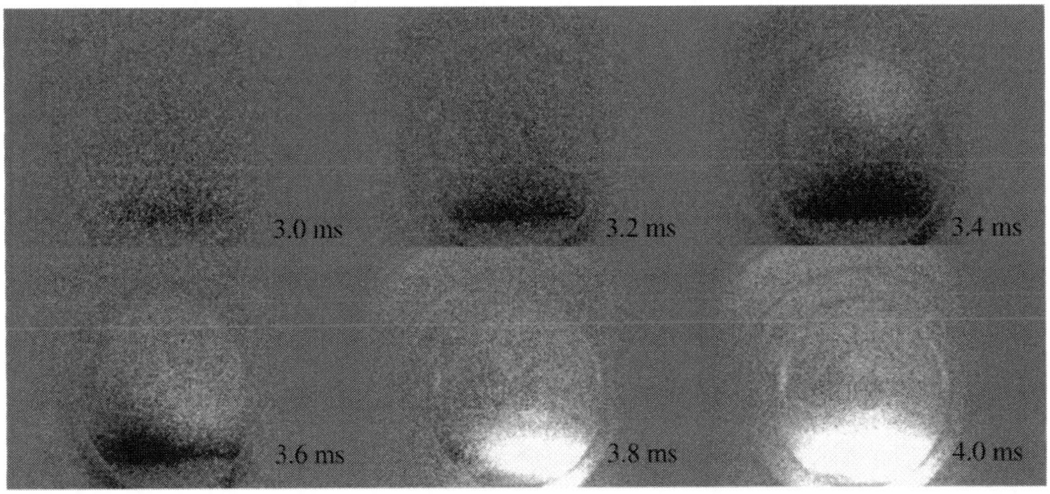

Figure 2. Time sequence of change in emission during N_2 injection into Ar plasma. Change in emission is determined via background subtraction. Here black represents a decrease in emission, white represents an increase and gray is no change. Time is given relative to initiation of injection.

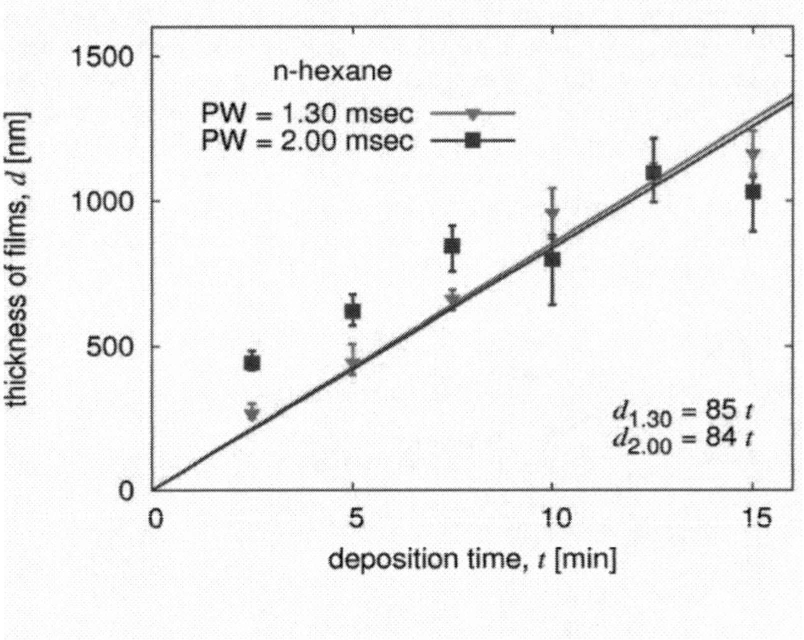

Figure 3. Typical film thicknesses for hexane injections.

Such process steps result in films such as those shown in figure 5. Figure 6 shows a film produced from Hexane with 2 nm Fe nano-particles. Here the Fe represents ~4.3% of the film by volume. Process conditions are as in Figure 3, with the 1.3 ms injector pulse widths. Deposition times are ~10 minutes.

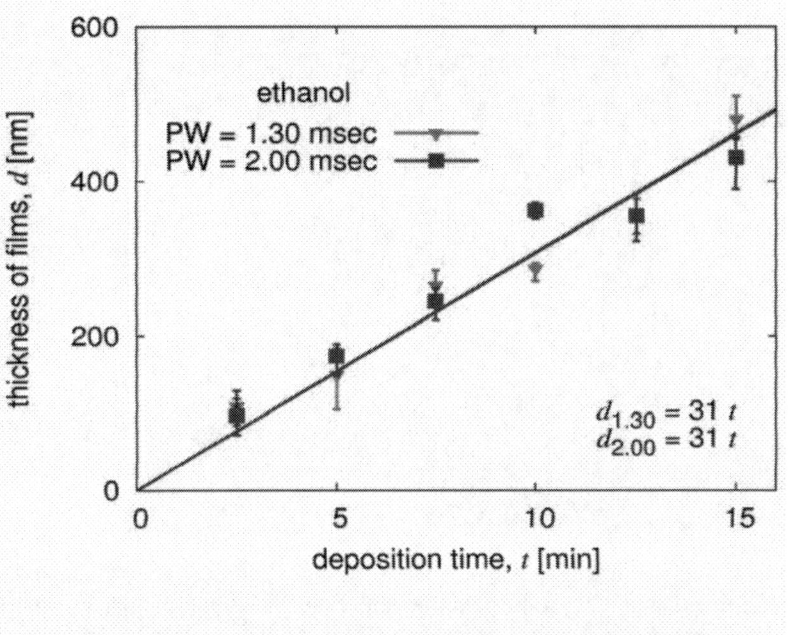

Figure 4. Typical film thicknesses for ethanol injections.

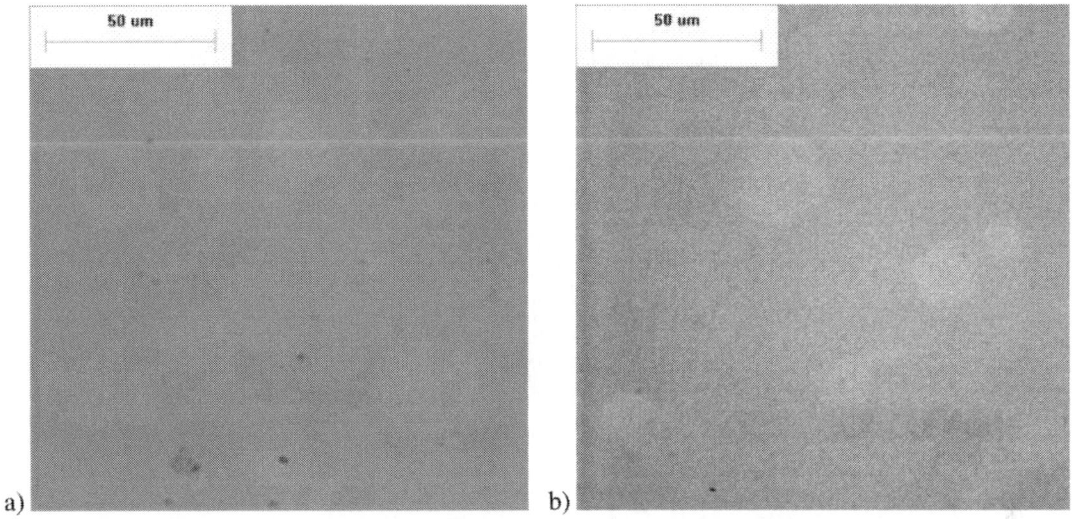

Figure 5. Typical films for a) hexane b) ethanol injections.

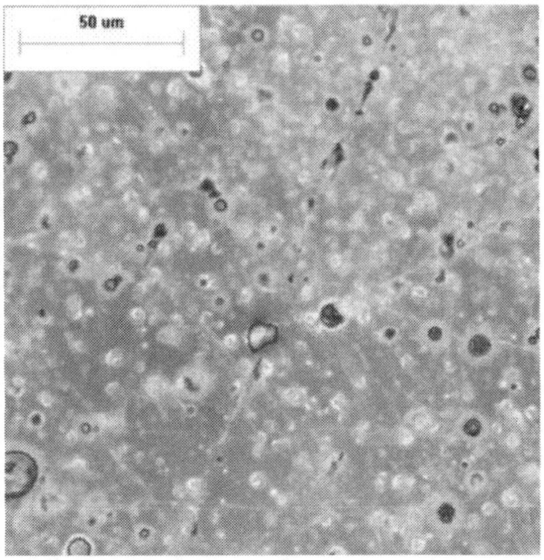

Figure 6. Typical films for hexane with 2 nm Fe nano-particles.

3. Modeling efforts

Models of the time evolution of the injected liquid have been developed based on the processes shown in figure 7[3]. Without a plasma, the model indicates that the time required for the droplet to fully evaporate is a function of the background pressure, initial (wall) temperature, the number of droplets inserted simultaneously and initial size, see figure 8.

It is evident that the evaporation falls has two distinct temporal periods. The first period is set by evaporation of the solvent, along with a rapid cooling, see figure 9. For example both water and benzene rapidly reach freezing temperatures and the liquid evaporation process becomes a sublimation process.

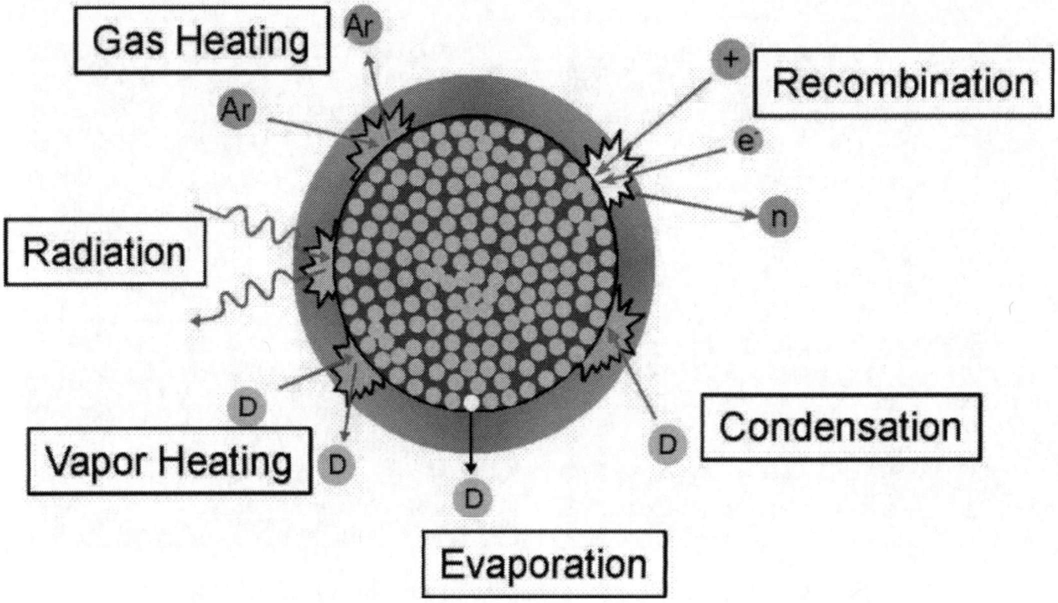

Figure 7. Parameters considered in the model of the evolution of liquid droplets in a low-pressure, low-temperature plasma discharge. Note that we have not yet included surface chemical reactions in the model.

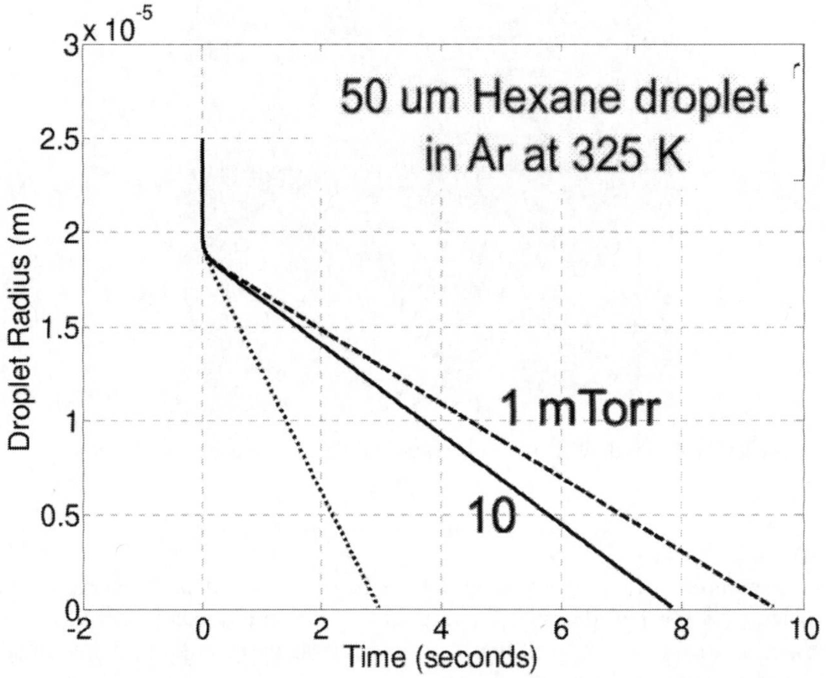

Figure 8. Time sequence of 50 μm n-hexane droplet injected into 1 and 10 mTorr Ar. Plasma greatly speeds this evaporation rate.

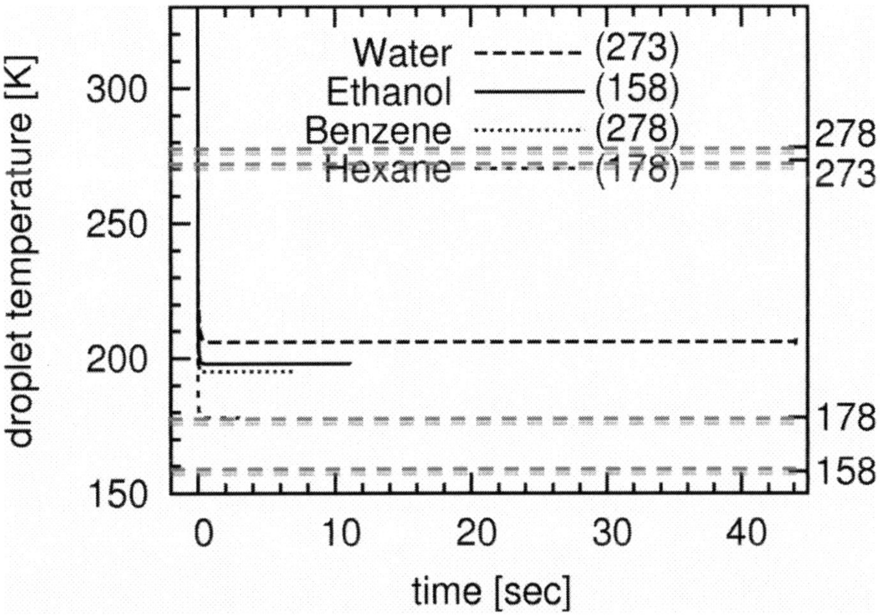

Figure 9. Droplet temperature as a function of time. We note that certain droplets will rapidly fall below the freezing point, e.g. benzene at 278 °C and water at 273 °C. Once this happens the evaporation is via sublimation.

Adding a plasma brings additional heat to the surface through ion impact and electron-ion recombination. This can result in a significant increase in the evaporation – and a corresponding decrease in the total evaporation time. (This time is the duration it takes to go from an initial size of 50 μm to 100 nm.) Figure 10 shows the influence of a plasma on a hexane droplet. At low densities the plasma has very little impact. At very high plasma densities, the evaporation is almost instantaneous. At more typical densities, e.g. ~10^{16} m^{-3}, the temperature of the electrons has a major influence. The electron temperature impacts the energy flux to the surface via the floating potential of the droplet in the plasma. The higher the temperature the higher the potential and hence the higher the impact energy of the ions. At very high densities, the recombination of the ions and electrons on the surface drives the evaporation. Here the energy flux is set by the overall flux of ions and electrons to the surface.

Journal of Physics: Conference Series **162** (2009) 012014 doi:10.1088/1742-6596/162/1/012014

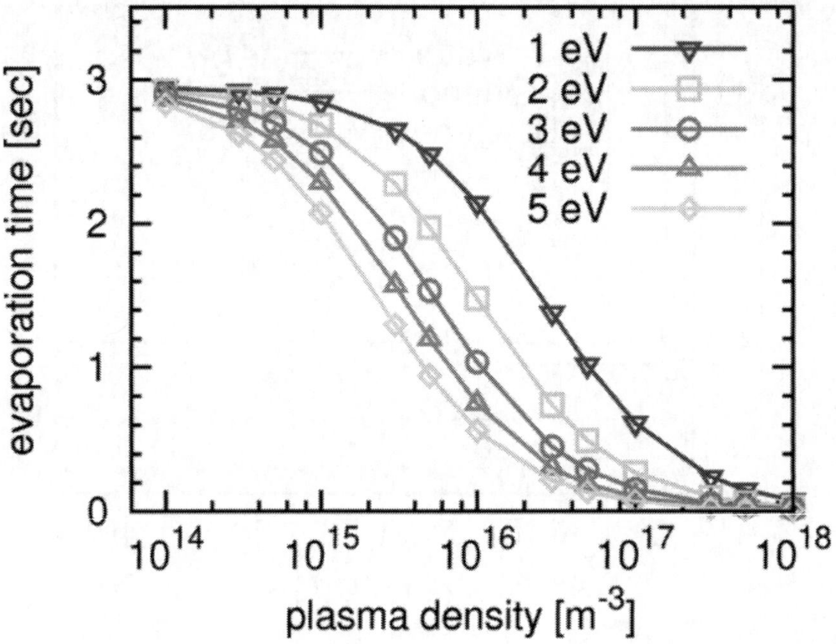

Figure 10. Evaporation times for 50 μm hexane droplets under various plasma conditions.

In addition to single component droplets, we have also examine two component droplets. In figure 11 we examine the temperature on a multi-component droplet as a function of time. Here there are a number of stages. In the first stage, the droplet temperature decreases rapidly as the initial evaporation occurs. It then reaches a stage in which the heat loss due to evaporation is balanced by plasma heating. When all of the hexane has evaporated, the droplet begins to heat up. It eventually reaches a steady state temperature that is set by a balance between radiative cooling and plasma heating.

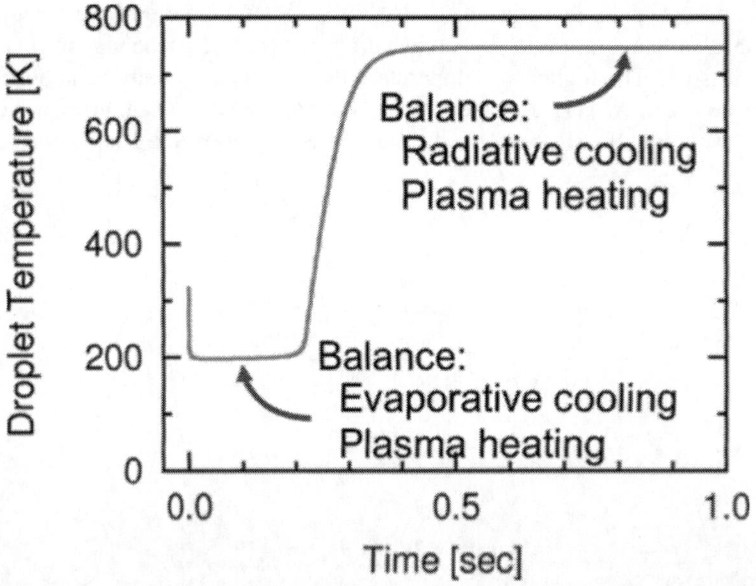

Figure 11. Modeled droplet temperature for hexane + 1% solid solute in Ar.

Journal of Physics: Conference Series **162** (2009) 012014 doi:10.1088/1742-6596/162/1/012014

When both the solute and the solvent have non-zero vapor pressures the droplet responds in a slightly different fashion. Figure 12 and 13 show the predicted droplet radius and droplet temperature as functions of time for a mixture of 1% ferrocene in hexane. First it observed that the droplet undergoes the typical rapid drop in radius as it also undergoes a rapid drop in temperature. It then reaches a quasi steady state temperature regime set by evaporative cooling (hexane) balancing plasma heating. When the hexane is exhausted, there is a brief period during which the droplet temperature increases rapidly, while the droplet does not shrink in size. After this period the droplet will have reached a new balance temperature, again set by a balance of evaporative cooling and plasma heating, with the evaporation now consisting of ferrocene.

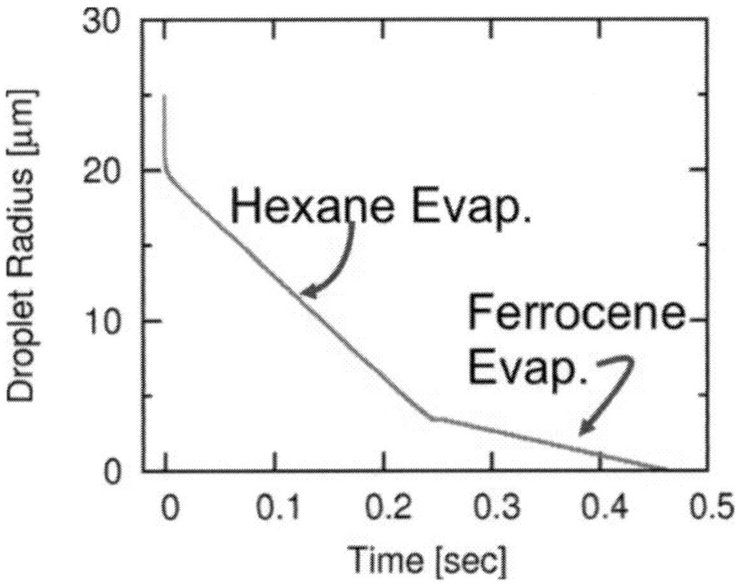

Figure 12. Modeled droplet radius for hexane + 1% ferrocene in Ar.

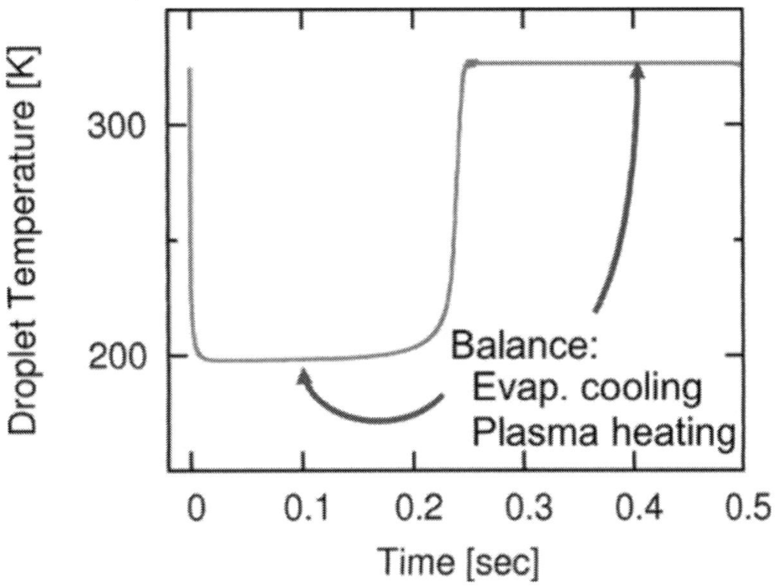

Figure 13. Modeled droplet temperature for hexane + 1% ferrocene in Ar.

Journal of Physics: Conference Series **162** (2009) 012014 doi:10.1088/1742-6596/162/1/012014

4. Conclusions

We have shown that it is possible to directly inject liquids (solvent and solute) into low-pressure plasmas and use them to produce films that are not possible through other methods. Our models indicated that the droplet undergo significant temperature and radial shifts. There is an initial temperature drop due to evaporative cooling followed by steady state temperatures set by energy balances. In some cases the droplet even freeze. The nature of the energy input, plasma vs. neutral bombardment, plays a major role in determining the lifetime of the droplet. Under just neutral bombardment, droplets can take many seconds to evaporate. Moderate plasma densities significantly reduce the evaporation times to a few tenths of second or less.

Acknowledgements

This work was funded in part by a grant from AFOSR through the SPRING consortium, contract FA 9550-05-1-0393

References

[1] Sénateur J-P et al. 2000 *Adv. Mater. Opt. Electron.* **10**, 137
[2] Ishigaki T et al. 2005 *Sci. Technol. Adv. Mater.* **6** 111
[3] Ogawa D, Saraf I, Sra A, Timmons R, Goeckner M, and Overzet L, "The Direct Injection of Liquid Droplets into Low Pressure Plasmas," Submitted to *J Vac. Sci. Technol.* A 2008

Journal of Physics: Conference Series **162** (2009) 012015 doi:10.1088/1742-6596/162/1/012015

Surface Functionalization of Organic Materials by Weakly Ionized Highly Dissociated Oxygen Plasma

Alenka Vesel[1] and Miran Mozetic[1]

[1]Jozef Stefan Institute, Jamova cesta 39, 1000 Ljubljana, Slovenia

E-mail: alenka.vesel@ijs.si, miran.mozetic@guest.arnes.si

Abstract. A review on surface modification of different polymers by treatment in oxygen plasma is presented. Plasma is created in a high frequency inductively coupled gaseous discharge at the power of about 200 W. In such discharge created in pure oxygen, plasma with the following parameters is obtained: the electron temperature of about 50.000 K, the charged particle density around $1x10^{16}$ m^{-3}, and the neutral oxygen atom density of the order of $10^{21}m^{-3}$. A huge flux of neutral oxygen atoms on the surface of samples exposed to plasma assures for rapid interaction with polymer materials. The modification of surface properties of the following polymers was studied: polyethyleneterephthalate (PET), polyethersulphone (PES), polyphenylenesulfide (PPS), Nylon 6 polyamide (PA6), polytetrafluoroethylene (PTFE), polystyrene (PS), polypropylene (PP) and cellulose (ink-jet paper and textile). The polymer samples were treated for 3 s in oxygen plasma at a pressure of 75 Pa where the O-atom density was the largest at $4x10^{21}$ m^{-3}. The appearance of the functional groups on the surface of the samples was monitored by high resolution X-ray photoelectron spectroscopy (XPS). The results show that oxygen plasma treatment is an effective tool for surface modification. On all polymer surfaces increased concentration of oxygen is detected. The high resolution C1s peaks indicate formation of several new oxygen-containing functional groups. On all polymers groups like C-O, C=O and O=C-O are observed. The concentration of these groups depends on the type of polymer. The highest uptake of oxygen by the polymer was found for cellulose and the lowest for polypropylene. The only exception was polymer PTFE where practically no chemical changes were observed after plasma treatment.

1. Introduction

1.1. Surface energy and adhesion

Surface properties of materials depend on the properties of bulk materials as well as surface contaminants. The surface energy is largest for metals, followed by ceramics and glasses, and is lowest for non-polar polymers. Even carefully cleaned materials are often covered with very thin films of different impurities such as native oxide films and traces of hydrocarbons that definitely modify the original surface energy of pure materials. The surface energy often determines the adhesion of different coatings on the substrates. Apart from the roughness, the surface energy is usually the key parameter that affects the quality of coatings on polymers and polymer matrix composites. Modern industry requires components that are made from materials with desired mechanical, chemical, optical and electrical properties, and are coated with different films often incompatible with the substrates. In such cases, the surface properties should be modified without modification of the bulk properties.

Several techniques have appeared to reach this goal: they include a variety of wet chemical treatments, modification by energetic particle beams and application of gaseous plasma. Although atmospheric plasmas are increasingly popular, low pressure plasmas have some advantages. The main advantage of low pressure plasma is spreading over a large volume that assures for rather uniform treatment of three dimensional objects with a complex shape.

1.2. Plasma treatment of selected polymers – state of the art

Many materials including most polymers and polymer matrix composites have poor adhesion properties and wettability. This is especially the case of hydrophobic polymers (such as polypropylene, polystyrene and polytetrafluoroethylene). The surface energy should be increased in order to assure a rather good adhesion of various coatings. The process that allows for increasing the surface energy is called surface activation. The surface of organic materials should be activated before printing, painting, coating, for improving biocompatibility etc. One of the best methods for modifying the surface properties of polymer materials is plasma treatment. Plasma treatment is an ecologically benign method and is gradually replacing traditional wet chemical techniques, which usually involve application of ecologically unfriendly chemicals. A broad range of surface energy can be achieved by using appropriate plasma for polymer treatment, from extreme hydrophilicity to significant hydrophobicity. Increased hydrophobicity is usually achieved by application of plasma rich with fluorine, while hydrophilicity is often achieved using oxygen plasma. In some applications, especially to increase the biological compatibility of organic materials, nitrogen or ammonia plasma gives better results [1]. It should be noted that plasma treatment usually does not produce only one type of a functional group on a polymer surface. Typically, several different functional groups appear on the surface of polymer during plasma treatment. It is therefore necessary to apply such plasma that facilitates formation of the functional groups that are most important for a given application and to attempt to shift the distribution in favour of a specific functionality by changing plasma parameters [2]. Oxygen plasma treatment usually causes formation of different functional groups such as C-O, C=O, O=C-O. Sometimes, more exotic groups can be produced on the surface of some polymers [2,3].

Literature reports different treatment times used for surface modification of polymers ranging from milliseconds [4,5] to several minutes [6]. A millisecond treatment often causes oxidation and removal of surface contaminants which is often sufficient to improve surface wettability. A further treatment usually causes chemical bonding of oxygen atoms at active sites on the polymer surface, leading to formation of various functional groups that modify the surface wettability. With prolonged treatment time excessive change scission may appear leading to a layer of low-molecular-weight fragments on the surface [3]. Also, prolonged treatment often causes chemical etching of polymer materials, increased sample temperature and eventual irreversible damage of the bulk properties. In many cases, the longer exposure of the surface, not only of polymers, leads to creation of various nanostructures [7-12].

An important drawback of plasma treatment for activation of polymers is ageing. Functional groups formed on the plasma treated surface are not stable with time, as the surface is in non-equilibrium condition and tends to approach more energetically favorable state, often recovering to its untreated state. Thus, the surface keeps loosing its hydrophilic character spontaneously. The ageing effect is obviously more pronounced at higher substrate temperature. Two different processes have been reported to be responsible for surface ageing: the first one is the reorientation of the polar groups into the bulk polymer, and the second one is mobility of small polymer chain segments into the matrix, both leading to different free surface energy. It was also reported that the chain mobility mainly occurs in the amorphous phase of polymers while the mobility in the crystalline phase is fairly limited due to orderly packed structure. Therefore more crystalline polymers are ageing slower. Since plasma treatment can increase the surface crystallinity due to preferential etching of the amorphous phase, the polymers treated for longer times are usually ageing slower [3-5,13]. This is not always true, however – too long treatment times may again lead to faster ageing due to formation of small fragments loosely bound on the surface. Such surface has a greater tendency to ageing because of migration of small fragments to the bulk. Also, as mentioned above, prolonged treatment often causes chemical etching and thus increasing temperature, that definitely affects the characteristics of plasma treated materials.

Plasma treatment usually affects the surface layers of polymer, often only first few nanometers of material. The bulk properties remain fairly intact [14]. The quickest method to check the effect of plasma treatment on the surface properties is a wettability test by measuring the contact angle of a suitable liquid drop with a well known surface energy – often distilled water. This method is extremely easy to apply athough the results are sometimes dubious. As it is a macroscopic method, it does not say anything about the local distribution of surface functional groups, let alone the type of the groups. A suitable method for determination of the type of functional groups on a polymer surface is X-ray Photoelectron Spectroscopy (XPS), often called Electron Spectroscopy for Chemical Analysis (ESCA) [14-16]. Unfortunately, commercial XPS instruments have poor lateral resolution so a microbeam ECSA at an appropriate synchrotron is required for determination of the distribution of functional groups on micro scale. The interpretation of XPS spectra may be quite difficult, too. A fundamental problem in polymer analysis is surface charging due to a loss of surface electrons by X-ray irradiation. This effect is often more pronounced when using a monochromatic source of X-rays, which improves the energy resolution otherwise. To avoid charging effects, neutralization with a low energy electron flux is used. A common practice is manual shifting of unfunctionlized C1s peak (C-C) to 284.8 eV. In some cases, however, almost all carbon atoms in the surface layer of polymer are chemically shifted – an example is cellulose, where all carbon atoms are bound to at least one oxygen atom. For analyses of these materials, a peak which is assigned to hydrocarbon contamination can be used as a reference (if such contaminants are presented). Obviously, it is not possible for polymers treated with oxygen plasma since reactive particles often remove such contaminants even at short treatment time as mentioned above.

Polymers are usually quite stable during typical analysis times. Prolonged exposure to X-rays, however, may cause radiation damage of the sample. In such cases, the photoelectron spectrum slowly changes. A visual evidence of this effect is sample discolouration [17]. For example, this can be very easily observed on paper substrates. Halogen containing polymers can be also quite sensitive to X-ray induced sample degradation. The result is a depletion of halogen atoms with the exposure time [18].

1.3. Oxygen plasma

Plasma is usually created in a gaseous discharge. Electrons are accelerated in an electric field and quickly thermallize at elastic collisions. Their energy distribution function is therefore roughly Maxwellian with the temperature of several eV. Electrons in the high-energy tail of the distribution function have enough energy for direct ionization of gaseous molecules, while those in the low energy part of the distribution function are only capable to excite rotational and vibrational excited states of molecules. In the case of oxygen molecules, the excitation energy for metastable molecules $O_2{}^1\Delta$ and $O_2{}^1\Sigma$ is about 1 and 2 eV, respectively, the dissociation energy is 5.2 eV, while the ionization energy is 12 eV. Electrons with the average energy of few eV are therefore likely to excite molecules into metastable states and ro-vibrational states, while dissociation and ionization in less probable.

Excited particles tend to de-excite. There are many channels for de-excitation of oxygen molecules, some of them occur in the gas phase, while the others take place on the surface of the discharge chamber. In any case, the conservation of energy and momentum as well as rules of quantum mechanics should be obeyed. Vibrationally excited states are de-excited in the gas phase primarily by vibration interchanges (V-V transitions) and super-elastic collisions with atoms (V-T transitions) [19]. Neutral oxygen atoms in the ground state can recombine to molecules only at three-body collisions that are unlikely to occur at low pressure (say below few mbar) so they are rather stable in the gas phase.

Surface de-excitations often play a dominant role in low-pressure plasmas. The probability for ion recombination is close to 1 [20], while the recombination of neutral oxygen atoms depends largely on the type of material facing plasma as well as its temperature and morphology. The probability for heterogeneous surface recombination ($O + O \rightarrow O_2$) for many glasses and some ceramics is often low

Journal of Physics: Conference Series **162** (2009) 012015 doi:10.1088/1742-6596/162/1/012015

(the typical order of magnitude in 10^{-4}), while for many metals and some porous ceramics it is of the order of 10^{-1} [21-22].

The density of different excited species in plasma depends on excitation and de-excitation probabilities. The excitation probabilities depend mostly on electron density and temperature, while the de-excitation probabilities depend on gas pressure and surface properties. By choosing smooth materials with low recombination coefficients for recombination of O atoms it is often possible to achieve plasma with a low density of ions but a very high density of neutral atoms. It is often possible to obtain oxygen plasma with the ion density below 10^{16} m^{-3} and the neutral atom density above 10^{21} m^{-3} [23-26].

Reactive particles created in oxygen plasma react with organic materials at certain rates. The rates depend on the type of polymer and particle as well as the sample temperature and particles kinetic temperature. High oxidation selectivity for different types of organic materials can only be obtained with cold plasmas. Cold plasma is a state of gas with a low kinetic energy of heavy particles, i.e. all particles except electrons. The low kinetic energy of heavy particles is assured if their heating is minimized. There are some channels for heating heavy particles in plasmas. Elastic collisions between fast electrons and heavy particles do not lead to substantial kinetic energy exchange due to a small mass of the electrons. The major channel for heating heavy particles at collisions with energetic electrons is a dissociation event. At direct electron impact dissociation, the excessive kinetic energy of the electron can be observed as the kinetic energy of newly formed atoms, which can move apart with a substantial kinetic energy. Such energetic atoms get effectively thermallized at elastic collisions with other heavy particles. A good way to avoid this sort of plasma heating is application of plasma with a rather low electron temperature.

Another mechanism of heating heavy particles is super-elastic collisions between vibrationally excited molecules and oxygen atoms. The reaction cross-section is large [20]. The only way to avoid such collisions is application of fully dissociated plasma where such collisions are unlikely to occur.

An important channel for heating heavy particles may be acceleration of ions in an electric field. The ions are accelerated in electric field as are the electrons. As long as the electric field frequency is low, the ions can follow any change of the local electric field. The energy an oxygen ion can pick from the electric field is

$$W = (e^2E^2)/(2m\omega^2). \quad (1)$$

Here, e is the ion charge, E peak electric field, m ion mass and ω electric field frequency. The kinetic energy an oxygen ion in a high vacuum can gain in the electric field therefore depends on the frequency. As long as the frequency is low (say below 100 kHz), the ions are well accelerated in the field. But as the frequency is increased, the ions are not able to follow the field. At the frequency of about few MHz, the kinetic energy an ion can pick from the field is less than the average thermal energy of an ion at room temperature. This means that ions in the electric field with a frequency above 10 MHz cannot pick energy worth mentioning. As long as the electric field frequency is higher than 10 MHz, the ions thus cannot be accelerated in the field and cannot contribute to neutral gas heating.

Upper considerations lead to the conclusion that best oxygen plasma for selective oxidation of different materials is created in a radio-frequency discharge. A high degree of dissociation of oxygen molecules is obtained in a discharge chamber made from glass, which has a smooth surface and a low coefficient for recombination of oxygen atoms. There are two extreme modes for RF generator coupling: i) capacitive, and ii) inductive. In practice, the coupling is often a mixture of both extremes. Inductive coupling is often obtained using a coil wounded around a glass tube. In this case, electrons are accelerated in induced electric field sustained due to alternating magnetic field in the coil. The electric field at the axis is rather low, and is increasing towards the edge of the glass tube.

Another extreme is capacitive coupling. In this case, the charged particles are accelerated in the alternating electric field between two parallel electrodes. A sheath with a substantial potential is established next to the powered electrode (which often has a smaller area than the grounded chamber). As long as the sheath is almost collisionless (i.e. the mean free path is larger than the sheath thickness) the ions entering the sheath from the gas phase are accelerated towards the electrode and do not transfer kinetic energy to other particles. They bombard the electrode and some are reflected as neutral fast atoms or molecules. These fast particles do heat the neutral gas. The heat exchange between positive ions and other heavy particles is increased in the case the sheath is not collisionless. In such cases there are more channels for kinetic energy exchange in the gas phase. This often occurs at elevated pressure, say above 0.1 mbar.

As the electric field frequency increases towards the microwave range, the ions can gain practically no energy from the field, and also electrons cannot pick as much energy as in the case of radio-frequency discharges. As a general rule, the electron temperature in simple microwave discharges is always lower than in radio-frequency discharges with comparable power. More energy is transferred to neutral gas heating so the microwave plasma is never as cold as the RF plasmas.

Parameters of low pressure plasma created by inductively coupled RF discharge in a glass tube depend on discharge power and pressure. Typical values are as follows: neutral gas kinetic temperature is often equal to ion kinetic temperature and is a bit more than room temperature - values between 300 and 500 K are common. The electron temperature is often about 50.000 K or more. At the pressure of few 10 Pa, the density of electrons and ions is often between 10^{15} and 10^{16} m^{-3}, while the density of neutral oxygen atoms is of the order of 10^{21} m^{-3}. The plasma potential is often of the order of 10 V and the Debye length about 10^{-4} m. The density of neutral oxygen atoms is certainly the most important parameter.

Several methods have appeared to measure the O density in highly dissociated oxygen. The methods include optical spectroscopy [27-36], mass spectrometry [37,38], gas titration [25,39] and catalytic probes [40-55]. The latter was found to have some advantages over other techniques, as catalytic probes enable real/time measurements and do not disturb the original concentration of O atoms. The disadvantages include a poor understanding of surface recombination phenomena and sensitivity to high-frequency interferences. From the latter point of view, fiber optics catalytic probes (FOCP) have a definite advantage: as any connection is made optical, they are completely immune to stray effects caused by a high frequency electromagnetic field [51-55]. On the other hand, the FOCPs cannot measure low densities of O atoms.

The interaction of inductively coupled oxygen plasma with solid materials is almost entirely potential. As shown in upper text, the ion density is usually below 10^{16} m^{-3} and their kinetic energy at the sample surface about 10 eV. On the other hand, we have neutral atoms with the density often exceeding 10^{21} m^{-3}. The dissociation degree therefore exceeds the ionization fraction by 5 orders of magnitude if not more so. A large flux of O atoms onto a sample surface assures rich surface chemistry if the samples are organic materials. The first effect of interaction is removal of surface organic impurities, followed by surface functionalization with oxygen-rich functional groups. Next effect is slow etching of the organic material. [56-58] Since the interaction is almost purely potential, the etching largely depends on the nature and structure of organic materials.

2. Experimental

2.1. Experimental setup

Experiments were performed with different polymers including PP, PS, PET, PES; PPS; PA6, PTFE and cellulose materials like ink-jet paper and textile. The samples of these materials were treated in the experimental system shown in figure 1. The system is pumped with a two-stage oil rotary pump with a pumping speed of 16 m^3/h. The discharge chamber is a borosilicate glass (Schott 8250) cylinder with a length of 200 mm and an inner diameter of 36 mm. A narrow glass tube with an inner diameter of 5 mm and a length of 6 cm leads to the afterglow chamber, which is also a borosilicate glass cylinder, with a length of 400 mm and an inner diameter of 36 mm. Plasma is created inside the discharge chamber with an inductively coupled RF generator [59-62], operating at a frequency of 27.12 MHz and an output power of about 200 W. The plasma parameters are measured with a double Langmuir probe and a catalytic probe. The Langmuir probe is placed into the discharge chamber, while the catalytic probe is mounted in the afterglow chamber. Commercially available oxygen is leaked into the discharge chamber, as shown in figure 2. The pressure is measured with an absolute vacuum gauge. The pressure is adjusted during continuous pumping using a precise leak valve.

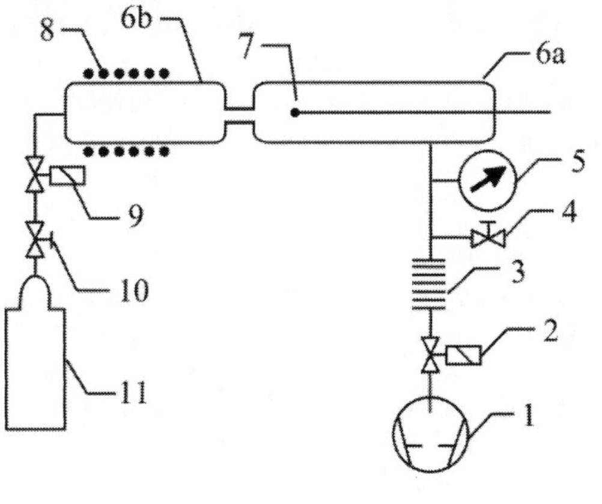

Figure 1. Schematic of the experimental setup: 1 – rotary pump, 2 – gate valve, 3 – Hopkins trap, 4 – air inlet valve, 5 – vacuum gauge, 6a – discharge chamber, 6b – post-discharge chamber, 7 – catalytic probe, 8 – RF coil, 9 – leak valve, 10 – high-pressure valve, 11 – oxygen.

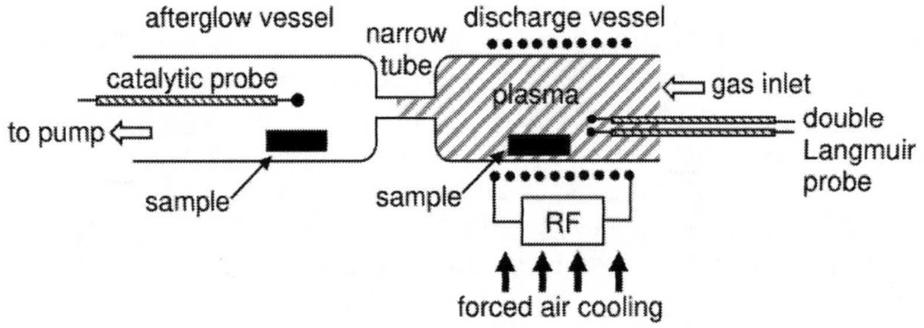

Figure 2. The discharge chamber.

2.2. XPS characterization

The samples were exposed to air for a few minutes after the plasma treatment and then mounted in an XPS instrument (TFA XPS Physical Electronics). The base pressure in the XPS analysis chamber was about 6×10^{-10} mbar. The samples were excited with X-rays over a 400-μm spot area with monochromatic Al $K_{\alpha 1,2}$ radiation at 1486.6 eV. The photoelectrons were detected with a hemispherical analyzer positioned at an angle of 45° with respect to the normal to the sample surface. The energy resolution was about 0.6 eV. Survey-scan spectra were made at a pass energy of 187.85 eV, while C 1s, S 2p, N 1s, F 1s and O 1s individual high-resolution spectra were taken at a pass energy of 23.5 eV and a 0.1-eV step. Since the samples are insulators, we used an additional electron gun to allow for surface neutralization during the measurements. The spectra were fitted using MultiPak v7.3.1 software from Physical Electronics, which was supplied with the spectrometer. The curves were fitted with symmetrical Gauss-Lorentz functions. The peak width (FWHM) was fixed during the fitting process.

3. Results and discussion

3.1. Plasma characterization

Figure 3 represents the density of electrons in the discharge tube as measured with a double electrical probe. The density reaches its highest value at the pressure of about 20 Pa. At lower pressure it decreases rapidly with decreasing pressure (below about 1 Pa plasma cannot be ignited any more) while at higher pressure it decreases slowly with increasing pressure. This phenomenon is explained

Journal of Physics: Conference Series **162** (2009) 012015 doi:10.1088/1742-6596/162/1/012015

by more frequent collisions of electrons at higher pressure. Namely, as mentioned above, the electrons gain maximal energy from the induced electric field when the mean free path is equal to the oscillation amplitude. As pressure becomes large, this is not the case any more: electrons suffer collisions before they can reach the maximal oscillating energy in the electric field. As the result of such collisions the electron temperature is lowered and so is the ionization probability. At pressure above 200 Pa plasma cannot be ignited any more.

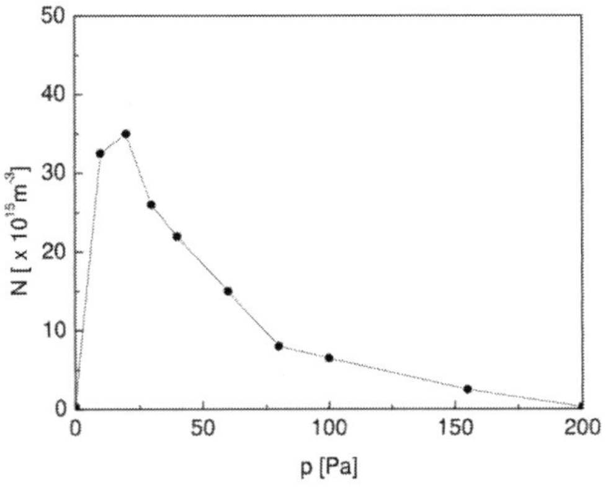

Figure 3. The plasma density versus pressure.

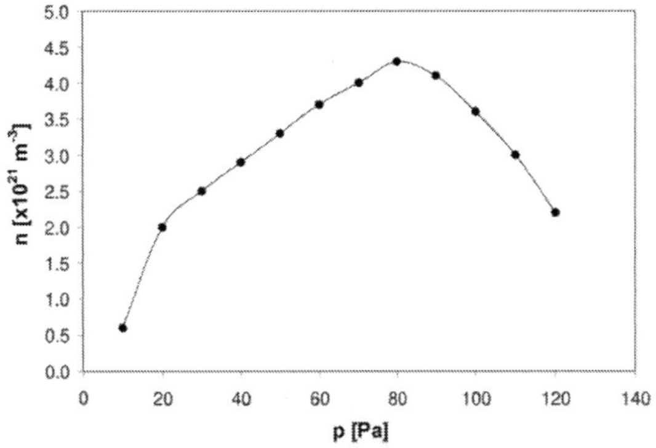

Figure 4. The density of neutral oxygen atoms versus pressure.

Figure 4 represents the density of neutral oxygen atoms in the early afterglow (at the catalytic probe position). In the plasma itself, the O-atom density is somewhat larger. At low pressure the O-atom density is rather low and slowly increases with increasing pressure. As the pressure is increased, the O-atom density increases monotonously until it reaches a broad maximum. At high pressure, the O-atom density decreases with increasing power. The appearance of maximum on the curve presented in figure 4 is explained by different mechanisms of oxygen atom production and loss. At low pressure, the O-atom density is limited by surface effects rather than the discharge power. At high pressure the limiting factor is the poor density of electrons as well as their temperature. At even higher pressure the gas-phase atom loss by three-body would become important if the power were increased. The optimal conditions for a large O-atom density are met at pressure between about 50 and 100 Pa. In this range, the O-atom density depends largely on the discharge power: a higher power would cause a higher density. The pressure at which the maximum appears depends on power, too: at higher power the maximum is shifted to a higher pressure. In any case, the theoretical limit of the O-atom density is full dissociation.

Since the neutral oxygen atom density exceeds the oxygen ion density by orders of magnitude, the experiments with polymer treatment were performed at the maximum of the O density, i.e. the pressure of 75 Pa.

3.2. Surface activation of polymers

Different polymers were exposed to oxygen plasma. They include:

- only carbon and hydrogen containing polymers: aliphatic polypropylene PP (figure 5a) and aromatic polystyrene PS (figure 5b)
- polymers containing oxygen: polyethyleneterephthalate PET (figure 5c) and cellulose CELL (figure 5d) like textile and ink-jet paper
- polymers containing sulphur: polyphenylenesulfide PPS (figure 5e) and polyethersulphone PES (figure 5f)
- a polymer containing nitrogen: Nylon 6 polyamide PA6 (figure 5g)
- a polymer containing flourine: polytetrafluoroethylene PTFE (figure 5h)

Journal of Physics: Conference Series **162** (2009) 012015 doi:10.1088/1742-6596/162/1/012015

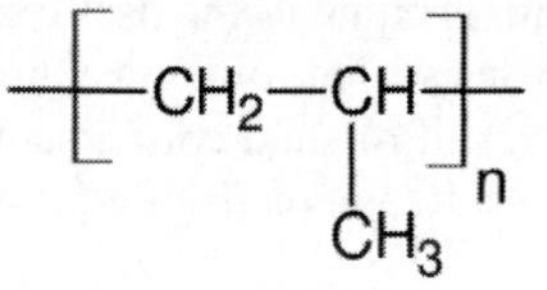

(a) PP

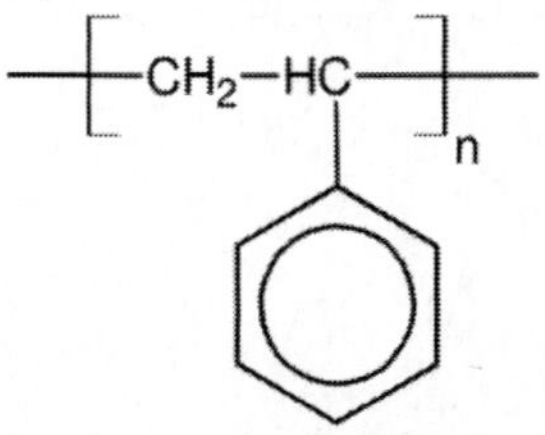

(b) PS

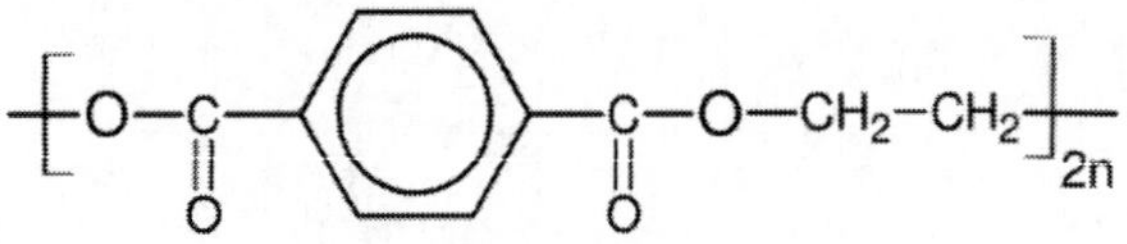

(c) PET

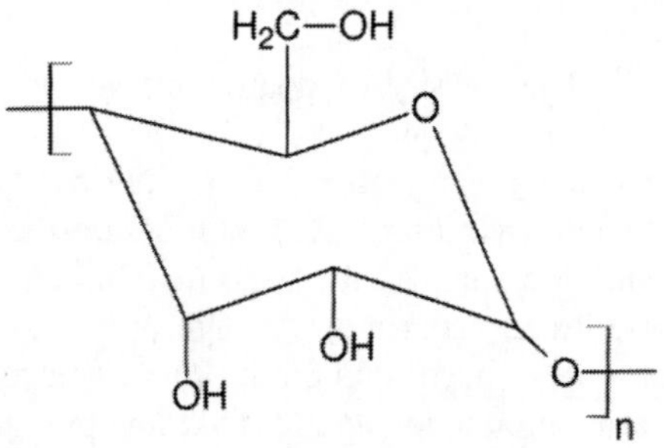

(d) cellulose

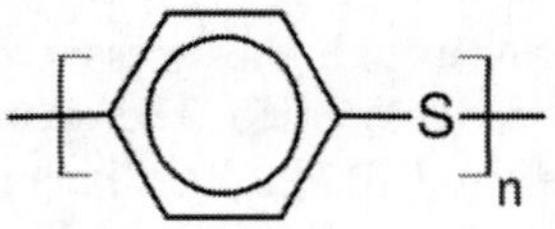

(e) PPS

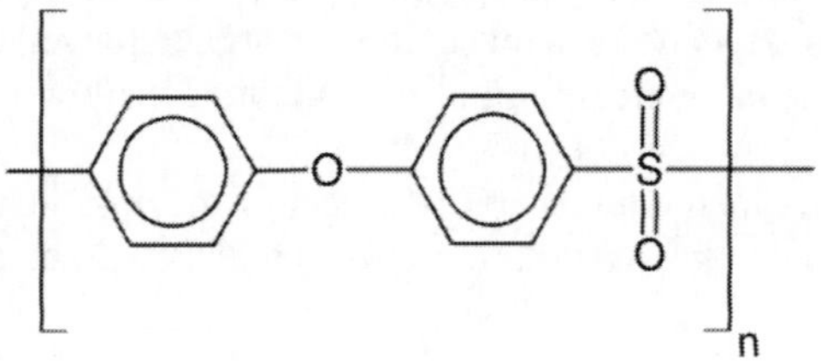

(f) PES

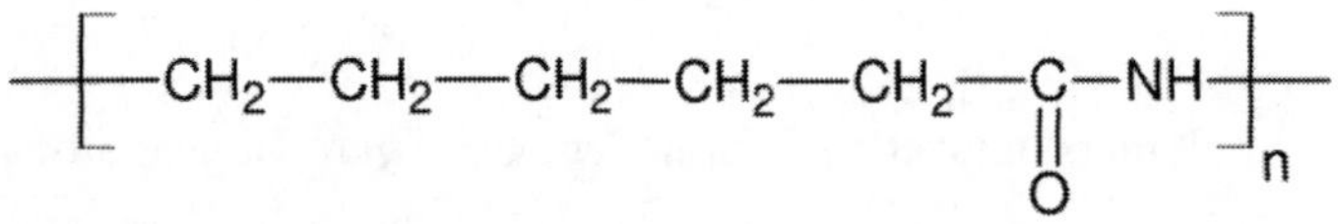

(g) PA6

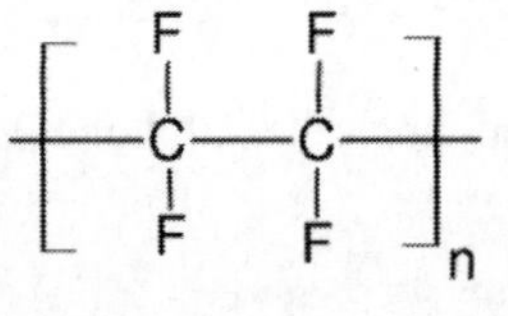

(h) PTFE

Figure 5. Structure of polymers used for plasma activation: (a) PP, (b) PS, (c) PET, (d) cellulose, (e) PPS, (f) PES, (g) PA and (h) PTFE.

3.2.1. *Carbon containing polymers.* Carbon containing polymers consist of carbon and hydrogen only. Therefore their high resolution XPS C1s peak (thereafter: carbon peak) is rather uniform and positioned at a binding energy of 285 eV which corresponds to C-C and C-H bonds. Since there is no oxygen in the original polymers, they are very good candidates for studding the effect of oxygen plasma treatment, because it is easily to observe new peaks due to oxygen incorporation into the surface after plasma treatment. One of such candidates is PP which consists of aliphatic chain containing carbon atoms (figure 5a). Figure 6a represents the carbon peak for the untreated PP and PP treated in oxygen plasma for 3 s. As already mentioned the C1s peak of untreated sample is uniform, while the peak after the treatment clearly reveals an appearance of new peaks resulting from plasma oxidation. A more detailed picture of these new functional groups is obtained using a curve fitting procedure (figure 6b). Besides the main peak that corresponds to C-C bonds and is assigned C1 in figure 6b, there is another peak C2 which corresponds to C-O bonds, a peak C3 which corresponds to C=O bonds and a peak C4 which corresponds to O-C=O bonds.

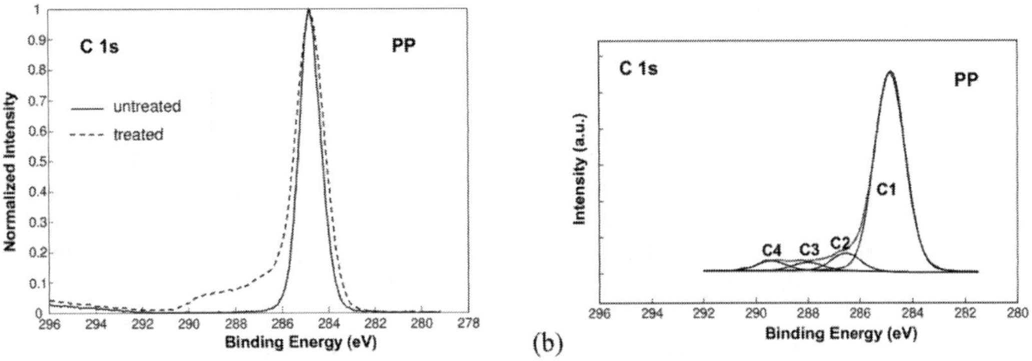

(a) (b)

Figure 6. High-resolution C 1s peak of untreated (a) and treated (b) PP surface.

Similar effect is observed for plasma treated PS (figure 7a) which is another example of a structurally simple polymer. Here, plasma induced modifications are more pronounced indicating a higher concentration of new functional groups at the surface. In this case, not only peaks C2, C3 and C4 are observed, but an additional peak C5 appears at a binding energy of 290 eV (figure 7b) which may correspond to -C(=O)-O-C(=O)- or to -O-C(=O)-O- functional groups on the surface [18].

Another important characteristic of untreated PS in comparison with untreated PP is a small peak at a binding energy of about 291.5 eV (figure 7a), which is not observed in the case of PP. This peak is due to the π-π^* shake-up transition and it is characteristic for the aromatic phenyl ring. Therefore, this peak is observed only at polymers having phenyl rings [2,18]. Changes in the intensity of this peak can provide information regarding the extent of ring-opening induced by plasma treatment. In our case, after the plasma treatment, the intensity of this peak decreased significantly indicating that plasma caused a destruction of the phenyl ring in PS.

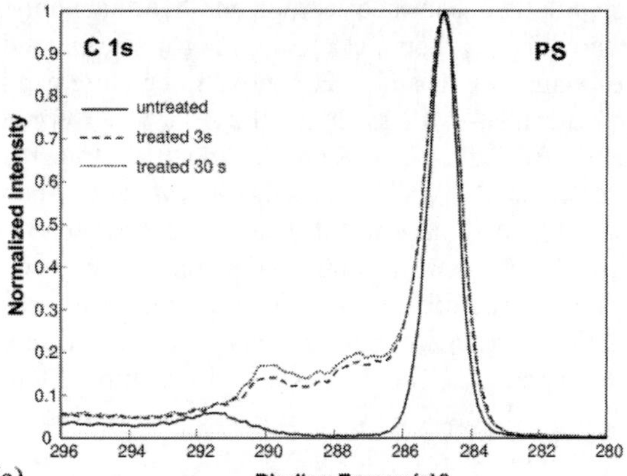

(a)

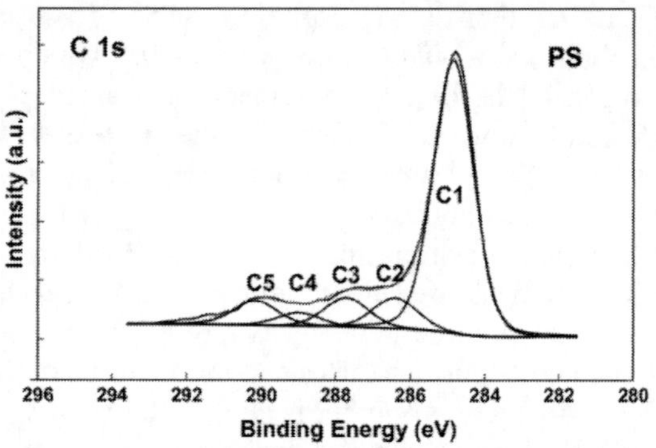

(b)

Figure 7. High-resolution C 1s peak of untreated (a) and treated (b) PS surface.

3.2.2. Oxygen containing polymers. The carbon C1s peak of oxygen containing polymers does not have such a simple shape like that of pure hydrocarbons. The interpretation of XPS spectra after oxygen plasma treatment of these polymers can be quite difficult due to problems in distinguishing between original and newly formed oxygen functional groups on the surface. A polymer which has been extensively studied is PET (polyethyleneterephthalate) [4,5,12,63-66]. Figure 8a shows the carbon peak for an untreated PET. We can observe three peaks: C1 corresponds to C-C bonds in phenyl ring, C2 corresponds to C-O bond (ether) and C3 corresponds to O=C-O bond (ester group) (figure 5c). After the plasma treatment the intensity of the peaks C2 and C3 increases remarkably (figure 8b) and a new peak C4 is observed. This peek has been attributed to C=O functional group [63-67]. The modifications of the functional groups on PET polymer are clearly visible indicating extremely strong surface activation.

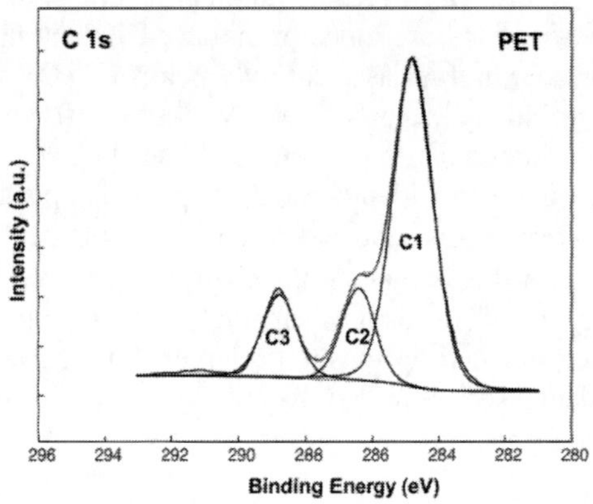

(a)

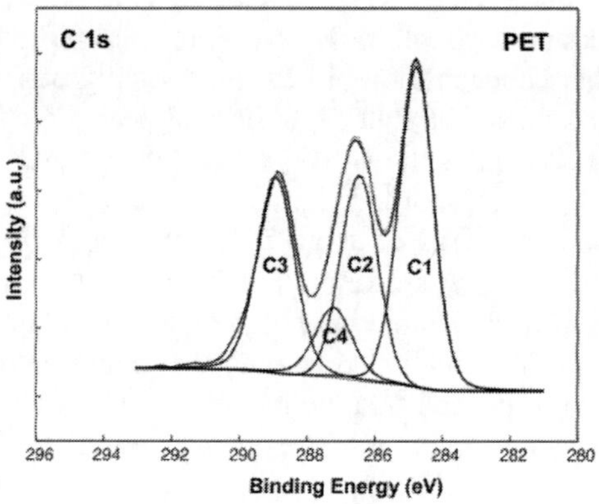

(b)

Figure 8. High-resolution C 1s peak of (a) untreated and (b) treated PET surface.

Interesting enough, such increase in the surface activation is not observed for cellulose [68]. In this material, all carbon atoms are bound to at least one oxygen atom (figure 5d): each cellulose unit contains five carbon atoms with a single bond to oxygen C-O (hydroxyl groups) and one carbon atom with two bonds to oxygen O-C-O. Thus, for pure cellulose one would expect just two peaks C2 and C3. In our case we study 2 different products made from cellulose: textile and paper. For the case of textile (figure 9a), C1 peak is clearly observed as well. This peak is probably due to the presence of surface contaminants that appeared on the surface during or after production. After the plasma treatment of textile (figure 6b), the contribution of C3 peak is increased, while the contribution of C2 and C1 peaks decreased. Furthermore, a new peak C4 with a binding energy of 289.2 eV appeared. An increase of the C3 peak is associated with either an increase of the surface concentration of O-C-O group, or a formation of a new functional group C=O, or both. The new peak C4 is attributed to a formation of the O=C-O group. Here, it is worth mentioning that, since carbon atoms in the cellulose are already bonded to at least one oxygen atom, the incorporation of more oxygen atoms from plasma causes strong degradation of the molecule. This can be a reason for the decrease of the relative intensity of the C2 peak.

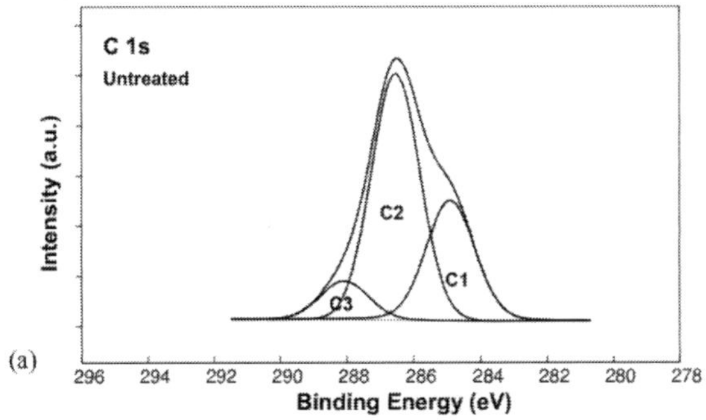

(a) textile 0s

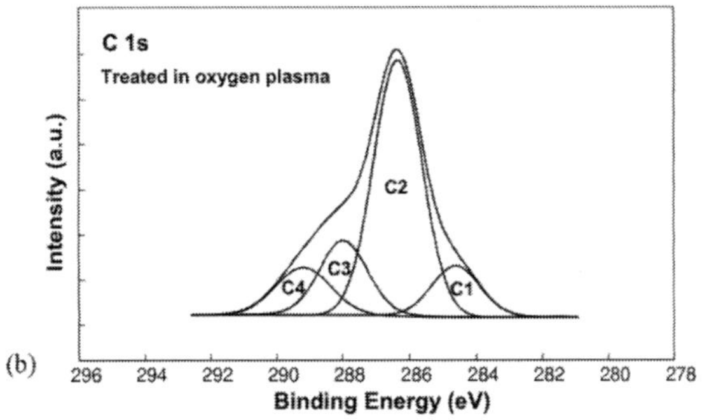

(b) textile 5s

Figure 9. High-resolution C 1s peak of untreated (a) and treated (b) textile surface.

Interesting results are obtained at plasma activation of another cellulose-made product, ink-jet paper. Commercially available paper was used for current experiments. Apart from cellulose, it contains about 10 weight % of alkyl ketene dimer (AKD) and few weight % of $CaCO_3$. In the untreated sample (figure 10a) we can see a carbon peak typical for cellulose. We can not observe any carbon peak associated with carbon in $CaCO_3$, although this material is present in the bulk. The situation after 3 s of treatment (figure 10b) is quite the same as for untreated sample – it is difficult to see any changes (apart from largely increased C4 peak), while after 200 s of treatment the situation is much different (figure 10c). After 200 s of plasma treatment the organic part (cellulose) was probably heavily oxidized and partially removed ("burnt"), while inorganic particles (calcium carbonate) remained. In the carbon peak of the ash we can observe peaks C1, C2, C3 and C4 like in the case of textile, and a new peak C5, which is associated to carbon atoms in $CaCO_3$. Such prolonged plasma treatment is known as plasma ashing and it allows for detection of inorganic material which is present in organic samples in so small quantities that cannot be detected with standard methods, since the concentration is below the detection limit of many techniques for surface characterization.

Journal of Physics: Conference Series **162** (2009) 012015 doi:10.1088/1742-6596/162/1/012015

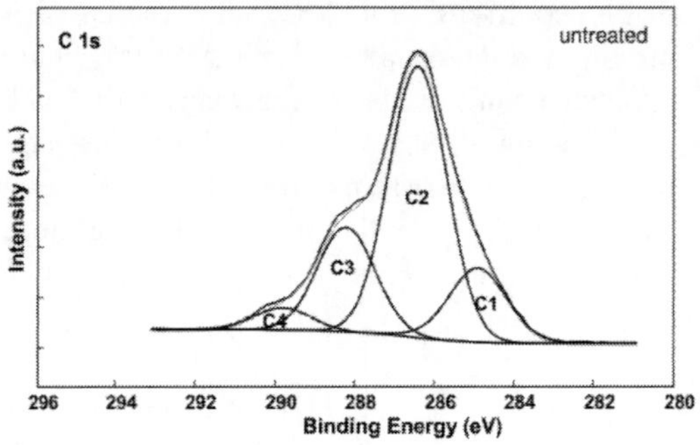

(a)

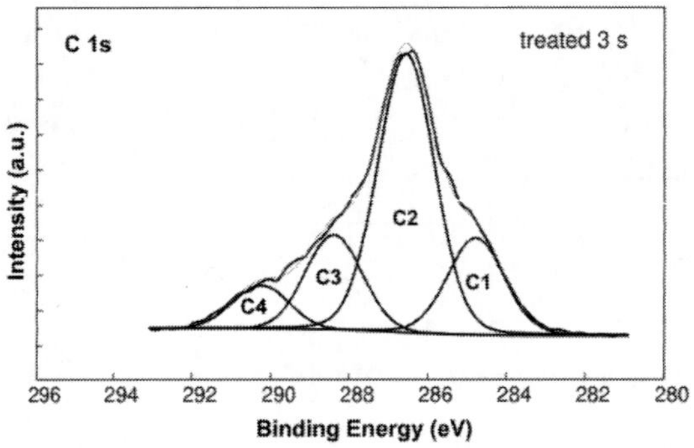

(b)

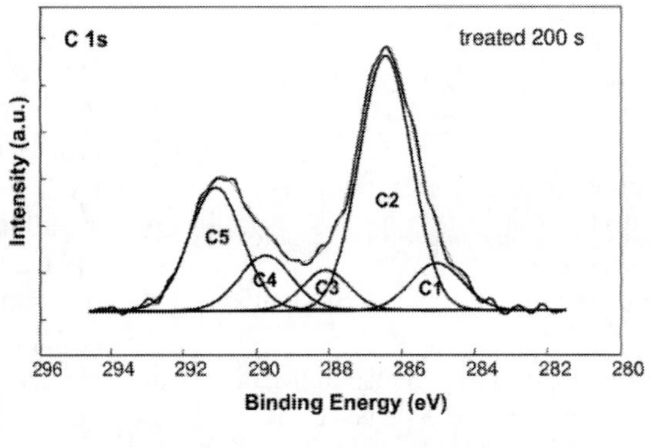

(c)

Figure 10. High-resolution C 1s peak of untreated (a) and treated (b) surface of ink-jet paper.

3.2.3. *Sulphur containing polymers.* In XPS characterization of sulphur containing polymers there is a well known problem with overlapping of the peaks due to C-C bond at a BE of 285 eV and C-S bond at a BE of 285.3 eV [69,70]. The carbon peak of an untreated PES consists of 2 peaks only (figure 11a); the larger one C1 at 285 eV and the smaller one C2 at 286.5 eV. The larger peak corresponds to C-C bonds, while the smaller peak corresponds to C-O bonds. As already mentioned, it is difficult to see the C-S peak since it is overlapping with the peak C1. Figure 11b shows the carbon C1s peak of the plasma treated sample. The carbon peak of the plasma-treated sample is different from the peak of untreated sample. Now, four separate peaks can be observed; a peak C3 that corresponds to C=O, and a peak C4 that corresponds to O=C-O. The peak C2 is somehow enlarged in comparison to the untreated sample [70]. Obviously, carbon atoms in PES polymer get well oxidized during plasma activation.

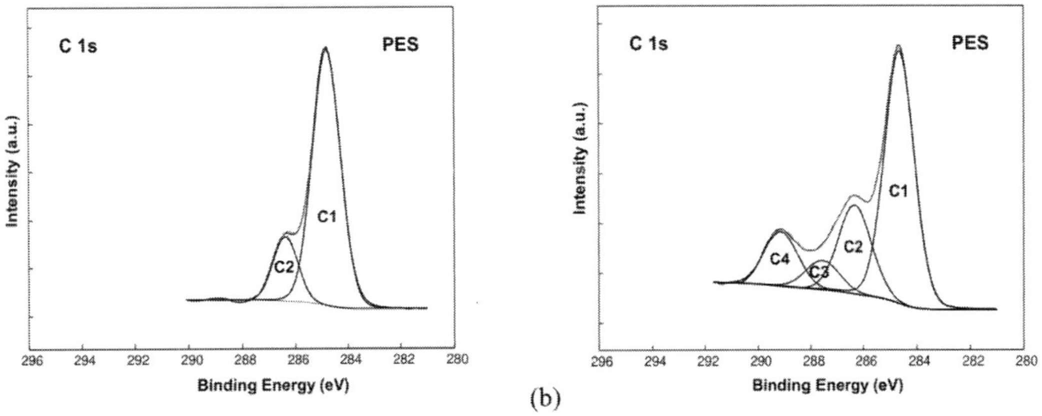

Figure 11. High-resolution C 1s peak of untreated (a) and treated (b) PES surface.

Interesting results are obtained for plasma activation of PPS polymer. As shown in figure 5, sulphur is not bonded to oxygen atoms as in the case of PES, but only to carbon atoms. Figure 12a represents the shape of the carbon peak before and after plasma treatment. Surprisingly enough, the carbon peak of oxygen plasma treated sample is not much different from the untreated sample. One can therefore conclude that the oxygen–carbon bonds are presented in rather small concentrations. Nevertheless, oxidation of carbon (figure 12c) results in formation of C-O, C=O and O-C=O functional groups on the surface like in the case of PES, but at lower concentration. The right mechanism for PPS activation by oxygen plasma can be deduced from figure 12b, however. This figure represents the high resolution S2p peak before and after plasma treatment. We can see that in this case not only carbon is oxidized but also sulphur [45]. Before plasma treatment, sulphur S2p peak (duplet) is positioned at a BE of 163.7 eV corresponding to oxidation state S^{2-} (C-S-C bond). After the oxidation, a new broad peak appears at a BE of about 169 eV. According to the literature, this peak corresponds to either of the double peaks S^{4+} or S^{6+}. Namely, the S^{4+} oxidation state is found at 168.2 and 169.4 eV, and S^{6+} at 169.1 and 170.2 eV, respectively. The oxidation state of sulphur therefore changes dramatically during the treatment with oxygen plasma.

From the XPS results of plasma activation of PPS we can conclude that sulphur is more easily oxidized in this material than carbon. We can not observe this in the case of polymer PES because sulphur atoms in virgin PES (figure 5f) are already bonded to oxygen and further oxidation is not possible. Prolonged oxygen plasma treatment of PES can cause formation of SO_3^{2-} (leading to polymer degradation) which is desorbed from the surface as shown by other authors [69].

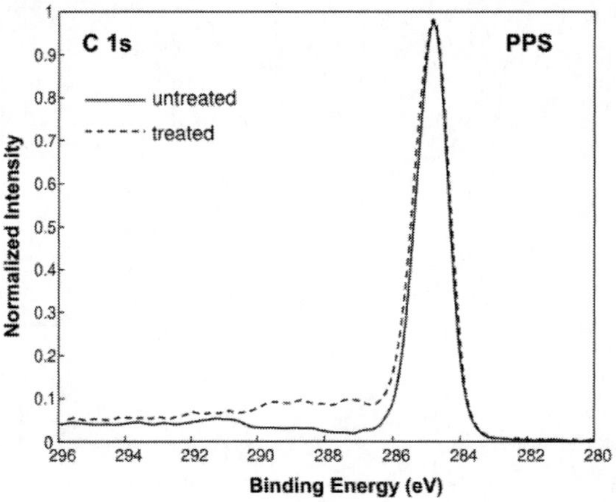

(a)

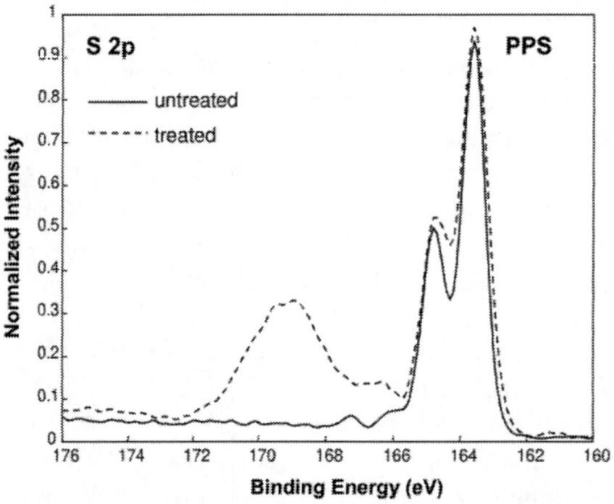

(b)

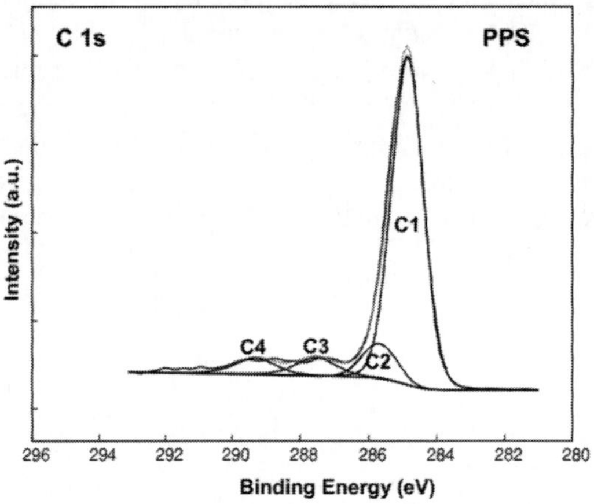

(c)

Figure 12. High-resolution C 1s peak of untreated (a) and treated (b) PPS surface.

3.2.4. Nitrogen containing polymers. At plasma treatment of nitrogen-containing polymers we have never observed oxidation of nitrogen atoms (what one might have expected from the results with PPS). Only carbon atoms get oxidized during treatment with oxygen plasma.

One problem associated with XPS characterization of nitrogen containing polymers is a difficult determination of the exact type and concentration of nitrogen functional groups, since there is a problem with strong overlapping of oxygen- and nitrogen-containing functionalities. They appear at very similar binding energies (figure13) [18]. Moreover, relevant literature reports different data for binding energies of different nitrogen peaks which are positioned quite close together: C-N (285.5 eV – 286.3 eV), C=N (285.5 eV – 286.6 eV), C≡N (286.7 eV – 287.0 eV) [72-77] and this makes the interpretation of XPS high resolution C1s peak extremely difficult.

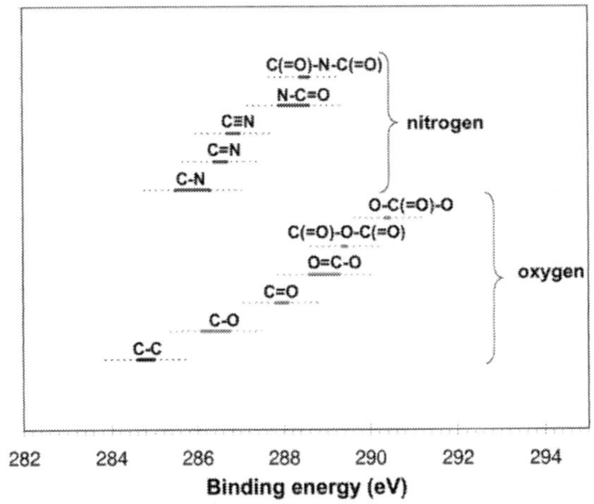

Figure 13. Expected positions of oxygen and nitrogen functional groups: full line – position of the peak maximum, dotted line – peak width (FWHM).

Also, the N1s peak can not give a decisive answer about the nitrogen containing functionalities [78]. The N1s peak is always composed of a single relatively broad symmetric peak that could correspond to different nitrogen states. According to the literature we can find several carbon-nitrogen species (like amines, amides, imides, nitriles, etc.) in the range between 399.1 eV and 400.2 eV [18,78]. As reported by Morent at al, it is very difficult to incorporate nitrogen into polymer surfaces [78]. Therefore, he assumed that during plasma treatment only C-N groups are usually formed on the surface. Amide groups (N-C=O) can be also present on the surface, while the presence of the groups where nitrogen is bounded to oxygen (nitro, oxime and nitrate groups) can be definitively excluded, since they should appear at energies 406-408 eV [18,78] and this is never observed for nitrogen containing polymers treated with oxygen plasma.

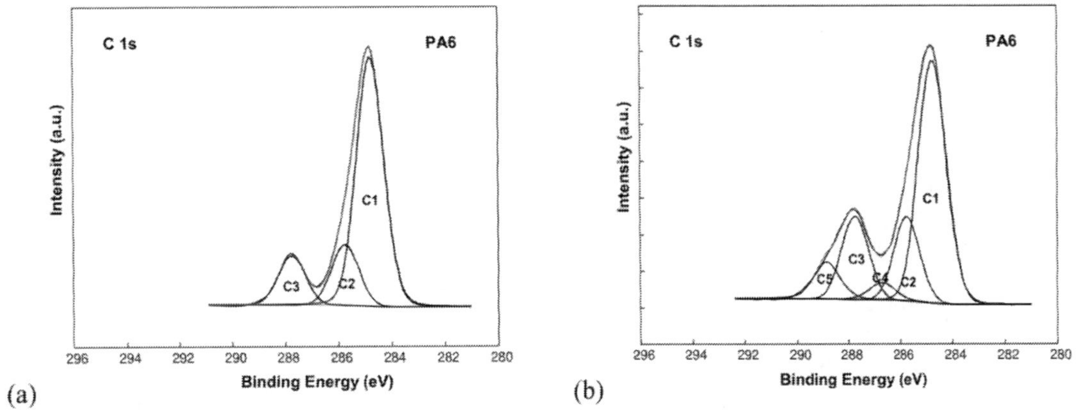

Figure 14. High-resolution C 1s peak of untreated (a) and treated (b) PA6 surface.

As an example of nitrogen-containing polymer we used a commercial polyamide PA6. Figure 14a shows the carbon peak of the untreated PA6 sample. The carbon peak is composed of three peaks: C1 corresponding to C-C bond, C2 at a BE of 286.1 eV corresponding to C-N bond and C3 at a BE of 287.9 eV corresponding to O=C-N bond (amide group). The carbon peak of a PA6 sample treated in oxygen plasma (figure 14b) shows that the peaks C2 and C3 are increased, and another peak C4 appears as well,. This peak corresponds to the O=C-O functional group. The increase of the C2 component can be explained by formation of C-O groups, that appear at a similar binding energy as C-N group, while the increase of the C3 peak can be explained by the formation of the C=O functional group.

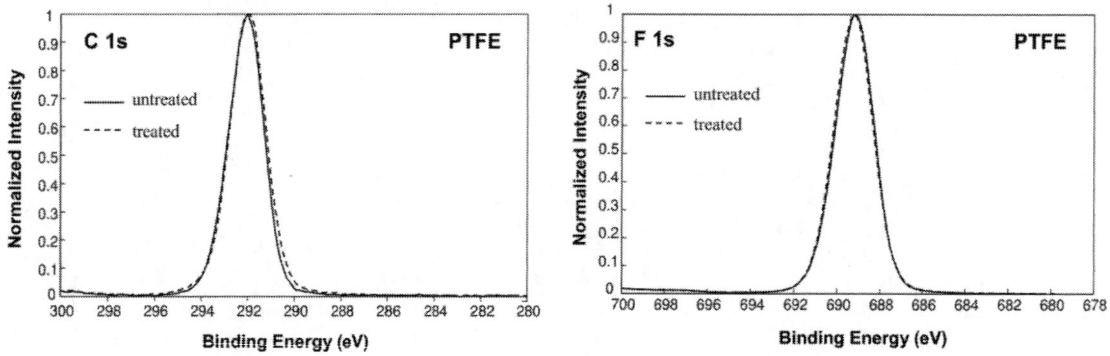

(a) (b)

Figure 15. High-resolution C 1s peak of untreated (a) and treated (b) PTFE surface.

Table 1. Surface composition of the polymer samples before and after plasma treatment.

Sample	C	O	N	S	F	Ca	Na	K	O/C
Untreated PA6	76.2	13.1	10.7						0.17
Treated PA6	63.0	25.3	11.7						0.40
Untreated PET	73.4	26.6							0.36
Treated PET	60.0	40.0							0.67
Untreated PES	74.6	20.4		5.0					0.27
Treated PES	57.3	37.8		5.0					0.66
Untreated PTFE	33.3	0.6			66.1				0.02
Treated PTFE	31.7	0.5			67.8				0.02
Untreated textile	64.2	35.8							0.56
Treated textile	49.6	50.4							1.02
Untreated paper	75.3	23.0	1.4			0.4			0.31
Treated paper	48.9	45.2	2.3			2.4	1.2		0.92
Treated paper-ash	29.8	47.6				8.5	2.9	11.2	1.60

3.2.5. *Halogen containing polymers.* PTFE is a chemically inert polymer. It is very difficult to activate its surface by plasma treatment. Figure 15a shows a comparison of the carbon peak for untreated and plasma-treated PTFE surface. There is practically no difference in the shape of the carbon peak [79]. Furthermore, the comparison of the fluorine peaks (figure 15b) does not show any difference, either. The survey-scan measurements showed no changes in oxygen concentration at the surface (table 1). For the case of PTFE we performed prolonged treatment in oxygen plasma but no modifications were observed. Even after 10 min of treatment the XPS peaks remained quite the same. So oxygen plasma treatment seems to be ineffective for the case of this polymer. Here it is also worth mentioning that halogen-containing polymers are known to be sensitive to X-ray irradiation, resulting in a decrease in the halogen peak intensity and an increase in the C1s peak intensity [17,18,80]. In our case we did not observe this effect even after several hours of exposure to X-rays.

Table 2. Comparison of concentration of different functional groups for different polymers.

	C-C [3]C-S	C-O [4]C-N	C=O [4]O=C-N	O=C-O	[1]-C(=O)-O-C(=O)- [2] CO_3^{2-}
PP untreated	100%	/	/	/	/
PP treated 3s	78.2%	12.1%	5.3%	4.5%	/
PS untreated	100%	/	/	/	/
PS treated 3s	70.7%	8.9%	8.6%	4.1%	7.8%[1]
PET untreated	75.8%	13.0%	/	11.2%	/
PET treated 3s	34.0%	30.4%	3.9%	31.7%	/
Paper untreated	16.3%	57.0%	22.0%	4.7%	/
Paper treated 3s	19.0%	54.9%	18.0%	8.2%	/
Paper treated 200s	8.9%	48.3%	7.1%	11.1%	24.6%[2]
Textile untreated	30.7%	60.0%	9.3%	/	/
Textile treated 3s	12.2%	57.7%	17.7%	12.4%	/
PES[3] untreated	82.1%	17.9%	/	/	/
PES[3] treated 3s	52.4%	22.3%	9.8%	15.6%	/
PPS[3] untreated	100%	/	/	/	/
PPS[3] treated 3s	78.9%	9.0%	6.4%	5.7%	/
PA[4] untreated	70.2%	16.5%	13.3%	/	/
PA[4] treated 3s	55.8%	17.7%	19.0%	7.5%	/

4. Conclusions

Oxygen plasma was found as an effective method for surface modification of different polymers. The only exception was PTFE, which is known as chemically very inert material and it was not possible to functionalize it by exposure to our plasma. On all other samples, a higher oxygen concentration was detected on the surface after the plasma treatment (Table 1). Plasma treatment did not produce only one unique functionality on a polymer surface but different functional groups was created. On all polymers except PTFE new groups like C-O, C=O and O=C-O were observed; only their concentration was different depending on polymer type (Table 2). For example, the rate of incorporation of new species after oxygen plasma treatment at identical conditions was found to be greater for PS than for PP, although both polymers contain carbon and hydrogen only. The surface incorporation of new species depends on the number of available carbon atoms that can bond with the reactive species in the plasma. The primary available carbon atoms are those that are only bonded to other carbon atoms or hydrogen. Secondary sites would be carbon atoms with only a single bond to oxygen or nitrogen [2,77]. Plasma treatment of sulphur containing polymers resulted in oxidation of sulphur atoms (unless they are already oxidized in the original material), while this is not the case for oxidation of nitrogen containing polymers, where no groups with nitrogen bounded to oxygen were found. The surface chemistry of plasma-treated polymers containing carbon–oxygen or carbon–nitrogen functionalities is more complex and is significantly different from polymers containing only carbon–carbon bonds. Characterization of plasma-induced changes on oxygen- or nitrogen-containing polymers is more difficult due to the more complex chemical structure of the polymer and the number of possible chemical species that can be produced.

References

[1] Meyer-Plath A A, Schröder K, Finke B, Ohl A 2003 *Vacuum* **71** 391
[2] Gerenser L J 1996 *Surface Chemistry of Plasma-Treated Polymers, in Handbook of Thin Film Process Technology*, ed D A Glocker and S I Shah (IOP, Bristol)
[3] Strobel M, Lyons C S, Mittal K L 1994 *Plasma Surface Modification of Polymers: Relevance to Adhesion* (VSP, Utrecht)
[4] Borcia G, Anderson C A, Brown N. M. D 2003 *Plasma Sources Sci. Technol.* **12** 335
[5] Borcia G, Anderson C A, Brown N. M. D 2004 *Appl. Surf. Sci.* **225** 186
[6] Feng J, Wen G, Huang W, Kang E, Neoh K G 2005 *Polym. Degrad. Stabil.* **91** 12
[7] Cvelbar U, Pejovnik S, Mozetič M, Zalar A 2003 *Appl. surf. sci.* **210** 255
[8] Chen Z, Cvelbar U, Mozetič M, He J, Sunkara M.K 2008 *Chem. mater.* **20** 3224
[9] Mozetič M, Cvelbar U, Sunkara M K, Vaddiraju S 2005 *Adv. mater.* **17** 2138
[10] Ostrikov K, Murphy A B 2007 *J. Phys. D: Appl. Phys.* **40** 2223
[11] Tam E, Levchenko I, Ostrikov K 2006 *J. Appl. Phys.* **100** 036104
[12] Xu S Y, Long J D, Sim L N *et al.* 2005 *Plasma Proces. Polymers* **2** 373
[13] Morent R, De Geyter N, Leys C, Gengembre L, Payen E. 2007 *Surf. Coat. Technol.* **201** 7847
[14] Chan C-M, Ko T-M, Hiraoka H 1996 *Surf. Sci. Rep.* **24** 1
[15] Briggs D, Grant J T 2003 *Surface Analysis by Auger and X-ray Photoelectron Spectroscopy* (IMP & Surface Spectra, Trowbridge)
[16] Briggs D 2005 *Surface Analysis of Polymers by XPS and static SIMS* (Cambridge University Press, Cambridge)
[17] Briggs D, Seah M P 1994 *Practical surface Analysis, Auger and X-rays Photoelectron Spectroscopy* (Wiley, Chichester)
[18] Beamson G, Briggs D 1992 *High Resolution XPS of Organic Polymers – The Scienta ESCA300 Database* (Wiley, Chichester)
[19] Mozetic M, Zalar A 2003 *Mater. Sci. Forum* **437** 81
[20] Ricard A 1996 *Reactive plasmas* (SFV)
[21] Cvelbar U, Vujošević D, Vratnica Z, Mozetič M 2006 *J. Phys. D: Appl. Phys.* **39** 3487
[22] Cvelbar U, Mozetič M, Ricard A 2005 *IEEE Trans. Plasma Sci.* **33** 834
[23] Ricard A, Monna V 2002 *Plasma Sourc. Sci. Technol.* **11** A150
[24] Vesel A, Mozetic M 2001 *Vacuum* **61** 373
[25] Villeger S, Cousty S, Ricard A, Sixou M 2003 *J. Phys. D Appl. Phys.* **36** L60
[26] Drenik A, Cvelbar U, Vesel A, Mozetic M 2005 *Inf. MIDEM* **35** 85
[27] Robbins J J, Alexander R T, Xiao W, Vincent T L, Wolden C A 2002 *Thin Solid Films* **406** 145
[28] Kim M C, Yang S H, Boo J H, Han J G 2003 *Surf. Coat. Technol.* **174** 839
[29] Granier A, Vervloet M, Aumaille K, Vallee C 2003 *Plasma Sourc. Sci. Technol.* **12** 89
[30] Biederman H, Stelmashuk V, Kholodkov I, Choukourov A, Slavinska D 2003 *Surf. Coat. Technol.* **174** 27
[31] Cvelbar U, Krstulović N, Milošević S, Mozetič M 2007 *Vacuum* **82** 224
[32] Vujošević D, Mozetič M, Cvelbar U, Krstulović N, Milošević S 2007 *J. Appl. Phys.* **101** 103305
[33] Miyauchi M, Miyoshi Y, Petrović Z Lj, Makabe T 2007 *Solid-State Electronics* **51** 1418

[34] Miyoshi Y, Petrovic Z Lj, Makabe T 2002 *IEEE Trans. Plasma Sci.* **30** 130
[35] Hioki K, Itazu N, Petrovic Z Lj, Radmilovic M 2001 *Jpn. J. Appl. Phys.* **40** L1183
[36] Ostrikov K N, Xu S, Azam A B M S 2002 *J. Vac. Sci. Technol. A*, **20** 251
[37] Leu G F, Brockhaus A, Engemann J 2003 *Surf. Coat. Technol.* **174** 928
[38] Hong Y C, Uhm H S 2003 *Phys. Plasmas* **10** 3410
[39] Scott G B I, Fairley D A, Milligan D B, Freeman C G, McEwan M J 1999 *J. Phys. Chem.* A **103** 7470
[40] Elias L, Ogrizlo E A, Schiff H I 1959 *Can. J. Chem.* **37** 1680
[41] Carruth M R, DeHaye R F, Norwood J K, Whitaker A F 1990 *Rev. Sci. Instr.* **61** 1211
[42] Sorli I, Rocak R J 2000 *Vac. Sci. Technol. A* **18** 338
[43] Mozetic M, Vesel A, Gaillard M, Ricard A 2001 *Plasmas Polym.* **6** 41
[44] Babič D, Poberaj I, Mozetič M 2001 *Rev. Sci. Instr.* **72** 4110
[45] Poberaj I, Babič D, Mozetič M 2002 *J. Vac. Sci. Technol. A* **20** 189
[46] Mozetic M, Ricard A, Babic D, Poberaj I, Levaton J, Monna V, Cvelbar U 2003 *J. Vac. Sci. Technol. A* **21** 369
[47] Balat-Pichelin M, Vesel A 2006 *Chem. Phys.* **327** 112
[48] Vesel A, Mozetic M, Balat-Pichelin M 2007 *Vacuum* **81** 1088
[49] Mozetic M, Vesel A, Cvelbar U, Ricard A 2006 *Plasma Chem. Plasma Process.* **26** 103
[50] Mozetic M, Vesel A, Monna V, Ricard A 2003 *Vacuum* **71** 201
[51] Mozetic M, Cvelbar U, Vesel A, Ricard A, Babic D, Poberaj I 2005 *J. Appl. Phys.* **97** 103308
[52] Cvelbar U, Ostrikov K, Drenik A, Mozetič M 2008 *Appl. Phys. Lett.* **92** 133505
[53] Drenik A, Cvelbar U, Ostrikov K, Mozetič M 2008 *J. Phys. D: Appl. Phys.* **41** 115201
[54] Cvelbar U, Mozetič M 2007 *J. Phys.D: Appl. Phys.* **40** 2300
[55] Cvelbar U, Mozetič M, Poberaj I, Babič D, Ricard A 2006 *Thin Solid Films* **475**
[56] Cvelbar U, Mozetič M, Klanjšek Gunde M 2005 *IEEE Trans. Plasma. Sci.* **33** 236
[57] Klanjšek Gunde M, Kunaver M, Horvat A, Cvelbar U 2005 *Prog. Org. Coat.* **54** 119
[58] Klanjšek Gunde M, Kunaver M, Cvelbar U, Barle N 2005 *Vacuum* **80** 189
[59] Ostrikov K N, Xu S, Yu M Y 2000 *J. Appl. Phys.* **88** 2268
[60] Ostrikov K, Long J D, Rutkevych P P, Xu S 2006 *Vacuum* **80** 1126
[61] Puac N, Petrovic Z Lj, Radetic M, Djordjevic A 2005 *Mater. Sci. Forum* **494** 291
[62] Osiac M, Schwarz-Selinger T, O'Connell D, Hell B, Petrovic Z Lj, Malović G N, Gans T, Czarnetzki U 2007 *Plasma Sources Sci. Technol.* **16** 355
[63] Krstulović N, Labazan I, Milosevic S, Cvelbar U, Vesel A, Mozetic M 2006 *J. Phys. D: Appl. Phys.* **39** 3799
[64] Vesel A, Mozetic M, Zalar A 2008 *Vacuum* **82** 248
[65] Vesel A, Mozetic M 2008 *J. Phys. Conf. Ser.* **100** 012027
[66] Cvelbar U, Markoli B, Poberaj I, Zalar A, Kosec L, Spaić S 2006 *Appl. Surf. Sci.* **253** 1861
[67] Vesel A, Junkar I, Cvelbar U, Mozetic M, Kovac J 2008 *Surf. Interface Anal.* (in press)
[68] Vesel A, Mozetic M, Hladnik A, Dolenc J, Zule J, Milosevic S, Krstulovic N, Klanjsek-Gunde M, Hauptman N 2007 *J. Phys. D: Appl. Phys.* **40** 3689
[69] Feng J, Wen G, Huang W, Kang E, Neoh K G 2005 *Polym. Degrad. Stabil.* **91** 12
[70] Vrlinic T, Vesel A, Cvelbar U, Krajnc M, Mozetic M 2007 *Surf. Interface Anal.* **39** 476
[71] Cvelbar U, Mozetic M, Junkar I, Vesel A, Kovac J, Drenik A, Vrlinic T, Hauptman N, Klanjsek-Gunde M, Markoli B, Krstulovic N, Milosevic S, Gaboriau F, Belmonte T 2007 *Appl. Surf. Sci.* **253** 8669
[72] Larrieu J, Clement F, Held B, Soulem N, Luthon F, Guimon C, Martinez H 2005 *Surf. Interface Anal.* **37** 544
[73] Charpentier P A, Maguire A, Wan W-K 2006 *Appl. Surf. Sci.* **252** 6360
[74] Deslandes Y, Pleizier G, Poiré E, Sapieha S, Wertheimer M R, Sacher E 1998 *Plasmas Polym.* **3** 61
[75] Wilson D J, Williams R L, Pond R C 2001 *Surf. Interface Anal.* **31** 385

[76] Wang M J, Chang Y I, Poncin-Epaillard F 2005 *Surf. Interface Anal.* **37** 348
[77] Grace J M, Gerenser L J 2003 *J. Disp. Sci. Technol.* **24** 305
[78] Morent R, De Geyter N, Gengembre L, Leys C, Payen E, Van Vlierberghe S, Schacht E 2008 *Eur. Phys. J. Appl. Phys.* in press
[79] Vesel A, Mozetic M, Zalar A 2008 *Surf. Interface Anal.* **40** 661
[80] Artyushkova K, Fulghum J E *Surf. Interface Anal.* **31** 352

Journal of Physics: Conference Series **162** (2009) 012016 doi:10.1088/1742-6596/162/1/012016

Two-dimensional dusty plasma crystals and liquids

Z. Donkó[1,2], P. Hartmann[1,2], G. J. Kalman[2]

[1] Research Institute for Solid State Physics and Optics of the Hungarian Academy of Sciences, Budapest, Hungary

[2] Physics Department, Boston College, Chestnut Hill, MA, USA

E-mail: donko@mail.kfki.hu

Abstract. Strongly coupled plasmas – in which the average potential energy per particle dominates over the average kinetic energy – appear in a wide variety of physical systems. Among these systems, dust plasma crystals and liquids realized in low-pressure gas discharges by dispersing mesoscopic grains into the plasma have attracted a lot of attention during the past years. We describe the experimental realization of the quasi-two-dimensional dust system, summarize the basics of the computer simulation and theoretical approaches capable of their description in the liquid and solid phases. We discuss the properties of the dynamical density and current correlation spectra, generated by molecular dynamics simulations, and address the issues associated with the existence of different phases and transport coefficients (e.g. superdiffusive behavior) in the low-dimensional systems under study.

1. Introduction

In the majority of types of plasmas the kinetic energy of the particles is significantly higher than the potential energy arising from the interaction of the charged particles. Such plasmas are conventionally termed as "ideal" ones. In contrast with these, in "non-ideal plasmas" the interaction energy already plays a role in the physics of the systems. The (less frequent) cases when the potential energy dominates over the kinetic energy are termed as "strongly coupled plasmas". As examples of strongly coupled plasmas those appearing in some astrophysical objects (white dwarf interiors, neutron star crusts, supernova cores, and giant planetary interiors), semiconductor devices (degenerate electron or hole liquids in two-dimensional or layered nanostructures), colloidal suspensions, as well as dusty plasmas, can be mentioned [1].

The ratio of the interparticle potential energy to the kinetic energy per particle can be expressed by the *coupling parameter* defined as:

$$\Gamma = \frac{1}{4\pi\varepsilon_0}\frac{Q^2}{ak_\mathrm{B}T}, \qquad (1)$$

where T is the temperature and a is the Wigner-Seitz radius. In the case of ideal plasmas $\Gamma \approx 0$, in non-ideal plasmas $\Gamma \sim 1$, while in strongly coupled plasmas $\Gamma \gg 1$. In this paper we focus on this latter, less common type of plasmas.

Strongly coupled plasmas are known to undergo crystallization at high enough density and/or low enough temperature, i.e. at high enough Γ. In the case of Coulomb systems, in three dimensions this (first order) phase transition [2] takes place at a coupling parameter value $\Gamma \cong 175$ [3, 4].

In two dimensions, for Coulomb systems the liquid phase is limited to coupling values $\Gamma \lesssim 137$, where the system is known to undergo a transition to crystallized phase [5]. Regarding Yukawa systems systems ($\kappa > 0$), it has been shown theoretically that exact long range order cannot exist at finite temperatures $T > 0$ [6]. Thus infinite (Yukawa) single crystals do not exist in 2D. Nonetheless, crystal-like and fluid-like behavior has been observed in 2D systems and pronounced changes of certain characteristics (e.g. bond angular order parameter) have been detected in the transition region between these two "phases" both in simulations [7] and experiments [8]. In many papers on 2D systems (most frequently 2D dusty plasma configurations) the words "phase" and "phase transition" have routinely been used. One should, however, be aware of the above mentioned arguments against the existence of true long range order and real phase transitions taking place in low dimensional Yukawa systems.

In this paper we focus on laboratory dusty plasmas, where the main constituents of the systems are micron sized dielectric particles. These particles are charged up electrically due to the electron and ion currents from the background plasma, which is provided typically by a low pressure gas discharge (see later). The interaction potential for the dust grains can be well approximated by the Yukawa type, which describes the Coulomb repulsion and the screening by the background plasma.

$$\phi(r) = \frac{Q}{4\pi\varepsilon_0}\frac{\exp(-r/\lambda_{\rm D})}{r}. \tag{2}$$

Here Q is the charge of the particles and $\lambda_{\rm D}$ is the Debye length. The dimensionless *screening parameter* is given by:

$$\kappa = \frac{a}{\lambda_{\rm D}}, \tag{3}$$

where $a = 1/\sqrt{n\pi}$ in two dimensions (n is the areal density of particles). The (nominal) plasma frequency is defined as

$$\omega_0 = \sqrt{\frac{n_{\rm 2D} Q^2}{2\varepsilon_0 m a_{\rm 2D}}}. \tag{4}$$

It is noted that in 2D the plasma frequency may also have different definitions, and some of the authors use the lattice constant instead of the WS radius as a length scale.

The paper intends to give a brief account of basic features of two-dimensional and quasi-two-dimensional dusty plasmas in the strongly coupled domain [9]. The work is organized as follows. Section 2 gives a short description of the methods of our studies: the experimental realization of quasi-2D dusty plasma layers, the molecular dynamics simulation technique as well as the theoretical tool applicable for the derivation of wave dispersion relations, the Quasilocalized Charge Approximation. In section 3 representative results for the static properties (section 3.1) and collective excitations (section 3.2) are presented, and the transport properties are briefly discussed (section 3.3). A short summary is given in Section 4.

2. Methods

2.1. Experimental realization of dust plasma layers

Complex (dusty) plasmas [10, 11, 12, 13] can be created in laboratory experiments by dispersing micron-sized particles into gas discharges (see figure 1). The (typically noble gas) glow discharge can be driven either by a direct current (dc) or by a radio-frequency (rf) source, and serves primarily as a charging medium for the dielectric particles. The dust particles are exposed to electron and ion currents from the discharge plasma, and a dynamic equilibrium is rapidly reached. Their electric charge can be in the order of $\sim 10^3 - 10^4$ electron charges, depending on the diameter of the particles. The high charge of the particles gives rise to a strong electrostatic repulsion between them, and in a confined system, may lead to crystallization, see e.g. [14].

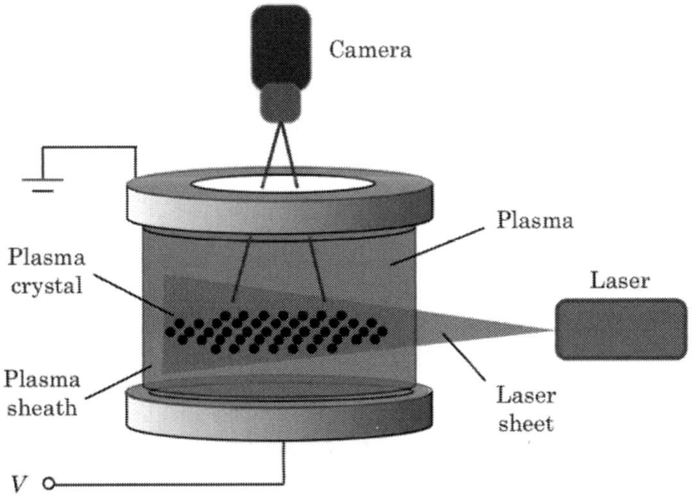

Figure 1. Experimental realization of a quasi-two-dimensional dust plasma layer.

The particles interact with their environment via several forces: gravity, ion drag force, neutral drag force, thermophoretic force, and the Lorentz force. The dominance of the different force contributions can be tuned by adjusting the experimental conditions, including choice of the particle size. In this paper we focus our attention on systems where the particles are levitated in a glow discharge with a horizontal plane-parallel electrode configuration. In this case the dominant forces are gravity and the electrostatic force arising from the "vertical" electric field of the sheath adjacent to the "lower" electrode of the discharge. These two main forces balance each other and the particles settle in a single 2D layer near the electrode sheath.

Figure 2. Snapshot (negative image) of particle configuration in a dusty plasma layer in the crystallized phase.

The number of particles which can be accommodated within a single layer is limited. If this limit is exceeded, additional layers form. The structure of these layers, especially the vertical correlations of the particle arrangements, are largely influenced by drag forces and the wake potential originating from directed ion flow. In this case particles may line up into vertical chains [15]. In this paper we restrict our studies to cases when a single layer of dust particles is formed. This layer is quasi-two-dimensional, as small displacements of the particles, perpendicular to the plane, are allowed.

The upper electrode of the discharge can be made of a mesh, or can be of solid material with a large diameter hole machined into it, this way allowing observation of the dust suspension with a camera. The video frames can be stored by a computer, which can reconstruct the phase space trajectories of the particles. For precise determination of particle positions algorithms providing subpixel resolution have been routinely used. As an illustration a snapshot of a single-layer particle cloud is shown in figure 2. The picture was taken in an experiment using 9.16 μm diameter melamine-formaldehyde particles.

2.2. Molecular dynamics simulations

Molecular dynamics (MD) simulations follow the motion of particles by integrating their equations of motion while accounting for the pairwise interaction of the particles, as well as for the forces originating from any external field(s), see e.g. [16]. The "core" of the MD codes describes the time evolution of phase space trajectories of the ensemble of particles, while "measurements" implemented in the code provide information (from the phase space coordinates) about the quantities of interest: about density and current fluctuations, transport coefficients, as well as the pair correlation function, which gives insight into the structure of the systems and is the basis of the calculation of thermodynamic quantities.

MD simulations are applicable for both finite and infinite systems. In the case of finite systems a confinement potential keeps the system together. Infinite systems are simulated by using periodic boundary conditions; whenever a particle leaves the primary computational cell (of finite volume), it enters the same box from the opposite direction. In the simulation of quasi-two-dimensional systems periodic boundary conditions are applied in the unconfined (x and y) directions, while a confinement force acts on the particles along the z direction.

In the simulations presented here we apply the Newtonian equation of motion (in which we consider the interparticle forces and well as external confinement forces) to trace the particles in the simulation. It is noted that in dusty plasmas friction forces and randomly fluctuating forces also act on the dust particles, due to the plasma / gas background environment (see e.g. [17]). Such forces can also be considered in the simulations to improve similarity with experimental conditions, see e.g. [18, 19, 20].

In the case of Yukawa potential the forces acting on the individual particles can be calculated by summation of pairwise interaction forces, limited to a finite radius around the particles. This "cutoff" radius can be set according to the value of the screening parameter κ. In the case of small values of κ, when the cutoff radius becomes comparable with the size of the simulation box, periodic images of the computational box have to be considered in the force calculation, as well. In the limit of $\kappa \to 0$, when the Coulomb potential in recovered, periodic images up to infinity have to be included in the summation in all the principal directions. In such cases special techniques, like the Ewald summation [21, 22, 23], the fast multipole method or the particle-particle particle-mesh method (PPPM, or P3M), can be used in MD simulations [24, 25, 26, 27].

2.3. Theoretical approaches

Many body systems can be treated theoretically in a straightforward way in the extreme limits of both weak interaction and very strong interaction. In the first case, one is faced with a gaseous system, or a Vlasov plasma, where correlation effects can be treated perturbatively ($\Gamma \ll 1$). Sophisticated theoretical approaches, like diagrammatic expansions [28] make it possible to extend standard methods to obtain thermodynamic results in the moderately coupled regime. The random phase approximation (RPA) [29] method, based on the linear response theory, is a useful tool to calculate dynamical properties (wave dispersions) in the case when correlation effects are negligible.

In the case of very strong interaction, the systems crystallize, the particles are completely localized and phonons are the principal excitations. For such conditions lattice-summation techniques serve as solid basis to obtain wave dispersion information.

In the intermediate regime – in the strongly coupled liquid phase – the localization of the particles in the local minima of the potential surface still prevails, however due to the diffusion of the particles themselves the time of localization is finite [30]. A successful theoretical approach for calculating structural properties, like the $S(k)$ static structure function, is the Hyper-Netted-Chain (HNC) method, see e.g. [31].

The localization of the particles (which may typically cover a period of several plasma oscillation cycles) serves as the basis of the *Quasi-Localized Charge Approximation* (QLCA) method [32, 33], which uses structural information (in the form of $S(k)$ static structure function or $g(r)$ pair correlation function) as input for the calculations of the dispersion relations of collective modes. The conceptual basis for the QLCA has been a model that implies the following assumptions about the behavior of strongly coupled Coulomb or Yukawa liquids [32]: (i) in the potential landscape deep potential minima form that are capable of trapping (caging) charged particles; (ii) a caged charge oscillates with a frequency that is determined both by the local potential well and the interaction with the other (caged) particles in their instantaneously frozen positions; (iii) the potential landscape changes slowly to allow the charges to execute a fair number of oscillations; (iv) the escape from the cages of the particles is caused by the gradual disintegration of the caging environment; the timescale of this process is determined by the coupling strength Γ; (v) the (time and velocity dependent) correlation between a selected pair of particles is well approximated by the (time and velocity independent) equilibrium pair correlation; (vi) the frequency spectrum calculated from the averaged (correlated) distribution of particles represents, in a good approximation the average of the distribution of frequencies originating from the actual ensemble. These assumptions have been confirmed to be reasonable in a series of studies (e.g. [30, 33]).

The main concern of the QLCA theory is the analysis of the collective behavior in strongly coupled many-particle systems. The formal tools for this are the dielectric function $\varepsilon_{\mu\nu}^{AB}(\mathbf{k},\omega)$ [having a tensor character (subscripts) in real space and a matrix character (superscripts) in species space] and the dynamical structure function $S^{AB}(\mathbf{k},\omega)$ or more generally, the dynamical current-current correlation function $T_{\mu\nu}^{AB}(\mathbf{k},\omega)$. The principal approximation of the QLCA method is to replace the fluctuating microscopic densities and their products by their ensemble averages, making use of the $S(\mathbf{k})$ static structure function of the system.

All the information pertaining to the mode structure is contained in the dielectric matrix that has a longitudinal and a transverse element:

$$\varepsilon_{\mathrm{L/T}}(\mathbf{k},\omega) = 1 - \frac{\omega_0^2(\mathbf{k})}{\omega^2 - D_{\mathrm{L/T}}(\mathbf{k})}. \tag{5}$$

Here the $D_{\mathrm{L}}(\mathbf{k})$ and $D_{\mathrm{T}}(\mathbf{k})$ local field functions are the respective projections of the QLCA dynamical matrix $D_{\mu\nu}(\mathbf{k})$ [32], which is a functional of the equilibrium pair correlation function (PCF) $h(r) \equiv g(r) - 1$ or its Fourier transform $h(\mathbf{k})$:

$$D_{\mu\nu}(\mathbf{k}) = -\frac{n}{m}\int d^2r\, M_{\mu\nu}(r)[e^{i\mathbf{k}\cdot\mathbf{r}} - 1]h(r) \tag{6}$$

with $M_{\mu\nu}(r) = \partial_\mu\partial_\nu\phi(r)$ being the dipole-dipole interaction potential associated with $\phi(r)$.

The dispersion relations of the longitudinal and transverse modes are determined from

$$\varepsilon_{\mathrm{L}}(\mathbf{k},\omega) = 0, \qquad \varepsilon_{\mathrm{T}}^{-1}(\mathbf{k},\omega) = 0. \tag{7}$$

3. Characteristics of strongly coupled Yukawa systems

Here we present results concerning both ideal 2D Yukawa liquids [34] and quasi-2D Yukawa liquids confined (in one direction) by a parabolic potential [35]. In the latter case the particles can freely move in the (x, y) plane, while a confinement potential $V_c(z) \propto z^2$ acts upon them when they are displaced from the $z = 0$ plane. The confinement force is linear with respect to the "vertical" displacement,

$$F_z = -f_0 \frac{Q^2}{4\pi\varepsilon_0 a^3} z, \tag{8}$$

where the strength f_0 (besides Γ and κ) is the third characteristic parameter of the system. At $f_0 = 1$ the confinement force at a vertical displacement $z = a$ is equal to the magnitude of the force between two particles separated by a, interacting via Coulomb potential.

3.1. Structural properties

Figure 3 illustrates the "raw" results (particle snapshots) of MD simulations of 2D Yukawa systems at two different values of the coupling parameter Γ, belonging to the liquid phase. In the case of Γ=5 – which corresponds to a warm liquid – little correlation is seen between the particle positions, while at Γ=120 – which corresponds to a very cold liquid — the separation of the first neighbors is seen to scatter within a relatively narrow range, although no quasi-two-dimensional order is present.

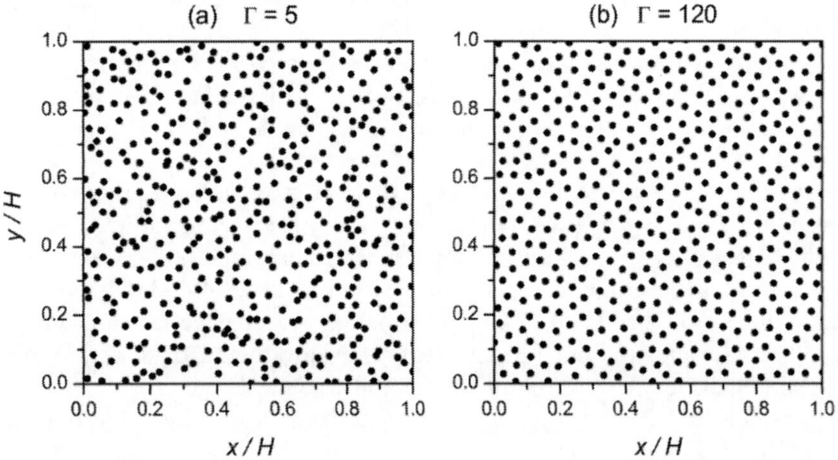

Figure 3. Snapshot of particle positions obtained in MD simulations of a 2D Yukawa liquids at (a) Γ=5 and (b) Γ=120 (κ=1). H is the edge length of the simulation box.

The degree of ordering can be quantitatively studied via the analysis of the pair correlation functions (PCFs). Figure 4 shows a series of PCFs for $\Gamma = 120$ and different κ values; all these conditions belong to the strongly coupled liquid phase. It can be seen that the pronounced order found at $\kappa = 0$ rapidly diminishes with increasing κ. (The amplitude of the first peak of the PCF can be, however, reestablished if Γ is also increased together with κ [34].) Here $\bar{r} = r/a$.

In the case of parabolic confinement the number of layers formed in the system depends on the strength of the confinement. Figure 5(a) shows the distribution of particle density along the z direction, $n(z)$, for different strengths of confinement (f_0), at $\Gamma = 100$ and $\kappa = 0.27$ (belonging to the liquid phase). We observe a transition from two-layered configuration to a single layer at $f_0 \approx 1.4$. Due to the finite temperature of the system the layers do not resemble crystal planes (like in ion trap experiments) but they are rather broad. At $f_0 = 1.4$ the width of the layer is $\approx 0.86a$ (at half of the maximum of the distribution). Doubling the amplitude of the potential results in a width $0.17a$. When the repulsion between the particles decreases at higher κ values, the potential with the same amplitude results in a stronger confinement. Thus, with increasing

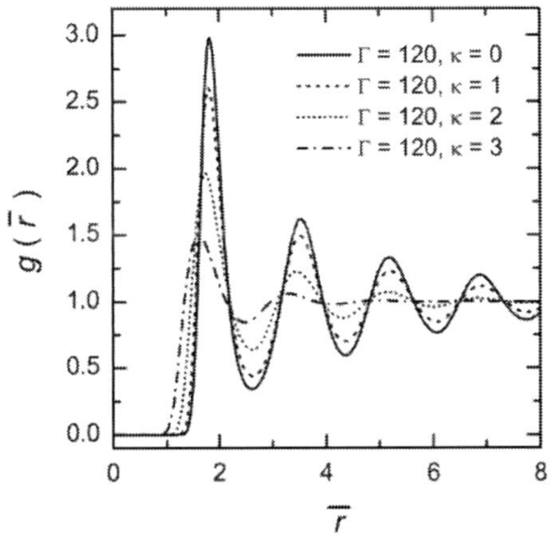

Figure 4. Pair correlation functions of 2D Yukawa liquids at $\Gamma = 120$, for different values of the screening parameter. Reproduced from Ref. [34]; copyright (2005) by the American Physical Society.

κ the number of layers, or the width of the single layer decreases, as illustrated in Fig. 5(b). Similar studies of layer formation in quasi-2D systems have been presented in [36, 37, 38].

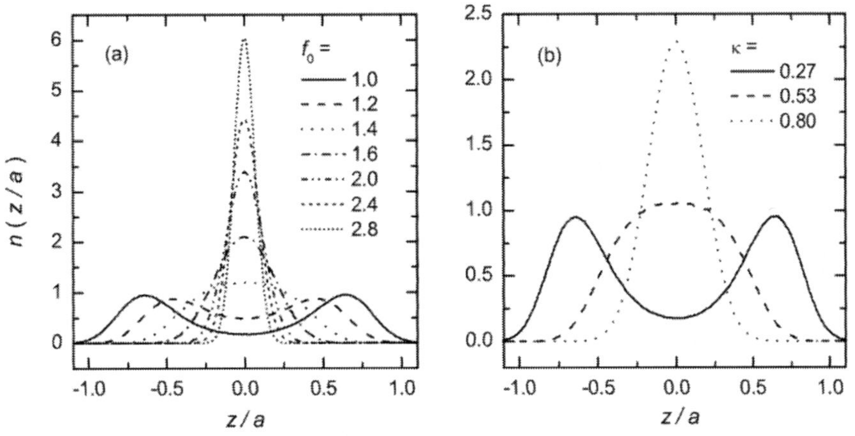

Figure 5. Density distribution of particles along the z direction in the quasi-2D system (normalized as $\int n(z/a)d(z/a) = 1$). (a) Dependence on the strength of confinement at $\Gamma = 100$ and $\kappa = 0.27$; (b) Dependence on κ at constant f_0. Reproduced from Ref. [35]; copyright (2004) by the American Physical Society.

Our further studies of the collective excitations (to be presented in the forthcoming section) concern the domain of parameters when a *single layer* if formed. At higher number of layers the mode structure is expected to be more complicated [39] but this is not within the scope of the present paper.

3.2. Collective excitations

The different waves that can build up in a quasi-two-dimensional dust plasma layer are shown in figure 6. In an ideally narrow layer only the compressional ($\mathcal{L}$) and in-plane transverse modes ($\mathcal{T}$) exist, while in the case of a layer of finite width (confined e.g. by a parabolic potential) a third collective mode, the out-of-plane transverse mode ($\mathcal{P}$) can also develop.

In the MD simulation information about the (thermally excited) collective modes and their dispersion is obtained from the Fourier analysis of the correlation spectra of the density and

Journal of Physics: Conference Series **162** (2009) 012016 doi:10.1088/1742-6596/162/1/012016

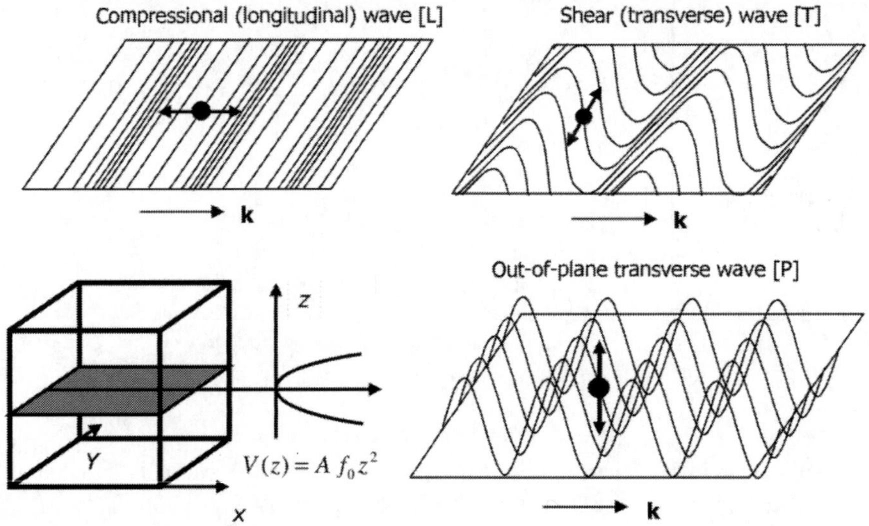

Figure 6. Types of collective excitations in a quasi-two-dimensional dust plasma layer.

current fluctuations. Taking as an example the microscopic density defined as

$$\rho(k,t) = \sum_j \exp\big[ikx_j(t)\big], \tag{9}$$

the dynamical structure function is obtained by the following expression [40]:

$$S(k,\omega) = \frac{1}{2\pi N} \lim_{\Delta T \to \infty} \frac{1}{\Delta T} |\rho(k,\omega)|^2, \tag{10}$$

where $\rho(k,\omega) = \mathcal{F}\big[\rho(k,t)\big]$ is the Fourier transform of (9). Similarly, the spectra of the longitudinal and transverse current fluctuations, $L(k,\omega)$, as well as $T(k,\omega)$ and $P(k,\omega)$, respectively, can be obtained from Fourier analysis of the quantities

$$\lambda(k,t) = k \sum_j v_{jx}(t) \exp\big[ikx_j(t)\big],$$
$$\tau(k,t) = k \sum_j v_{jy}(t) \exp\big[ikx_j(t)\big],$$
$$\pi(k,t) = k \sum_j v_{jz}(t) \exp\big[ikx_j(t)\big], \tag{11}$$

where x_j and v_j are the position and velocity of the j-th particle. Here we assume that **k** is directed along the x axis (the system is isotropic) and accordingly omit the vector notation of the wave number. The spectra are obtained for a series of wave numbers, which are multiples of $k_{\min} = 2\pi/H$, where H is the edge length of the simulation box. The dimensionless wave number is defined as: $\bar{k} = ka$.

Figure 7 shows the energy density of the different type current-current fluctuations on the $\omega - k$ plane.

The dispersion relations of the quasi-2D Yukawa system are displayed in Figure 8 for different values of f_0 and κ. $\Gamma = 100$ for all the cases, and thus the systems are in the liquid phase, where

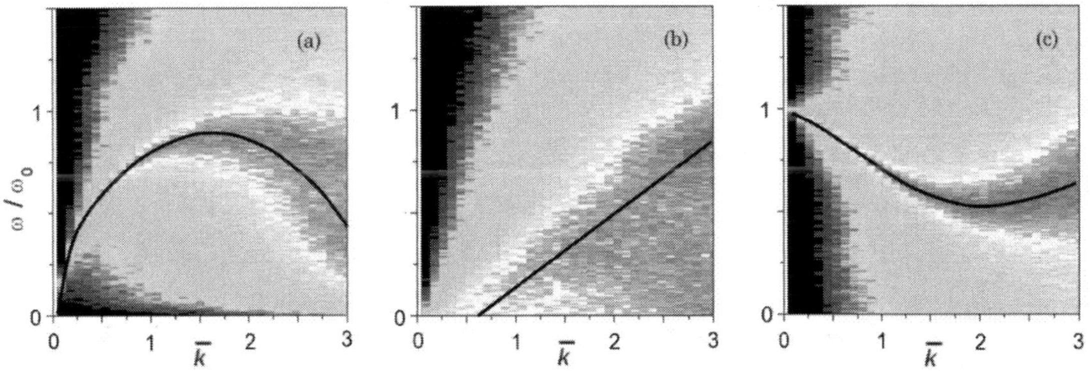

Figure 7. (Color online) Spectral composition of the energy of thermally excited waves in a quasi-2D dusty plasma system in the liquid phase at $\Gamma = 100$, $\kappa = 0.27$, and $f_0 = 2$. (a) Longitudinal, (b) in-plane transverse, and (c) out-of-plane transverse modes. The color scales are logarithmic, they intend to illustrate qualitative features only. Red color indicates concentration of the energy, the heavy black lines are approximate dispersion relations.

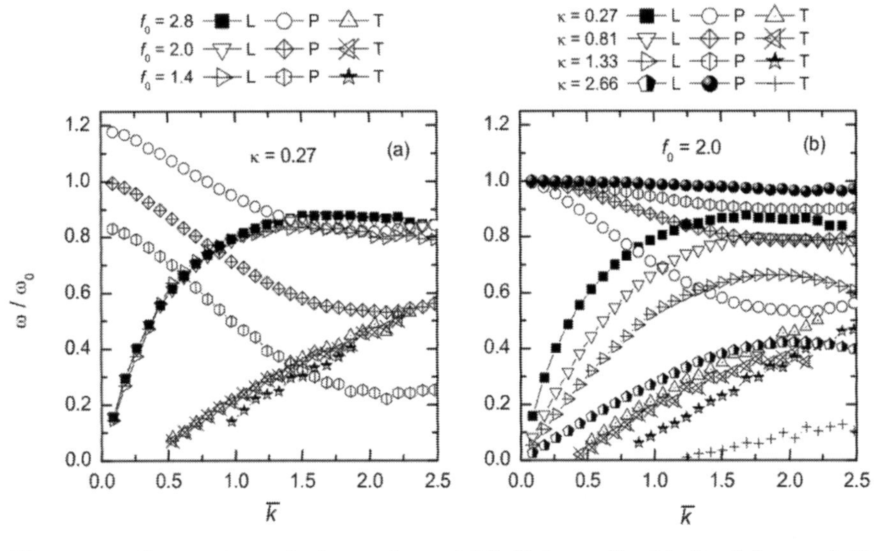

Figure 8. Dispersion relations of quasi-2D Yukawa liquids for (a) $\kappa = 0.27$ and different values of the amplitude f_0 of the confining potential, and (b) for fixed $f_0 = 2$ and different values of the screening parameter κ. Reproduced from Ref. [35]; copyright (2004) by the American Physical Society.

all directions are equivalent and the dispersion relations are isotropic. At constant κ, as shown in Figure 8(a), the frequency of the out-of-plane transverse mode changes significantly as the strength of the confinement force, f_0, is varied. At $k = 0$ the whole layer oscillates in unison in the potential well with a frequency $\omega(k = 0) = \sqrt{f_0 c/m}$, which gives $\frac{\omega(k=0)}{\omega_0} = \sqrt{f_0/2}$. A decreasing confinement force amplitude results in a smaller $\omega(k = 0)$ and $\omega(k \to \infty)$.

The $\mathcal{L}$ and $\mathcal{T}$ modes are only slightly affected by the value of f_0. The frequency of these modes is somewhat smaller at $f_0 = 1.4$, which is near the lower bound of f_0 for the formation of a single layer [35]. The $\mathcal{L}$ mode exhibits a quasi acoustic behavior, while the $\mathcal{T}$ mode shows an acoustic, $\omega \propto k$ dispersion, with a cutoff at a finite wave number. For the $\mathcal{P}$ mode we find $\mathrm{d}\omega/\mathrm{d}k < 0$ in the $\overline{k} \lesssim 2.1$ domain. Beyond this wave number the frequency of the mode slightly increases. This observation on the liquid system agrees well with that on the crystallized system [41], where the same behavior was found, except that in the latter system the critical wave number (at which the group velocity $\mathrm{d}\omega/\mathrm{d}k$ changes from negative to positive) also depends on the direction of the propagation in the lattice. At a constant f_0 the value of $\omega(k = 0)$ does not change when κ is varied, but – as shown in Figure 8(b) – $\omega(k > 0)$ increases with increasing κ. This is explained by the decreased interparticle force (at an average particle separation) at higher κ.

The dispersion relations of 2D Yukawa liquids generated by MD simulations have been cross checked with theoretical (QLCA) results in [42]. Dispersion characteristics can also be derived from experimental studies by tracking the phase space trajectories of the particles, which allows the calculation of the fluctuation spectra just like in the case of the simulations. Such studies have been presented e.g. in [43] and [44] for the solid phase and in [45] for the liquid phase of Yukawa systems.

3.3. Transport properties

MD methods offer two basic ways to study transport processes. In *non-equilibrium* MD simulation methods an external perturbation is applied to the system and the system's response (linked to the perturbation through a transport coefficient) is measured. In *equilibrium* simulations time correlation functions of certain microscopic quantities are measured and the macroscopic transport coefficients are obtained through the Green-Kubo (GK) relations. For the self-diffusion (D_s), shear viscosity (η), and thermal conductivity (λ) the GK relations are given as [46]:

$$D_\mathrm{s} = \frac{1}{\mathcal{D}} \int_0^\infty C_v \mathrm{d}t, \qquad C_v \equiv \langle \mathbf{v}(t) \cdot \mathbf{v}(0) \rangle, \tag{12}$$

$$\eta = \frac{1}{VkT} \int_0^\infty C_\eta \mathrm{d}t, \qquad C_\eta \equiv \langle P_{ij}(t) P_{ij}(0) \rangle, \tag{13}$$

$$\lambda = \frac{1}{VkT^2} \int_0^\infty C_\lambda \mathrm{d}t, \qquad C_\lambda \equiv \langle J_{Qi}(t) J_{Qi}(0) \rangle. \tag{14}$$

In the above formulae C_v, C_η, and C_λ are the velocity (VACF), shear stress (SACF) and energy current (EACF) autocorrelation functions, respectively. The calculation of these correlation functions requires time series for the particle velocity $\mathbf{v}$, the off-diagonal element of the pressure tensor P_{ij}, and the energy current J_{Qi}. Here, $\mathcal{D}$ is the dimensionality of the system and $i \neq j$ are space coordinates.

For the existence of transport coefficients the integrands in Eqs. (12)-(14) must converge. This may not always be the case in some lower dimensional ($\mathcal{D} < 3$) systems. Non-exponential long-time tails in the VACF of hard sphere and hard disk systems were first reported by Alder and Wainwright [47] based on their simulation studies. They have found a $t^{-\mathcal{D}/2}$ decay in these systems. According to this finding, for $\mathcal{D}$=2 the VACF is non-integrable, and consequently the diffusion coefficient does not exist. Further, it has been shown [48] that the kinetic contributions to the autocorrelation functions of shear stress and energy current – which are related to velocity correlations – exhibit the same behavior. Their findings were also confirmed by the calculations in [49]. Regarding systems with continuous potentials, a power law decay of the VACF was observed in 3D soft-repulsive and Lennard-Jones liquids [50], as well as a t^{-1} tail of the SACF was detected in the molecular dynamics simulations of a 2D soft disk fluid [51]. For the case of Coulomb interaction (2D classical electron liquid) the existence of self-diffusion coefficient has been a topic of controversy [52, 53, 54].

While the $t^{-3/2}$ decay in case of $\mathcal{D} = 3$ allows for an integrable VACF and thus for a meaningful diffusion coefficient to exist, in two dimensions, the corresponding $t^{-\mathcal{D}/2} = t^{-1}$ decay makes the VACF non-integrable. Thus the analysis of transport properties in 2D systems is of great interest and has already been the topic of some recent papers. As an example the equilibrium MD simulations of 2D Yukawa liquids by Liu and Goree [55] may be mentioned. In this study it was demonstrated that *superdiffusion* rather than normal diffusion occurs over a wide range of temperatures (at $\kappa = 0.56$).

The emergence and vanishing of superdiffusion in quasi-two-dimensional Yukawa systems ("bridging the gap" between 2D and 3D systems) have been investigated by molecular dynamics simulations by Ott *et al.* [56]. In this work the systems have been confined by two different potentials: (i) a parabolic potential and (ii) a soft box potential. Both of these potentials allowed a smooth change of the character of the system between a three-dimensional one and a two-dimensional one. The existence of a transition from normal diffusion to superdiffusion was identified in [56] by analyzing both the asymptotic behavior of the mean-squared displacement of the particles and the long-time tail of the velocity autocorrelation function.

Self-diffusion has also been experimentally studied in a two-dimensional underdamped laboratory liquid complex (dusty) plasmas by Nunomura *et al.* [57]. No superdiffusion has been observed in this work, but subdiffusion occurred at temperatures close to melting. Anomalous diffusion and non-Gaussian statistics have, on the other hand, been detected experimentally in a two-dimensional driven-dissipative Yukawa system by Liu and Goree [58]. In their experiment a single-layer dusty plasma suspension with a Yukawa interaction and frictional dissipation has been heated with laser radiation pressure to yield a structure with liquid ordering. Analyzing the time series for mean-square displacement, superdiffusion has been detected at a low but statistically significant level over a wide range of temperatures. Mass transfer processes have recently thoroughly been analyzed by Vaulina and coworkers both experimentally [59] and through numerical simulation [18].

Clarification of the behavior of the shear stress and energy current autocorrelation functions (SACF and EACF) of 2D Yukawa systems is a topic of current work [60]. The analysis of the long-time behavior of these functions is considerably more difficult as compared to the VACF, since, while the calculation of the VACF includes an averaging over a high number of particles and different initial times, for the SACF and EACF only the latter averaging is involved. Consequently the signal to noise ratio of the data obtained in a single simulation run for the SACF and the VACF may be orders of magnitude inferior compared to that of the VACF, and reaching of acceptable signal to noise levels requires numerous simulation runs with different initial conditions. Nonetheless, with the advance of computational resources details of the long-time behavior of the SACF and EACF of 2D and quasi-2D Yukawa liquids are also expected to be uncovered.

4. Summary

This paper intended to review some prominent features of strongly coupled 2D and quasi-2D Yukawa systems. The combination of theoretical and simulation methods on one hand, and the possibility of experimental realization of such many-particle systems, on the other hand, provides their fairly complete characterization.

We have presented here static properties, illustrated collective effects and briefly reviewed transport characteristics of such systems. While the first two of these are fairly well understood, further work is needed to shed light on the transport properties in the low-dimensional systems considered here.

5. Acknowledgment

This work has been supported by the Hungarian Fund for Scientific Research (OTKA-T-048389, OTKA-IN-69892, OTKA-PD-75113), by a bilateral grant between the Hungarian Academy of Sciences and the National Science Foundation of USA (MTA/NFS-102), by the National Science Foundation (PHY-0514619) and by the Janos Bolyai Research Scholarship of the Hungarian Academy of Sciences (PH).

References

[1] Kalman G J, Rommel J M and Blagoev K 1998 *Strongly Coupled Coulomb Systems (New York: Plenum)*; Special Issue 2003 *J. Phys. A: Math. Gen.* **36** no. 22; Special Issue 2006 *J. Phys. A: Math. Gen.* **39** no. 17.
[2] Brush S G, Sahlin H L and Teller E L 1966 *J. Chem. Phys.* **45** 2102
[3] Stringfellow G S, DeWitt H E and Slattery W L 1990 *Phys. Rev. A* **41** 1105
[4] Farouki R T and Hamaguchi S 1993 *Phys. Rev. E* **47** 4330
[5] Grimes C C and Adams G 1976 *Phys. Rev. Lett.* **36** 145
[6] Peierls R E 1935 *Ann. Inst. Henri Poincaré* **5** 177; Landau L D 1937 *Phys. Z. Sowjet.* **11** 26; Hohenberg P C 1966 *Phys. Rev.* **158** 383; Mermin N D 1968 *Phys. Rev.* **176** 250
[7] Totsuji 2001 *Physics of Plasmas* **8** 1856 Hartmann P, Donkó Z, Bakshi P, Kalman G J and Kyrkos S 2007 *IEEE Trans. Plasma Sci.* **35** 332
[8] Melzer A, Homann A and Piel A 1996 *Phys. Rev. E* **53** 2757; Quinn R A and Goree J 2001 *Phys. Rev. E* **64** 051404
[9] Donkó Z, Kalman G J and Hartmann P 2008 *J. Phys. Cond. Matter* **20** 413101
[10] Mendis D A 2002 *Plasma Sources Sci. Technol* **11** A219
[11] Merlino R L and Goree J A 2004 *Physics Today* **57** 32
[12] Fortov V E, Ivlev A V, Khrapak S A, Khrapak A G and Morfill G E 2005 *Phys. Rep.* **421** 1
[13] Ishihara O 2007 *J. Phys. D: Appl. Phys.* **40** R121
[14] Thomas H, Morfill G E, Demmel V, Goree J, Feuerbacher B and Möhlmann D 1994 *Phys. Rev. Lett.* **73** 652 Chu J H and Lin I 1994 *Phys. Rev. Lett.* **72** 4009; Hayashi Y, Tachibana K 1994 *Japan. J. Appl. Phys.* **33** L814 Melzer A, Trottenberg T and Piel A 1994 *Phys. Lett. A* **191** 301
[15] Lampe M, Joyce G, Ganguli V Gavrishchaka V 2000 *Phys. Plasmas* **7** 3851; Hebner G A and Riley M E 2004 *Phys. Rev. E* **69** 026405
[16] Frenkel D and Smit B 2001 *Understanding Molecular Dynamics Simulations* (Academic Press).
[17] Baimbetov F B, Ramazanov T S, Dzhumagulova K N, Kadyrsizov E R, Petrov O F and Gavrikov A V 2006 *J. Phys. A: Math. Gen.* **39** 4521
[18] Vaulina O S, Adamovich X G, Petrov O F and Fortov V E 2008 *Phys. Rev. E* **77** 066403
[19] Feng Y, Liu Bin and Goree J 2008 *Phys. Rev. E* **78** 026415
[20] Hou L-J and Piel A 2008 arXiv:0810.1623v1 [physics.plasm-ph]
[21] Ewald P P 1921 *Ann. Phys.* **64** 253
[22] Salin G and Caillol J-M 2002 *Phys. Rev. Lett.* **88** 065002; 2003 Phys. Plasmas **10** 1220.
[23] Johnson R E and Ranganathan S 2001 *Phys. Rev. E* **63** 056703
[24] Sagui C and Darden T A 1999 *Ann. Rev. Biophys. Biomol. Struct.* **28** 155
[25] David N and Hooker S M 2003 *Phys. Rev. E* **68** 056401
[26] David N, Spence D J and Hooker S M 2004 *Phys. Rev. E* **70** 056411
[27] Eastwood J W, Hockney R W and Lawrence D N 1980 *Comput. Phys. Commun.* **19** 215; Hockney R and Eastwood J 1981 *Computer Simulation Using Particles* (New York: McGraw-Hill)
[28] Hansen J P, McDonald I R 1976 *Theory of Simple Liquids* (Academic Press, London)
[29] Ehrenreich H and Cohen M H 1959 *Phys. Rev.* **115** 786
[30] Donkó Z, Kalman G J and Golden K I 2002 *Phys. Rev. Lett.* **88** 225001
[31] Lado F 1978 *Phys. Rev. B* **17** 2827; Rogers F J, Young D A, DeWitt H E and Ross M 1983 *Phys. Rev. A* **28** 2990
[32] Golden K I and Kalman G J 2000 *Phys. Plasmas* **7** 14
[33] Kalman G J, Golden K I, Donkó Z and Hartmann P 2005 *J. Phys. Conf.* **11** 254
[34] Hartmann P, Kalman G J, Donkó Z and Kutasi K 2005 *Phys. Rev. E* **72** 026409
[35] Donkó Z, Hartmann P and Kalman G J 2004 *Phys. Rev. E* **69** 065401
[36] Totsuji H, Kishimoto T, Inoue Y, Totsuji C and Nara S 1996 *Phys. Lett. A* **221** 215
[37] Bystrenko O 2003 *Phys. Rev. E* **67** 025401
[38] Qiao K and Hyde T W 2005 *Phys. Rev. E* **71** 026406

[39] Donkó Z, Kalman G J, Hartmann P, Golden K I and Kutasi K 2003 *Phys. Rev. Lett.* **90** 226804; Donkó Z, Hartmann P, Kalman G J and Golden K I 2003 *J. Phys. A: Math. Gen.* **36** 5877
[40] Hansen J P, McDonald I R and Pollock E L 1975 *Phys. Rev. A* **11** 1025
[41] Qiao K and Hyde T W 2003 *Phys. Rev. E* **68** 046403
[42] Kalman G J, Hartmann P, Donkó Z, Rosenberg M 2004 *Phys. Rev. Lett.* **92** 065001
[43] Nunomura S, Goree J, Hu S, Wang X and Bhattacharjee A 2002 *Phys. Rev. E* **65** 066402
[44] Nunomura S, Goree J, Hu S, Wang X, Bhattacharjee A and Avinash K 2002 *Phys. Rev. Lett.* **89** 035001
[45] Nunomura S, Zhdanov S, Samsonov D and Morfill G E 2005 *Phys. Rev. Lett.* **94** 045001
[46] Hansen J-P and McDonald I R 1976 *Theory of simple liquids* (Academic Press, New York)
[47] Alder B J and Wainwright T E 1970 *Phys. Rev. A* **1** 18
[48] Ernst M H, Hauge E H and van Leeuwen J M J 1970 *Phys. Rev. Lett.* **25** 1254
[49] Dorfman J R and Cohen E G D 1972 *Phys. Rev. A* **6** 776
[50] McDonough A, Russo S P and Snook I K 2001 *Phys. Rev. E* **63** 026109
[51] Morriss G P and Evans D J 1985 *Phys. Rev. A* **32** 2425
[52] Hansen J P, Levesque D and Weis J J 1979 *Phys. Rev. Lett.* **43** 979
[53] Baus M and Bosse J 1980 *Phys. Rev. A* **22** 2284
[54] Agarwal G K and Pathak K N 1983 *J. Phys. C: Solid State Phys.* **16** 1887
[55] Liu Bin and Goree J 2007 *Phys. Rev. E* **75** 016405
[56] Ott T, Donkó Z, Hartmann P and Bonitz M 2008 *Phys. Rev. E* **78** 026409
[57] Nunomura S, Samsonov D, Zhdanov S and Morfill G 2006 *Phys. Rev. Lett.* **96** 015003
[58] Liu Bin and Goree J 2008 *Phys. Rev. Lett.* **100** 055003
[59] Vaulina O S, Adamovich X G, Petrov O F and Fortov V E 2008 *Phys. Rev. E* **77** 066404
[60] Donkó Z, Goree J, Hartmann P and Liu Bin 2008 in preparation

Journal of Physics: Conference Series **162** (2009) 012017 doi:10.1088/1742-6596/162/1/012017

Effects of friction on modes in collisional multicomponent plasmas

J Vranjes[1,2], S Poedts[1], M Kono[3], M Y Tanaka[4]

[1]Center for Plasma Astrophysics, and Leuven Mathematical Modeling and Computational Science Center (LMCC), Celestijnenlaan 200B, 3001 Leuven, Belgium
[2]Faculté des Sciences Appliquées, avenue F.D. Roosevelt 50, 1050 Bruxelles, Belgium
[3]Faculty of Policy Studies, Chuo University, Hachioji, Tokyo 192-0393, Japan
[4]Department of High Energy Engineering Science, Kyushu University, Kasuga-koen 6-1, Kasuga, Fukuoka 816-8580, Japan

E-mail: jvranjes@yahoo.com; Stefaan.Poedts@wis.kuleuven.be; kono@fps.chuo-u.ac.jp; mytanaka@aees.kyushu-u.ac.jp

Abstract. The friction in multi-component plasmas is discussed, and its effects on ion acoustic (IA) and drift waves. In the case of the IA wave it is shown that the friction between the two species has no effect on the mode in a quasi-neutral plasma. Using the Poisson equation instead of the quasi-neutrality reveals the possibility for an instability driven by the collisional energy transfer. However, the different starting temperatures of the two species imply an evolving background. It is shown that the relaxation time of the background electron-ion plasma is, in fact, always shorter than the growth rate time. Therefore the instability is unlikely to develop. The drift wave instability is also reexamined taking into account the ion response in the direction parallel to the magnetic field lines, which appears due to friction with electrons and which can not be omitted in view of the momentum conservation. A modified instability threshold is obtained. In plasmas with dominant electron collisions with neutrals, the instability threshold is shifted towards higher frequencies, compared to the case of dominant electron collisions with ions. The difference between the two cases vanishes when the ion sound response is negligible, i.e., when the instability threshold disappears, and both ions and neutrals react to the electron friction in the same manner. The results obtained here should contribute to the definite clarification of some contradictory results obtained in the past.

1. Introduction

Plasmas both in the laboratory and in space are frequently in the state of partial thermodynamic equilibrium (i.e., with an initial temperature disparity of the plasma constituents). Collisions in such plasmas will after some time eventually result in equal temperatures of the species, implying an evolving plasma. There exists a long standing controversy in the literature, which deals with the effects of this temperature disparity on the perturbations in such an evolving plasma, more precisely on the ion acoustic (IA) waves. In Ref. [1] it is claimed that the corresponding energy transfer may result in the instability of the acoustic mode at large wavelengths (within the quasi-neutrality limit), and that this growth may be described within the fluid theory. The necessary condition for the instability obtained in Ref. [1] for an electron-ion plasma is, in fact, very easily satisfied because it requires only a very small temperature difference between the two species (electrons and ions), viz. $T_e > 4T_i/3$. This instability condition is obtained by using the energy equations including the source/sink terms originating from the collisional transfer, together with the corresponding friction force terms in the momentum equations. The sufficient instability condition is stronger because of additional dissipative effects, like viscosity and thermal conductivity. However, the current-less instability described in Ref. [1] is based on a model which disregards the same temperature disparity in the description of the spatially homogeneous background, which, due to the same reasons, must be time evolving. In other words, the effects of collisions in the background plasma have been explicitly neglected. Note that because of the time evolution, the term background is used instead of the equilibrium. These effects of collisions have been discussed in Ref. [2], published one year after Ref. [1], and for the same quasi-neutrality case. There, it is claimed that there is no instability for any temperature ratio of the two plasma components, and moreover, that this holds even in a current-carrying plasma, as long as the difference between the electron and ion equilibrium velocity remains below the sound speed. All that was needed to come to that conclusion was to let the background plasma evolve freely in the presence of the given temperature difference. However, we observe that Ref. [2] has apparently remained almost unnoticed by researchers, in contrast to the widely cited Ref. [1]. In the present work, this controversy is revisited for any two-component plasma. Essential for the problem is the energy equation describing the temperature variation. In the simplified form that we shall use, it contains only the collisional energy transfer source/sink term on the right-hand side. This simplified form is used for clarity only because, in the absence of currents, that term alone is supposed to yield an instability.

Regarding the drift wave, it is known that it can be excited by electron collisions. In this case, the usual relation between the electron perturbed density and the perturbed potential becomes modified due to the presence of the collisional term, so that the potential lags behind the density [3, 4]. The effect appears regardless whether the electrons collide, in their predominant motion along the magnetic field vector, with ions or with neutrals [5, 6]. Compared with the kinetic instability (which is due to the inverse electron Landau damping effect, that appears because the mode frequency is slightly below the diamagnetic frequency), the collisional instability is dominant provided that the electron parallel mean-free path is smaller than the parallel wavelength. Hence the interest in the drift modes with very large parallel wave-lengths and relatively short perpendicular wave-lengths. The modification in the perturbed electron number density is due to the friction, which is typically kept in the electron momentum equation only. However, the momentum conservation requires the presence of the corresponding term in the ion momentum too. Here we perform derivations with such a full friction force term for some simple cases in order to demonstrate the differences introduced by this friction-induced response of the heavier species.

2. Friction in ion acoustic mode

In view of the controversy mentioned above, here we give some details following Braginskii [7] where the energy equation for any species a is given in the form:

$$\frac{3}{2}n_a\frac{\partial T_a}{\partial t} + n_aT_a\nabla\cdot\vec{v}_a + \frac{3}{2}n_a(\vec{v}_a\cdot\nabla)T_a = Q_a. \tag{1}$$

The corresponding equation for the species b has the same shape, but with a minus sign on the right-hand side. We use the Landau formula for the energy transfer source/sink term [8] $Q_a = 3m_b\nu_{ba}n_b(T_b - T_a)/m_a$, where [9]

$$\nu_{ba} = 4\left(\frac{2\pi}{m_b}\right)^{1/2}\left(\frac{q_aq_b}{4\pi\varepsilon_0}\right)^2\frac{n_aL_{ba}}{3(T_b + T_am_b/m_a)^{3/2}}. \tag{2}$$

The Coulomb logarithm is given by $L_{ba} = \log[r_d/b_0]$, $r_d = r_{da}r_{db}/(r_{da}^2 + r_{db}^2)^{1/2}$, $r_{dj} = v_{Tj}/\omega_{pj}$, and $b_0 = [|q_aq_b|/(4\pi\varepsilon_0)]/[3(T_a + T_b)]$ is the impact parameter.

2.1. Non-evolving background

The two perturbed energy equations without the evolving background effects, corresponding to the model from Ref. [1], are:

$$\frac{\partial T_{(a,b)1}}{\partial t} + \frac{2}{3} T_{(a,b)0} \nabla \cdot \vec{v}_{(a,b)1} = \pm 2 \frac{m_b}{m_a} \nu_{ba} (T_{b1} - T_{a1}) \pm 2\nu_{ba} \frac{m_b}{m_a} (T_{b0} - T_{a0}) \frac{n_{b1}}{n_0}.$$

Here, the minus sign applies to the species b.

The momentum equation which we use throughout the text for the species a is of the form $m_a n_a \partial \vec{v}_a / \partial t = -q_a n_a \nabla \phi - \nabla(n_a T_a) - m_a n_a \nu_{ab}(\vec{v}_a - \vec{v}_b)$, and the continuity equation has its standard form. Similar equations are used for the species b, where the friction term is of the form $\vec{F}_{fb} = -m_b n_b \nu_{ba}(\vec{v}_b - \vec{v}_a)$.

In the case of quasi-neutral perturbations, the two number densities $n_{(a,b)1}$ are calculated from the continuity equations and are made equal. The dispersion equation in that case reads:

$$\left(\omega + \frac{i4 m_b \nu_{ba}}{m_a}\right)\left(\omega^2 - \frac{5k^2}{3} \frac{T_{a0} + T_{b0}}{m_a + m_b}\right) = 0. \tag{3}$$

Hence, even using the same model as in Ref. [1] there is neither an instability nor damping of the acoustic mode, regardless of the ratio T_{a0}/T_{b0}.

Observe that the momentum conservation condition $\nu_{ab} = m_b n_b \nu_{ba}/(m_a n_a)$ is nowhere used in the derivation of Eq. (3). This is because the friction terms vanish in any case. In fact, from the two continuity equations we have the velocities $v_{a1} - v_{b1} = (\omega/k)(n_{a1}/n_{a0} - n_{b1}/n_{b0})$. Using $q_a n_a = q_b n_b$ and assuming the constant charge on the two species a and b, we have $v_{a1} - v_{b1} = 0$ so that the assumption of quasi-neutrality cancels the friction completely. Otherwise, for a varying charge we would have $v_{a1} - v_{b1} = [\omega n_{a1}/(k n_{a0})][1 - z_{a1} z_{b0}/(z_{b1} z_{a0})]$, where z_j denotes the charge number, and therefore the friction effects would remain. Further in the text we assume singly charged species.

2.2. Isothermal perturbations

The derivations are now repeated for isothermal quasi-neutral perturbations. Keeping the full friction force $\vec{F}_f$ in both momentum equations, and within the same quasi-neutrality limit, yields a *real dispersion equation* $\omega^2 = k^2(T_{a0} + T_{b0})/(m_a + m_b)$. Hence, in the given limit the collisions (through friction) do not affect the isothermal ion acoustic mode.

This fact is usually overlooked in the literature where, in the limit $m_a \gg m_b$, a typical mistake is that the friction is kept for the lighter species only, in the incomplete simplified form $\vec{F}_{fb} = -m_b n_b \nu_{ba} \vec{v}_b$. For an electron-ion plasma this gives the incorrect result $\omega = \pm k(c_s^2 + v_{Ti}^2)^{1/2} - \nu_{ei}/2$. We stress again that the collisions appear in Eq. (3) only from the energy equations, yet they do not affect the IA mode.

Using the Poisson equation instead of quasi-neutrality, for isothermal perturbations we obtain coupled and damped IA and Langmuir waves

$$\omega^4 + i(\nu_{ab} + \nu_{ba})\omega^3 - \left[k^2\left(v_{Ta}^2 + v_{Tb}^2\right) + \omega_{pa}^2 + \omega_{pb}^2\right]\omega^2 - ik^2\left(\nu_{ab} v_{Tb}^2 + \nu_{ba} v_{Ta}^2\right)\omega$$
$$+ k^4 v_{Ta}^2 v_{Tb}^2 + k^2\left(v_{Ta}^2 \omega_{pb}^2 + v_{Tb}^2 \omega_{pa}^2\right) = 0. \tag{4}$$

In the low frequency limit $\omega \ll \omega_{p(a,b)}$ and for an e-i plasma, from Eq. (4) we have

$$\omega = \pm k v_s \left(1 - r_{de}^2 k^2 \frac{\nu_{ie}^2 r_{de}^2}{v_s^2}\right)^{1/2} - i\nu_{ie} r_{de}^2 k^2. \tag{5}$$

The IA mode is damped and the damping is k-dependent. Here, $v_s^2 = c_s^2 + v_{Ti}^2$ and we have used the momentum conservation $\nu_{ie} = m_e \nu_{ei}/m_i$.

2.3. Evolving plasma

From Eq. (1) it is seen that the background temperature is also evolving in time as

$$\frac{\partial T_{(a,b)0}}{\partial t} = \pm 2\frac{m_b}{m_a}\nu_{ba}(T_{b0} - T_{a0}). \tag{6}$$

Keeping the collision frequencies constant (the approximation discussed below), this gives the temperatures for the two species

$$T_{(a,b)0}(t) = [\widehat{T}_{(a,b)0}(1 + \exp(-4\nu_{ab}t)) + \widehat{T}_{(a,b)0}(1 - \exp(-4\nu_{ab}t))]/2.$$

It is seen that they evolve towards the common value $(\widehat{T}_{a0} + \widehat{T}_{b0})/2$.

On the other hand, solving Eq. (6) numerically, with time dependent collision frequencies (2) gives a slightly faster relaxation for the two temperatures. To get a feeling on the relaxation time scale, this is presented in Fig. 1 by taking $n_0 = 10^{18}$ m^{-3} and $\widehat{T}_{a0} = 0.1$ eV, $\widehat{T}_{b0} = 3\widehat{T}_{a0}$.

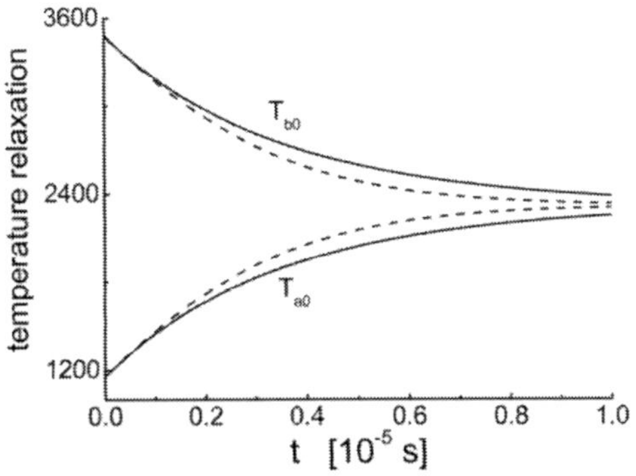

Figure 1. Approximative (full lines), and exact relaxation with time-dependent collision frequencies (dashed lines) of the background plasma temperatures (6).

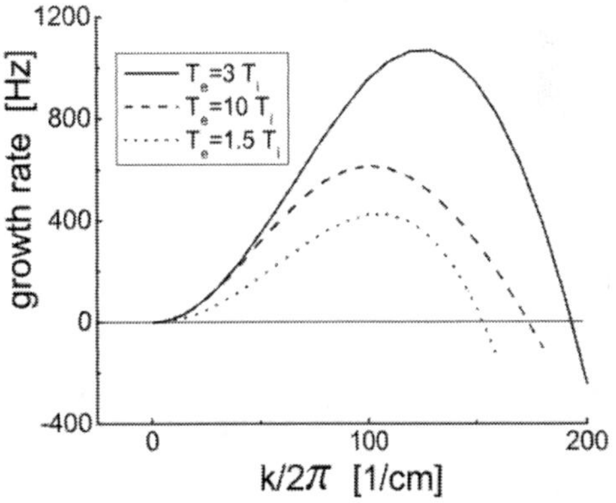

Figure 2. The growth rate of the IA mode in electron-proton plasma with $n_0 = 10^{18}$ m^{-3} and for several values of T_e/T_i.

Eq. (6) is to be used in the linearization of Eq. (1), which in the case $n_{a0} = n_{b0} = n_0$ yields:

$$\frac{\partial T_{a1}}{\partial t} + \frac{2}{3}T_{a0}\nabla\cdot\vec{v}_{a1} = +2\frac{m_b}{m_a}\nu_{ba}\left(T_{b1} - T_{a1}\right) - 2\nu_{ba}\frac{m_b}{m_a}\frac{n_{a1} - n_{b1}}{n_0}\left(T_{b0} - T_{a0}\right). \tag{7}$$

The corresponding equation for the component b is

$$\frac{\partial T_{b1}}{\partial t} + \frac{2}{3}T_{b0}\nabla\cdot\vec{v}_{b1} = -2\frac{m_b}{m_a}\nu_{ba}\left(T_{b1} - T_{a1}\right). \tag{8}$$

We stress that Eq. (3) is obtained also by using Eqs. (7, 8) in the quasi-neutral limit (implying that the last term in Eq. (7) is omitted). Hence, the IA mode appears unaffected by friction in the quasi-neutral limit even if the energy equations are used, and if the background is described correctly as evolving.

We now use the two energy equations (7, 8) *with the Poisson equation.* The dispersion equation becomes [10]

$$\omega^6 + i\nu_{ba}\left(1 + \frac{5m_b}{m_a}\right)\omega^5 - \omega^4\left[\frac{5}{3}k^2\left(v_{Ta}^2 + v_{Tb}^2\right) + \omega_{pa}^2 + \omega_{pb}^2 + 4\nu_{ba}^2\frac{m_b}{m_a}\left(1 + \frac{m_b}{m_a}\right)\right]$$

$$+i\omega^3\nu_{ba}\left[k^2v_{Tb}^2\frac{2m_b^2}{m_a^2}-k^2v_{Ta}^2\left(\frac{5}{3}+\frac{22m_b}{3m_a}\right)-4\omega_{pa}^2\left(1+\frac{m_b}{m_a}\right)-7k^2v_{Tb}^2\frac{m_b}{m_a}\right]$$

$$+\omega^2\left[\frac{25}{9}k^4v_{Ta}^2v_{Tb}^2+\frac{5k^2}{3}\left(v_{Ta}^2\omega_{pb}^2+v_{Tb}^2\omega_{pa}^2\right)+4k^2\nu_{ba}^2\frac{m_b}{m_a}\left(v_{Ta}^2\left(\frac{2}{3}+\frac{m_b}{m_a}\right)+v_{Tb}^2\frac{m_b^2}{m_a^2}\left(\frac{8}{3}-\frac{m_b}{m_a}\right)\right)\right]$$

$$+i10\omega\nu_{ba}k^2\left[k^2v_{Tb}^2\frac{m_b}{m_a}\left(v_{Ta}^2-\frac{v_{Tb}^2}{3}\frac{m_b}{m_a}\right)+\frac{2}{3}\omega_{pa}^2\left(v_{Ta}^2+v_{Tb}^2\frac{m_b}{m_a}\right)\right]$$

$$+4k^4\nu_{ba}^2\frac{m_b}{m_a}\left(v_{Ta}^2-v_{Tb}^2\frac{m_b}{m_a}\right)^2=0. \tag{9}$$

Solving for the IA mode yields approximately

$$\omega_{IA}^2=\frac{5k^2(\omega_{pb}^2v_{Ta}^2+\omega_{pa}^2v_{Tb}^2+5k^2v_{Ta}^2v_{Tb}^2/3)}{3\left[\omega_{pa}^2+\omega_{pb}^2+(5/3)k^2(v_{Ta}^2+v_{Tb}^2)\right]}. \tag{10}$$

The corresponding imaginary part of the frequency is:

$$\gamma=\frac{1}{2\left[\omega_{pa}^2+\omega_{pb}^2+(5/3)k^2(v_{Ta}^2+v_{Tb}^2)-2\omega_{IA}^2\right]}\left\{\nu_{ab}\left(\omega_{IA}^2-(5/3)k^2v_{Tb}^2\right)\left(1+(4/3)k^2v_{Ta}^2/\omega_{IA}^2\right)\right.$$

$$+\nu_{ba}\left(\omega_{IA}^2-(5/3)k^2v_{Ta}^2\right)\left[1+(4/3)\nu_{ab}k^2v_{Tb}^2/(\nu_{ba}\omega_{IA}^2)\right]+2(1-T_{a0}/T_{b0})\left(\nu_{ab}k^2c_s^2/\omega_{IA}^2\right)\left[\omega_{IA}^2\right.$$

$$\left.\left.-(5/3)k^2v_{Tb}^2-(8\nu_{ab}\nu_{ba}/\omega_{IA}^2)\left(\omega_{IA}^2-k^2v_{Ta}^2\right)+(8\nu_{ab}^2/\omega_{IA}^2)\left(\omega_{IA}^2-k^2v_{Tb}^2\right)\right]\right\}. \tag{11}$$

In principle, Eq. (11) reveals the possibility for a growing IA mode if the Poisson equation is used instead of the quasi-neutrality, in a time-evolving background plasma. For example, this can be easily demonstrated in the limit of negligible terms originating from the last term in Eq. (7). This is permissible on condition $|(T_{a1}-T_{b1})/(n_{a1}-n_{b1})|\gg|T_{a0}-T_{b0}|/n_0$, or in an alternative form, $|(T_{a1}-T_{b1})/(T_{a0}-T_{b0})|\gg r_{db}^2k^2|q_b\phi_1|/T_{b0}$. In that limit, the numerical solution of Eq. (9) yields the growth-rate of the IA mode in an electron-ion plasma that is presented in Fig. 2. Here, $n_0=10^{18}$ m^{-3} and we take several values of T_e/T_i, where $T_i=0.1$ eV. The growth rate increases with T_e/T_i but only up to $T_e/T_i\simeq 3$. For even higher values of the temperature ratio the instability ceases, this is represented by the dashed ($T_e/T_i\simeq 10$) line.

However, we stress that the system evolves in time. Thus, in order to have a reasonably fast growth of the perturbations, the following condition must be satisfied [cf. Eq. (6)] :

$$\gamma_r\equiv 2(m_b/m_a)\nu_{ba}\ll\gamma. \tag{12}$$

Here, γ_r^{-1} determines the relaxation time for the background. Taking the electron-ion case like in Ref. [1] and the corresponding self-evident conditions $m_b\ll m_a$, $T_{a0}<T_{b0}$, $k^2v_{Ta}^2<\omega_{IA}^2<\omega_{pa}^2<\omega_{pb}^2$, from Eq. (11) to the leading order terms we obtain

$$\gamma-\gamma_r\simeq-\frac{\nu_{ba}}{2\left(\omega_{pb}^2+5k^2v_{Tb}^2/3\right)}\left\{4\omega_{pa}^2-\omega_{IA}^2+\frac{2k^2c_s^2}{\omega_{IA}^2}\left[k^2v_{Ta}^2\left(\frac{5}{3}+\frac{8\nu_{ab}^2}{\omega_{IA}^2}\right)+8\nu_{ab}^2\right]\right\}. \tag{13}$$

Hence, because always $\omega_{pa}\geq\omega_{IA}$, here we have

$$\gamma<\gamma_r, \tag{14}$$

i.e., the system relaxes on a time scale that is (much) shorter than the eventual growth time of the perturbations. Consequently the assumed instability actually can not develop.

Journal of Physics: Conference Series **162** (2009) 012017 doi:10.1088/1742-6596/162/1/012017

3. Drift wave and friction

In the presence of electron collisions the drift wave can be growing. The friction force term in the electron parallel momentum equation is usually written in the form $-m_e n_e \nu_{ej} \vec{v}_e$, where $j = i, n$. As a result one obtains a standard phase shift in the electron Boltzmann distribution $n_1/n_0 = (e\phi_1/\kappa T_e)(1 - i\delta)$, that is responsible for the mode growth. Here and further in the text the temperature is in K.

However, the conservation of momentum implies that the friction term in the electron momentum equation should read $-m_e n_e \nu_{ej}(\vec{v}_e - \vec{v}_j)$ even when conventional criteria for a negligible parallel dynamics of heavier particles are fulfilled, so that the corresponding momentum component of the heavier species includes the friction term $-m_j n_j \nu_{je}(\vec{v}_j - \vec{v}_e)$. Below, we perform derivations with such a full friction force term for some simple cases [11] in order to demonstrate the differences introduced by this friction-induced response of the heavier species.

3.1. Plasma with dominant collisions between charged particles

We stress that this case may also include a rather weakly ionized plasma with $n_0 \ll n_n$ (the index 0 here and below denotes the electron or ion equilibrium quantities). This is because of the much larger cross section for collisions between charged particles. To have dominant collisions with protons in a plasma containing electrons, protons and neutral atoms, the electron number density should satisfy the condition

$$\frac{n_0}{n_n} > \frac{3\sigma_{en}(4\pi\varepsilon_0 \kappa T_e)^2}{(8\pi)^{1/2} e^4 L_{ei}}. \tag{15}$$

Here we used the standard notation, and L_{ei} is the Coulomb logarithm. Taking as an example $T_e = 10^4$ K, which gives [12, 13] $\sigma_{en} = 2.5 \cdot 10^{-19}$ m^2 (here $L_{ei} = 6$ for $n_0 = 10^{18}$ m^{-3}), it turns out that the electron-ion collisions are more frequent than the electron-neutral collisions provided that $n_0/n_n > 0.009$. For the given temperature this is close to well known estimate [14] showing that, in terms of electron collisions, an ion is equivalent to $3.4 \cdot 10^5 (300/T_e)^2 \simeq 300$ neutral atoms.

We use the continuity equation for electrons and ions placed in an external magnetic field $\vec{B}_0 = B_0 \vec{e}_z$

$$\frac{\partial n_j}{\partial t} + \nabla_\perp(n_j \vec{v}_{\perp j}) + \nabla_z(n_j \vec{v}_{zj}) = 0, \quad j = e, i, \tag{16}$$

The linearized perpendicular velocities of electrons and ions are given by

$$\vec{v}_{\perp e1} = \frac{1}{B_0}\vec{e}_z \times \nabla_\perp \phi_1 - \frac{v_{Te}^2}{\Omega_e}\vec{e}_z \times \frac{\nabla_\perp n_{e1}}{n_{e0}}, \quad v_{Te}^2 = \kappa T_e/m_e, \tag{17}$$

$$\vec{v}_{\perp i1} = \frac{1}{B_0}\vec{e}_z \times \nabla_\perp \phi_1 - \frac{1}{\Omega_i B_0}\frac{\partial}{\partial t}\nabla_\perp \phi_1. \tag{18}$$

The electron parallel velocity is determined from

$$0 = e n_0 \frac{\partial \phi_1}{\partial z} - \kappa T_e \frac{\partial n_1}{\partial z} - m_e n_0 \nu_{ei}(v_{ez1} - v_{iz1}), \tag{19}$$

and the ion velocity from

$$\frac{\partial v_{iz1}}{\partial t} = -\frac{e}{m_i}\frac{\partial \phi_1}{\partial z} - \nu_{ie}(v_{iz1} - v_{ez1}). \tag{20}$$

In 'standard' derivations the ion parallel velocity is only due to the parallel component of the perturbed electric field, while the friction induced term in (20) is omitted with the usual excuse of the huge difference in mass between the two species. Similarly, the corresponding ion term is neglected in Eq. (19).

In this case, the perturbed ion number density is described by [3, 4]

$$\frac{n_1}{n_0} = \left(\frac{\omega_*}{\omega} + \frac{k_z^2 c_s^2}{\omega^2} - k_y^2 \rho_s^2\right) \frac{e\phi_1}{\kappa T_e}. \tag{21}$$

The electron number density is calculated from Eqs. (16), (19) yielding the dispersion equation:

$$\omega^2(1 + k_y^2\rho_s^2) - \omega_*\omega - k_z^2 c_s^2 + \frac{i\omega}{k_z^2 D_z}\left[\omega(\omega - \omega_*) - k_z^2 c_s^2\right] = 0, \tag{22}$$

$$D_z = \frac{v_{Te}^2}{\nu_{ei}}, \quad \omega_* = -\frac{k_y \kappa T_e}{eB_0}\frac{n_0'}{n_0}, \quad \rho_s = \frac{c_s}{\Omega_i}.$$

Here, we have taken $\nabla n_0 = \vec{e}_x dn_0/dx$, the perturbations are assumed to be of the form $\sim f(x)\exp(-i\omega t + ik_y y + ik_z z)$, and we work in the frame of a local approximation. The imaginary part of the frequency is

$$\omega_i \simeq \frac{\omega_r^2}{k_z^2 D_z}\frac{\omega_r^2 k_y^2 \rho_s^2 - k_z^2 c_s^2}{\omega_r^2(1 + k_y^2\rho_s^2) + k_z^2 c_s^2}. \tag{23}$$

Hence, there appears to be a threshold for the instability, that appears due to the ion sound response in the parallel direction.

In deriving (23) we have used the limit

$$|\omega| \ll k_z^2 D_z, \tag{24}$$

and assume that ω and ω_* are of the same order. Used for convenience, the condition (24) is in fact not always easily satisfied. Physically it describes the condition of isothermal electrons along the field lines that has been assumed. It can be rewritten as $(\omega/k_z)/v_{Te} \ll k_z v_{Te}/\nu_{ei}$. The right-hand side gives the ratio of the electron mean free path v_{Te}/ν_{ei} and the parallel wavelength, that in fact must be much less than unity in order to remain within a proper fluid theory (i.e., for collisions being able to maintain Maxwellian distribution).

However, a self-consistent analysis should include the full friction force effect in both, i.e., the ion and electron, parallel equations. This is simply due to the fact that the two forces are necessarily equal by magnitude. Hence, we keep Eqs. (19, 20) as they are, *with the complete friction terms.*

Combining the two parallel momentum equations, and using the conservation of momentum $m_i\nu_{ie} = m_e\nu_{ei}$, yields $v_{iz1} = k_z c_s^2 n_1/(\omega n_0)$. Instead of Eq. (21), we now have

$$\frac{n_1}{n_0}\left(1 - \frac{k_z^2 c_s^2}{\omega^2}\right) = \left(\frac{\omega_*}{\omega} - k_y^2\rho_s^2\right)\frac{e\phi_1}{\kappa T_e}. \tag{25}$$

A procedure similar as earlier, now yields the electron number density

$$\frac{n_1}{n_0} = \frac{\omega_* + ik_z^2 D_z}{\omega - k_z^2 c_s^2/\omega + ik_z^2 D_z}\frac{e\phi_1}{\kappa T_e}. \tag{26}$$

Within the same approximations as earlier (i.e., $\omega^2, \omega_* \omega, k_z^2 c_s^2 \ll k_z^4 D_z^2$), the dispersion equation that we now have is:

$$\omega^2(1+k_y^2\rho_s^2) - \omega_*\omega - k_z^2 c_s^2 + \frac{i(\omega^2 - k_z^2 c_s^2)(\omega^2 - \omega_*\omega - k_z^2 c_s^2)}{\omega k_z^2 D_z} = 0. \tag{27}$$

The real frequency is the same as in the earlier, obtained from the real part of Eq. (27). The imaginary part of the frequency now becomes

$$\omega_i \simeq \frac{\omega_r^2 k_y^2 \rho_s^2}{k_z^2 D_z} \frac{\omega_r^2 - k_z^2 c_s^2}{\omega_r^2(1+k_y^2\rho_s^2) + k_z^2 c_s^2} = \frac{\omega_r^2 k_y^2 \rho_s^2}{k_z^2 D_z} \frac{\omega_* \omega_r - \omega_r^2 k_y^2 \rho_s^2}{\omega_r^2(1+k_y^2\rho_s^2) + k_z^2 c_s^2}. \tag{28}$$

We remark the obvious difference in the instability threshold in the expression (23), and the correct expression (28).

3.2. On electron-ion momentum transfer

The huge difference in mass is a typical excuse for neglecting the ion response to the friction. However, this difference in mass may be compensated by the frequent electron collisions with ions, so that sooner or later the ions start to move in the parallel direction due to the electron drag. To get a feeling on the effects of collisions and the corresponding time scales, we may discuss the following two separate cases.

a) Assuming that the following condition is satisfied $v_{Te} \gg \omega/k_z \gg c_s$, $c_s = (\kappa T_e/m_i)^{1/2}$, the ions respond in the parallel direction only through the friction. The electron velocity V_0 is assumed nearly constant due to the parallel electric field of the wave. From (20), assuming that the ions are initially at rest, the ion velocity, normalized to V_0, becomes $1 - \exp(-\nu_{ie} t)$. Taking $n_0 = 10^{18}$ m^{-3}, $T_e = 10^4$ K, $T_i = 2 \cdot 10^3$ K, we have $\nu_{ie} = 7 \cdot 10^3$ Hz. A simple plot of the ion velocity reveals that it becomes close to 1 already after about 0.0007 seconds.

b) Taking another extreme case where electrons initially, due to any external reason acquire a velocity $v_{e0} = V_0$, without any additional force, and where $v_{i0} = 0$. In this case the electron velocity is not kept constant, the interaction of the two fluids yields the evolution of the two velocities:

$$\vec{v}_e = \frac{\nu_{ie}\vec{V}_0}{\nu_{ei}+\nu_{ie}} + \frac{\nu_{ei}\vec{V}_0}{\nu_{ei}+\nu_{ie}} \cdot \exp[-(\nu_{ei}+\nu_{ie})t], \tag{29}$$

$$\vec{v}_i = \frac{\nu_{ie}\vec{V}_0}{\nu_{ei}+\nu_{ie}} - \frac{\nu_{ie}\vec{V}_0}{\nu_{ei}+\nu_{ie}} \cdot \exp[-(\nu_{ei}+\nu_{ie})t]. \tag{30}$$

Here, the electron and ion velocities monotonously change in time towards the common velocity (the first term on the right-hand side) $v_c \simeq V_0 m_e/m_i \ll V_0$ which, for the same parameters as above, is achieved within the time interval shorter than 10^{-6} sec. The characteristic time for the velocity relaxation is $\sim 1/(\nu_{ei} + \nu_{ie})$.

A real physical situation, as in the case of the drift wave discussed earlier, is expected to be somewhere in between the two extremes presented above. Hence, in spite of a huge mass difference, the collisions (friction) will force ions to move along the magnetic field lines, and, due to the same reason, the electron velocity amplitude associated with the drift wave is expected to be considerably smaller.

3.3. Dominant collisions with neutrals

The electron parallel momentum Eq. (19) now reads

$$0 = en_0 \frac{\partial \phi_1}{\partial z} - \kappa T_e \frac{\partial n_1}{\partial z} - m_e n_0 \nu_{en} (v_{ez1} - v_{nz1}). \tag{31}$$

The ion dynamics is the same as above, so we use Eq. (21). The dynamics of neutrals is completely described by

$$\frac{\partial v_{nz1}}{\partial t} = -\nu_{ne}(v_{nz1} - v_{ez1}). \tag{32}$$

This is used in Eq. (31), with the momentum conservation condition that now reads $m_n n_n \nu_{ne} = m_e n_0 \nu_{en}$, yielding

$$v_{ez1} = \frac{iak_z v_{Te}^2}{\nu_{en}} \left(\frac{e\phi_1}{\kappa T_e} - \frac{n_1}{n_0} \right), \quad a = 1 + \frac{i\epsilon \nu_{en}}{\omega}. \tag{33}$$

Here, $\epsilon = m_e n_0/(m_n n_n)$, and this is a small quantity for any plasma. For instance, for an electron-proton plasma in a hydrogen gas $m_n = m_i = m_p$, it is of the order 10^{-6} or less. Equation (33) is used in the electron continuity yielding

$$\frac{n_1}{n_0} = \frac{\omega_* + iak_z^2 D}{\omega + iak_z^2 D}, \quad D = \frac{v_{Te}^2}{\nu_{en}}. \tag{34}$$

In the case $\omega, \omega_* \ll k_z^2 D$, Eq. (34) can be written as

$$\frac{n_1}{n_0} = \left[1 + \frac{\epsilon^2 \nu_{en}^2}{\omega^2} - \frac{\epsilon \nu_{en}}{\omega} \frac{\omega + \omega_*}{k_z^2 D} + \frac{i(\omega - \omega_*)}{k_z^2 D} \right] \left(1 + \frac{\epsilon^2 \nu_{en}^2}{\omega^2} - \frac{2\epsilon \nu_{en}}{k_z^2 D} \right)^{-1}. \tag{35}$$

We have $1 \gg 2\epsilon \nu_{en}/k_z^2 D \gg \epsilon^2 \nu_{en}^2/\omega^2$, the real part is therefore very close to unity, while the imaginary part is simply $i(\omega - \omega_*)/k_z^2 D$. Consequently, combining Eq. (35) with Eq. (21), the real part of the frequency appears described as earlier, while the imaginary part becomes

$$\omega_i \simeq \frac{\omega_r^2}{k_z^2 D} \frac{\omega_r^2 k_y^2 \rho_s^2 - k_z^2 c_s^2}{\omega_r^2 (1 + k_y^2 \rho_s^2) + k_z^2 c_s^2}. \tag{36}$$

Compared to the previously discussed $e - i$ collision case (28), it is seen that i) in both cases the threshold is caused by the ion sound response, however, ii) the instability threshold in Eq. (36) is shifted towards higher frequencies (because $k_y \rho_s$ is usually less than unity).

Although the two cases describe two physically different plasma environments, the explanation for case ii) should be as follows. The lower threshold frequency in the $e - i$ case implies that electrons experience a larger amount of collisions with ions within a wave period, which is in fact necessary to compensate for the ion movement in the parallel direction (due to the parallel electric field). This is because moving ions (in the same direction as electrons) represent a less efficient barrier for electron parallel motion and, in order to have the necessary phase shift between the density and potential, the electrons should have more collisions for the instability to take place. On the other hand, in the $e - n$ case, a higher frequency (equivalent to a smaller amount of collisions) for the instability to develop is possible because neutrals are less movable in the parallel direction (they do not react to the parallel electric field), and therefore they represent a more effective barrier.

4. Summary

The results obtained for the IA case can be summarized as follows. i) The friction does not affect the IA mode in the limit of quasi-neutral perturbations. ii) Even using the non-evolving model equivalent to Ref. [1], there is no instability of the IA mode, contrary to claims from Ref. [1]. iii) When the background plasma is properly described as evolving in time, and as long as the quasi-neutrality is used, collisions do not produce a growth of the ion acoustic mode. iv) When the Poisson equation is used instead of quasi-neutrality, in principle there is a possibility for a positive growth-rate of the IA mode. It appears as a combined effect of the breakdown of the charge neutrality from one side (introduced by the Poisson equation), and the heat transfer (the compressibility and advection in energy equation) from the other side, all within the background of a time-evolving plasma. However, as the equilibrium plasma evolves in time, with the relaxation time t, the obtained growth time must be (much) shorter than the relaxation time. Yet, this shows to be impossible and we conclude that there is no instability in the electron-ion plasma with an initial temperature disparity if the plasma evolves freely.

In the case of the drift wave, the self-consistent inclusion of the momentum conservation in the ion and electron equations, which originates from the collisions between the two fluids, yields a different instability threshold that, to the best of our knowledge, has not been discussed in the literature so far. The correct expressions (28) and (36) should be used for estimates of the growth rate and the instability threshold for the collisional drift mode.

Acknowledgments

The results presented here are obtained in the framework of the projects G.0304.07 (FWO-Vlaanderen), C 90205 (Prodex), GOA/2004/01 (K.U.Leuven), and the Interuniversity Attraction Poles Programme - Belgian State - Belgian Science Policy.

References

[1] Rognlien T D and Self S A 1971 *Phys. Rev. Lett.* **27** 792
[2] Bateman G 1972 *Phys. Rev. Lett.* **29** 1499
[3] Weiland J 2000 *Collective Modes in Inhomogeneous Plasmas* (Bristol: Institute of Physics Pub.) p 32
[4] Bellan P M 2006 *Fundamentals of Plasma Physics* (Cambridge: Cambridge University Press) pp 253-258
[5] Vranjes J and Poedts S 2006 *Phys. Lett.* A **348** 346
[6] Vranjes J, Pandey B P and Poedts S 2006 *Planet. Space Sci.* **54**, 695
[7] Braginskii S I 1965 *Reviews of Plasma Physics* vol. 1 (New York, London: Consultants Bureau) p 211
[8] Landau L 1937 *J. Exp. Theor. Phys. (U.S.S.R.)* **7** 203
[9] Spitzer L 1984 *Physics of fully ionized gasses* (New York, London: Intersience Publishers) p 135
[10] Vranjes J, Kono M, Poedts S and Tanaka M Y 2008 *Phys. Plasmas* **15** 092107
[11] Vranjes J and Poedts S 2008 *Phys. Plasmas* **15** 034504
[12] Mitchner M and Kruger C H 1973 *Partially Ionized Gasses* (New York: John Willey and Sons) p 102
[13] Zecca A, Karwasz G P and Brusa R S 1996 *Riv. Nuovo Cim.* **19** 128
[14] Ratcliffe J A 1959 *The Magneto-Ionic Theory and its Applications to the Ionosphere* (Cambridge: Cambridge University Press) p 33.

Source Apportionment of Atmospheric Bulk Deposition in the Belgrade Urban Area Using Positive Matrix Factorization

M. Tasić[1], Z. Mijić[1], S. Rajšić[1], A. Stojić[1], M. Radenković[2], J. Joksić[2]
[1]Institute of Physics, Pregrevica 118, 10080 Belgrade, Serbia

[2]Institute of Nuclear science Vinča, Belgrade, Serbia

E-mail: tasic@phy.bg.ac.yu

Abstract. The primary objective of the present study was to assess anthropogenic impacts of heavy metals to the environment by determination of total atmospheric deposition of heavy metals. Atmospheric depositions (wet + dry) were collected monthly, from June 2002 to December 2006, at three urban locations in Belgrade, using bulk deposition samplers. Concentrations of Fe, Al, Pb, Zn, Cu, Ni, Mn, Cr, V, As and Cd were analyzed using atomic absorption spectrometry. Based upon these results, the study attempted to examine elemental associations in atmospheric deposition and to elucidate the potential sources of heavy metal contaminants in the region by the use of multivariate receptor model Positive Matrix Factorization (PMF).

1. Introduction

Clean air is considered to be a basic requirement for human health and well being. Various chemicals are emitted into the air from both, natural and anthropogenic sources. In spite of the introduction of cleaner technologies in industry, energy production and transport, air pollution remains a major health risk and tighter emission controls are being enforced by many governments. The studies of the transport and mobilization of trace metals up to now have attracted attention of many researchers [1,2,3]. Trace metals are persistent and widely dispersed in the environment and interacting with different natural components result in toxic effects on the biosphere.

The urban atmosphere is subjected to large inputs of anthropogenic contaminants produced by both stationary (power plants, industries and residential heating) and diffuse sources (road traffic) [4,5]. According to their different physical and chemical properties (vapor pressure, Henry's law constant), the size and composition of the source particles, these pollutants are partitioned between particulate and vapor phases and are subsequently transported to the Earth's surface through dry and wet deposition [6].

Most trace elements in terrestrial ecosystems originate from atmospheric wet and dry deposition, which are recognized as important processes determining pollutant dynamics in the atmosphere. The magnitude of atmospheric deposition and the ratio of wet to dry deposition is controlled among others by emission sources, distance to emission sources and the sampling site [7] and meteorological conditions, e.g. prevailing wind directions, and type, frequency and amount of precipitation. Because either mode, wet or dry, may contribute equally to or predominate in total deposition, it appears essential to include both wet and dry deposition to estimate loading from atmospheric deposition. From a biogeochemical perspective, the characterization of total atmospheric deposition is relevant in order to identify the variability and sources of the atmospheric pollutants [8]. Direct collection of atmospheric deposition using bulk sampling devices offers a practical approach to monitor atmospheric heavy metal deposition providing valuable information on the influences of atmospheric inputs of heavy metals on the surface environment [9,10]. During the past decades, wet and dry deposition of metals has been widely studied [10,11,12,13,14] and some authors have studied "bulk" (dry + wet) deposition [15,16,17,18].

Identification of air pollution source characteristics is an important step in the development of air quality strategies. Receptor modeling, using measurements of pollutant concentrations at one or more sample sites, is often a reliable way to provide information regarding source regions of pollution. Positive Matrix Factorization (PMF) modeling as source identification tool for atmospheric pollutants was already successfully tested in air quality studies. The objective of this study was to examine the use of PMF on elemental bulk atmospheric depositions data set for source apportionment purpose. Bulk atmospheric monthly deposits were sampled at three sites within the Belgrade urban area from June 2002 to December 2006 and analyzed for major (Fe and Al) and minor elements (Pb, Zn, Cu, Ni, Mn, Cr, V, As and Cd) using atomic absorption spectrometry. Seasonal trends were analyzed together with the spatial distribution and an assessment of the level of the trace element pollution was carried on.

2. Methodology

2.1. Studied sites and sample collection

The present study was performed in Belgrade (latitude N $44^0$49`14``, longitude E $20^0$27`44``), the capital of Serbia, situated at average height above sea level of 116.75 m, and on the junction of two rivers - Sava and Danube. It has a population of about 2 million in an area of about 3,222 km^2. The climate of Belgrade is moderate continental with fairly cold winters and warm summers. Mean annual air temperature is 11.7^0 C, mean annual rainfall of 669.5 l m^{-2} and mean atmospheric pressure of 1,001 millibars. The prevailing wind is N-NW, but a characteristic wind "Košava" (SE-ESE) blows with an annual frequency of 26% and an average speed of 4 m s^{-1}. In the winter, severe air pollution in the form of aerosol smog occurs frequently in the urban area of Belgrade, particularly during meteorologically calm (wind speed < 2 m s^{-1}) and stable conditions. Heavy automobile traffic is characteristic for Belgrade urban area. The average age of passenger cars is more than 15 years, which means that leaded gasoline (0.4 g l^{-1} Pb) is still widely used. There are 18 large heating plants with a total capacity of 2018 MW, run on natural gas or crude oil and 59 smaller plants run only on crude oil (approximately 193 MW). Fuel used for domestic heating consists mainly of coal or crude oil.

The total atmospheric deposition (TD) collection was performed using an open polyethylene cylinder (29 cm inner diameter and 40 cm height) fitted on a stand at about 2 m above the ground. The devices collected both dry deposition and precipitation continuously for one month periods from June 2002 to December 2006 at three sites in the urban part of Belgrade: a) on the roof of the Rector's Office building of Belgrade University in Student Square (RB), at a height of about 20 m, near a small city-park; b) Botanic Garden (BG), about 50 m far from heavy-traffic streets and c) the platform above the entrance steps to the Faculty of Veterinary Medicine (FVM) at a height of about 4 m from the ground, 5 m away from a street with heavy traffic and close to the big junction with the main state highway. The collection bottles were filled before each sampling period with 20 ml of 10% acidified (HNO_3 65% (Suprapure, Merck) ultra pure water. Precautions were taken to avoid contamination of samples in both the field and laboratory. Details on studies sites and sampling procedures are given elsewhere [19,20].

2.2. Analytical procedures

Total atmospheric deposition samples were evaporated to dryness, digested with 50 ml 0.1 N HNO_3 on ultrasonic bath, the digested solution was filtered using 0.45 μm porosity Sartorius membranes and analyzed using a flame and graphite furnace atomic absorption spectrometer. Laboratory blanks were analyzed in the same manner as field samples and the heavy metal concentration values were below the detection limit values for all analyzed metals.

The elemental composition (Al, V, Cr, Mn, Fe, Ni, Cu, Zn, As, Cd, and Pb) of total atmospheric deposition, was measured by the atomic absorption spectroscopy method (AAS). Depending on concentration levels, samples were analyzed for a set of elements by flame (FAAS) (Perkin Elmer AA 200) and graphite furnace atomic absorption spectrometry (GFAAS) using the transversely-heated graphite atomizer (THGA; Perkin Elmer AA 600) with Zeeman-effect background correction. The THGA provided a uniform temperature distribution over the entire tube length, rapid heating and an integrated L'vov platform, which gave an improved signal/interference ratio and high analytical sensitivity. Analyte injection (20 μl) and the atomization were done in five steps controlled by the appropriate software and auto-sampler.

For calibration, standard solutions containing all metals of interest were prepared using Merck certified atomic absorption stock standard solutions containing 1000 mg l^{-1} metal in 0.5 N HNO_3 and Milli-Q quality deionized water, with no matrix modifier addition. Detection limits for the trace elements were found to be: 0.04 ng ml^{-1} for Cd, 0.1 ng ml^{-1} for Cr, 0.2 ng ml^{-1} for Cu, 0.5 ng ml^{-1} for Pb, 2 ng ml^{-1} for Zn, 0.4 ng ml^{-1} for Ni, 0.2 ng ml^{-1} for Mn, 5 ng ml^{-1} for V, 0.5 ng ml^{-1} for Fe, 2 ng ml^{-1} for Al, and 0.5 ng ml^{-1} for As. Standard practices for the handling of trace metal samples were implemented. For quality assurance, NIST 2783 standard reference material was used. Altogether 140 samples were analyzed.

Data treatment included the calculation of Al, V, Cr, Mn, Fe, Ni, Cu, Zn, As, Cd, and Pb monthly deposition fluxes.

2.3. Source apportionment using PMF

Positive Matrix Factorization (PMF) has been shown to be a powerful receptor modeling tool and has been commonly applied to particulate matter data [21,22,23] and recently to VOC data [24,25,26]. To ensure that receptor modeling tools are available for use in the development and implementation of air quality standards, the United States Environmental Protection Agency's Office of Research and Development has developed a version of PMF with the name of EPA PMF1.1 that is freely available [27].

PMF solves the general receptor modeling equation using a constrained, weighted, least-squares approach [28,29,30]. The general model assumes there are p sources, source types or source regions (termed factors) impacting a receptor, and linear combinations of the impacts from the p factors give rise to the observed concentrations of the various species.

The model can be written as

$$x_{ij} = \sum_{k=1}^{p} g_{ik} f_{kj} + e_{ij} \tag{1}$$

where x_{ij} is the concentration at the receptor for the jth species on the ith sample, g_{ik} is the contribution of the kth factor to the receptor on the ith sample, f_{kj} is the fraction of k factor that is species j or chemical composition profile of factor k, and e_{ij} is the residual for the jth species on the ith sample. The objective of PMF is to minimize the sum of the squares of the residuals weighted inversely with error estimates of the data points. Furthermore, PMF constrains all of the elements of G and F to be non-negative. The task of PMF analysis can thus be described as to minimize Q, which is defined as

$$Q = \sum_{i=1}^{n}\sum_{j=1}^{p}\left(\frac{x_{ij} - \sum_{k=1}^{p} g_{ik} f_{kj}}{s_{ij}}\right)^2 \qquad (2)$$

where s_{ij} is uncertainty of the jth species measured in i^{th} sample.

In this study the robust mode has been used for analyzing element concentrations in bulk atmospheric deposition data set. The robust mode was selected to handle outlier values (that is any data that significantly deviates from the distribution of the other data in the data matrix) meaning that outliers are not allowed to overly influence the fitting of the contributions and profiles. This can be achieved by a technique of iterative reweighing of the individual data values, thus, the least-squares formulation becomes to

$$Q = \sum_{i=1}^{n}\sum_{j=1}^{m}\left(\frac{e_{ij}}{h_{ij} s_{ij}}\right)^2 \qquad (3)$$

where

$$h_{ij}^2 = \begin{cases} 1 & \text{if } \left|e_{ij} / s_{ij}\right| \le \alpha, \\ \left|e_{ij} / s_{ij}\right| / \alpha & \text{otherwise,} \end{cases}$$

The parameter α is called the outlier threshold distance and the value $\alpha = 4$ was used in this analysis. One of the most important advantages of PMF is the ability to handle missing and below detection limit data by adjusting the corresponding error estimates. In this analysis missing values were replaced with the geometrical mean of the measured concentrations for each chemical species, and large error estimates were used for them.

3. Results and discussion

3.1. Elemental atmospheric total deposition rates

A total of 140 atmospheric deposits were collected monthly from June 2002 to December 2006 in three sites in the urban area of Belgrade, RB, BG and FVM, and trace and other metal (Al, V, Cr, Mn, Fe, Ni, Cu, Zn, As, Cd, and Pb) monthly fluxes were calculated. For ease of comparison, seasonal and annual atmospheric heavy metal deposition rates were calculated based on the produced data. To calculate the daily atmospheric depositions of heavy metals per unit area (μg m^{-2} day^{-1}) in each season, total mass of trace metal was divided by the number of sampling days, exposed to the ambient environment and cross sectional sampling area of the container.

Summary statistics of the average daily elemental atmospheric depositions in the Belgrade urban area, for all sampling sites, for the period of 2002 to 2006 is presented in Table 1.

Table 1. Descriptive statistics of average daily elemental atmospheric depositions (μg m^{-2} day^{-1}) for all sampling sites from June 2002 to December 2006

Elements	Fe	Cd	Pb	Cu	Ni	Zn	Cr	Mn	Al	V	As
N	140	140	140	140	140	140	140	140	140	140	93
Average	1627.3	0.6	59.4	94.5	31.0	113.5	4.5	71.8	901.1	54.1	9.7
S.D.	1245.7	0.5	39.1	82.9	73.8	91.4	5.2	55.0	600.4	65.4	22.5
Max	6113.6	3.6	229.8	423.0	598.2	615.4	30.0	351.2	3722.9	340.7	154.8
Min	42.8	< d.l	2.4	10.9	0.7	16.1	< d.l	3.4	42.5	0.4	0.2
98th Percentile	5322.3	1.8	148.4	316.4	186.2	347.8	19.0	228.7	2270.6	272.0	81.9

N – Number of samples; S. D. – Standard arithmetic deviation; d.l. – Detection limit

The most abundant trace metal in bulk depositions was Zn, followed in order of abundance by V, Mn, Ni, Cu, Pb, As, Cr and Cd. This order is similar to those already obtained for bulk deposition (wet and dry) in some other studies [9,10,15]. A large variation in the monthly deposition rates of all the elements was observed, which explains the high standard deviations from the mean deposition rates. This is ascribed mainly to weather conditions that strongly affect the dispersion and deposition of air pollution, as well as to the variation of emission sources over the seasons.

3.2. Spatial variations

Measured bulk atmospheric deposition rates, TD (μg m^{-2} day^{-1}) of analyzed elements at three sampling sites are presented in Table 2. In general, deposition of some heavy metals showed large variations between different sampling locations and this variation may be ascribed to variations in traffic density, presence of industries and different land use properties. As expected, the TD values were highest at FVM site for almost all metals except Cu. High deposition of Cu in samples at Rector' Office building (RB) comparing to other locations could point to the presence of point sources, which are clearly site specific (foundry nearby) and could be related to roof covering nearby as well. High total deposition values (μg m^{-2} day^{-1}) of Pb, Ni, V and Zn in samples at FVM have been expected due to its traffic oriented point.

The spatial distribution of Al and Fe shows that the concentrations of Fe are highest at FVM site. Values of the Fe/Al ratio range from 1.41 to 1.62, from 1.33 to 1.69 and from 1.91 to 2.28 at RB, BB and FVM, respectively. Fe/Al ratios in the Earth's crust are variable, depending on chemical enrichments and different ratios for the upper continental crust are reported such as 0.40 [31] and 0.85 [32]. The Fe/Al ratios measured at RB, BB and FVM are higher than Fe/Al ratios of pure crustal sources, indicating that both elements contained in the acid-soluble fraction of particles coming from atmospheric deposition have an anthropogenic source. At the BB site, the influence of crustal particles containing Al and Fe is enhanced as the ratios obtained at this location are closer to the Earth's crustal ratio.

Table 2. Measured bulk atmospheric deposition rates, TD (μg m^{-2} day^{-1}) of analyzed elements at sampling sites for the period June 2002 to December 2006

	Rector's Office building			Botanic Garden			Faculty of Veterinary Medicine		
Elements	Average	Max	Min	Average	Max	Min	Average	Max	Min
Fe	776.2	1451.3	42.8	1292.1	3736.5	275.5	2753.4	6113.6	541.8
Cd	0.4	2.5	<d.l	0.4	2.0	<d.l	0.8	3.6	<d.l
Pb	39.3	87.5	4.8	48.5	140.4	2.4	90.1	229.8	6.4
Cu	166.5	423.0	10.9	43.2	94.5	18.5	75.3	131.6	11.9
Ni	11.0	49.2	0.7	15.7	79.6	1.3	65.8	598.2	1.6
Zn	60.8	119.9	16.1	122.9	615.4	38.4	155.5	345.5	32.0
Cr	1.5	6.4	0.1	4.3	14.8	<d.l	7.5	30.0	1.4
Mn	31.1	56.7	16.8	72.1	351.2	22.4	111.2	264.3	3.4
Al	494.7	1410.5	42.5	869.1	2051.0	106.9	1330.9	3722.9	477.5
V	25.7	107.2	0.4	30.7	191.0	0.8	105.3	340.7	20.0
As	3.7	75.2	0.2	4.4	43.9	0.5	20.4	154.8	1.2

d.l. – Detection limit

3.3. Seasonal variation

Seasonal variations of elemental atmospheric bulk depositions have been analysed. In general, a seasonal variation is likely associated with the nature of the sites and seasonal trend is difficult to recognise because of the number of variables that influence removal from the atmosphere during each sampling period (e.g., the number and intensity of rainfall events, hours of dry weather, wind direction, and other meteorological parameters such as mixing depths). The analysis of seasonal variation of elemental TD in Belgrade atmosphere has shown the pronounced seasonal dependence for V, Ni (Fig.1) and As, less pronounced but significant for Fe, while seasonal variations of the other elements were not evident. The higher depositions of V, Ni and As in winter periods are related to the combustion of fossil fuels (fuel oil, coal) as the urban site of Belgrade is highly influenced by domestic heating and automobile traffic.

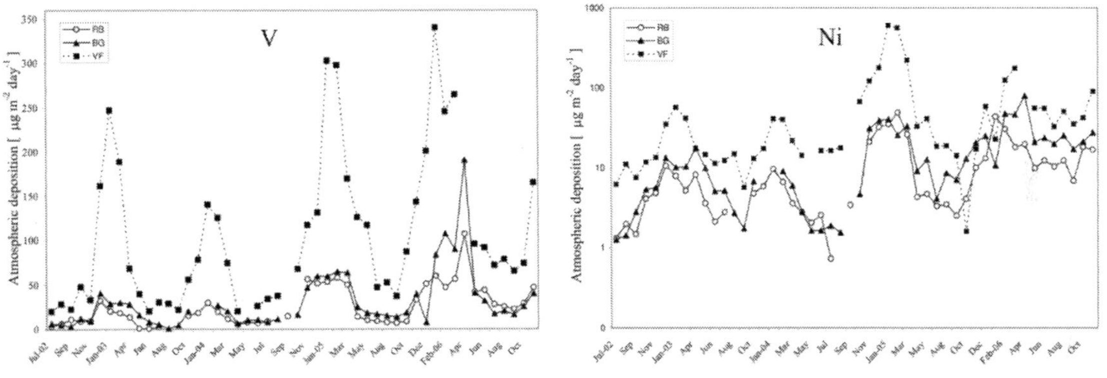

Figure 1. Seasonal variation in total atmospheric depositions of V and Ni in Belgrade urban area at Rector's Office building (RB), Botanic Garden (BG) and Faculty of Veterinary Medicine (FVM) for the period 2002 - 2006

Annual atmospheric deposition of heavy metals (Cd, Cu, Ni, Pb and Zn) in units of mg m^{-2} yr^{-1} in this study is compared with the data from other studies (Table 3).

Table 3. Annual atmospheric deposition fluxes (mg m^{-2} yr^{-1}) in different regions

Location	Period	Cd	Cu	Ni	Pb	Zn	References
Belgrade	2002-2006	0.20	34.49	11.30	21.69	41.42	This study
Paris	2001–2002	0.24	6.0	0.62	4.2	30	Motelay-Massei *et al* (2005)
Pearl River Delta, China	2001–2002		18.6		12.7	104	Wong *et al.* (2003)
Massachusetts Bay	1992–1993	0.27	2.50	1.50	1.80	7.80	Golomb *et al.* (1997)
Lake Superior	1993–1994	0.46	3.10	0.80	1.50	8.80	Sweet *et al.* (1998)
Lake Michigan	1993–1994	0.45	1.90	0.61	1.60	6.00	Sweet *et al.* (1998)
Fiorland, New Zealand	1993–1995	0.004	0.023		0.035	0.025	Halstead *et al.* (2000)
North Sea	1993–1994		1.24		3.52	6.50	Injuk *et al.* (1998)
Irish Sea	1993–1994		2.60		1.62		Williams *et al.* (1998)
Mediterranean Coast	1988–1993	0.31	2.60	0.57	3.80		Guieu *et al.* (1997)
Ligurian Sea	1997–1998	0.06	1.28	1.10	1.20	41.2	Sandroni and Migon (2002)

Journal of Physics: Conference Series **162** (2009) 012018 doi:10.1088/1742-6596/162/1/012018

Annual bulk deposition fluxes of Cu, Ni, and Pb exceeded the corresponding reported data from other sites. The largest contributor to the copper loading might be the specific local source in the vicinity of Rector`ş Office building, building siding and brake emissions from automobiles. Confirmation of the presence of nickel in elevated concentrations, in a great number of soil samples in Belgrade [40], at certain locations, points to its geographical origin followed by the emission of heavy fuel combustion. High Pb values are the consequence of the still wide use of leaded gasoline.

3.4. PMF modeling

A critical step in PMF analysis is determination of the number of factors. To determine the number of sources, it is reasonable to test different numbers of sources and use the one that both adequately fits the data and provides the most physically meaningful results.

The PMF analysis of the TD data set resulted in five factors for Rectorat`s Office building and Botanic Garden sites and six for Faculty of Veterinary Medicine site and they have been identified as possible sources at each site. Figs. 2, 3 and 4 present the identified source profiles and time series plots of estimated monthly contributions for bulk atmospheric depositions at Rector`s Office building, Botanic Garden and Faculty of Veterinary Medicine, respectively.

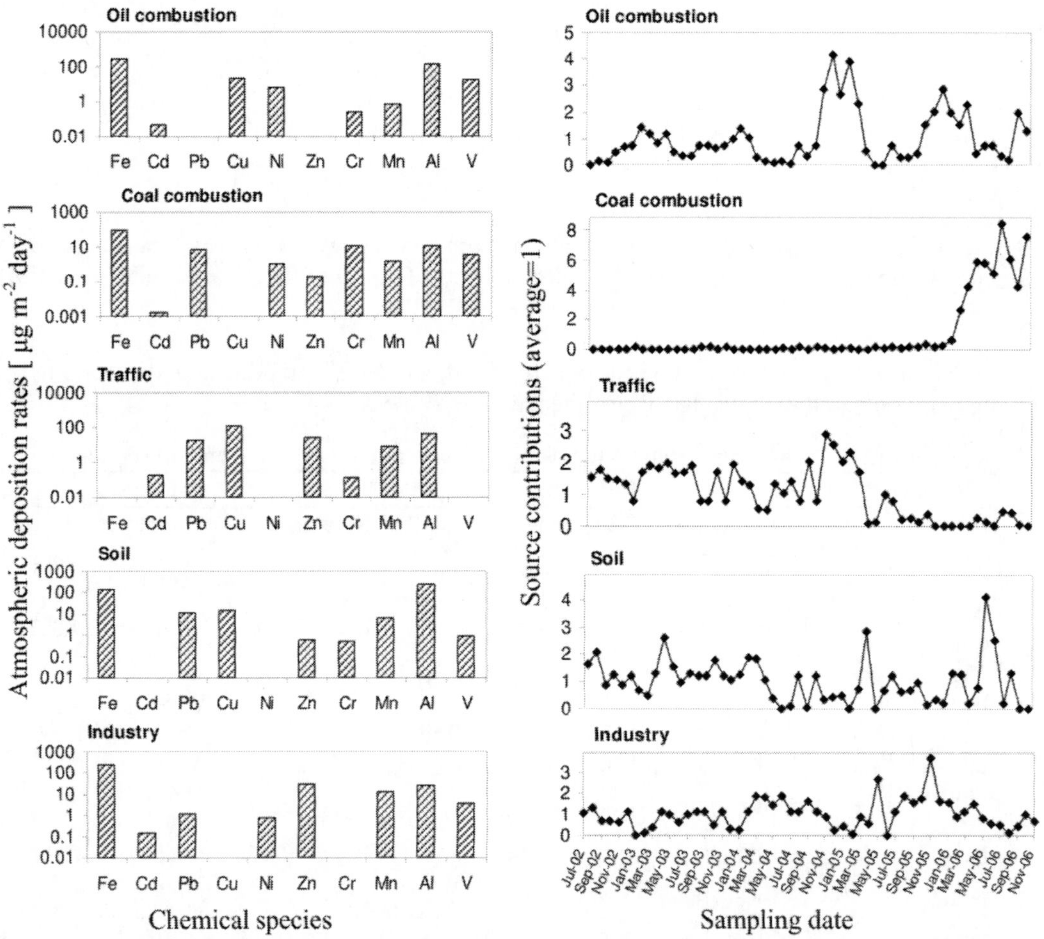

Figure 2. Source profiles and time series plot of source contributions resolved from the Rector's Office building

Among the resolved factors, five factors are common to RB and BB sites. They are traffic, soil, coal combustion, oil combustion, and industry.

Traffic factor at RB site has significant loadings of Pb, Cd, Mn, Fe, Zn, and Cu. Lead, Zn, and Cu are indicator elements of traffic emission; Cu, Fe and Zn are present in resuspended brake wear particles; tires are the major Zn source from road traffic; the high Mn concentrations, specially at FVM are related to motor vehicles that burn gasoline with the Mn additive. Fe and Mn are crustal elements, which may have been present in dust resuspended by traffic; Fe is related to heavy-duty diesel emissions. This factor related to traffic has median contribution of 27%, 21% and 26% for RB, BG and FVM sites respectively.

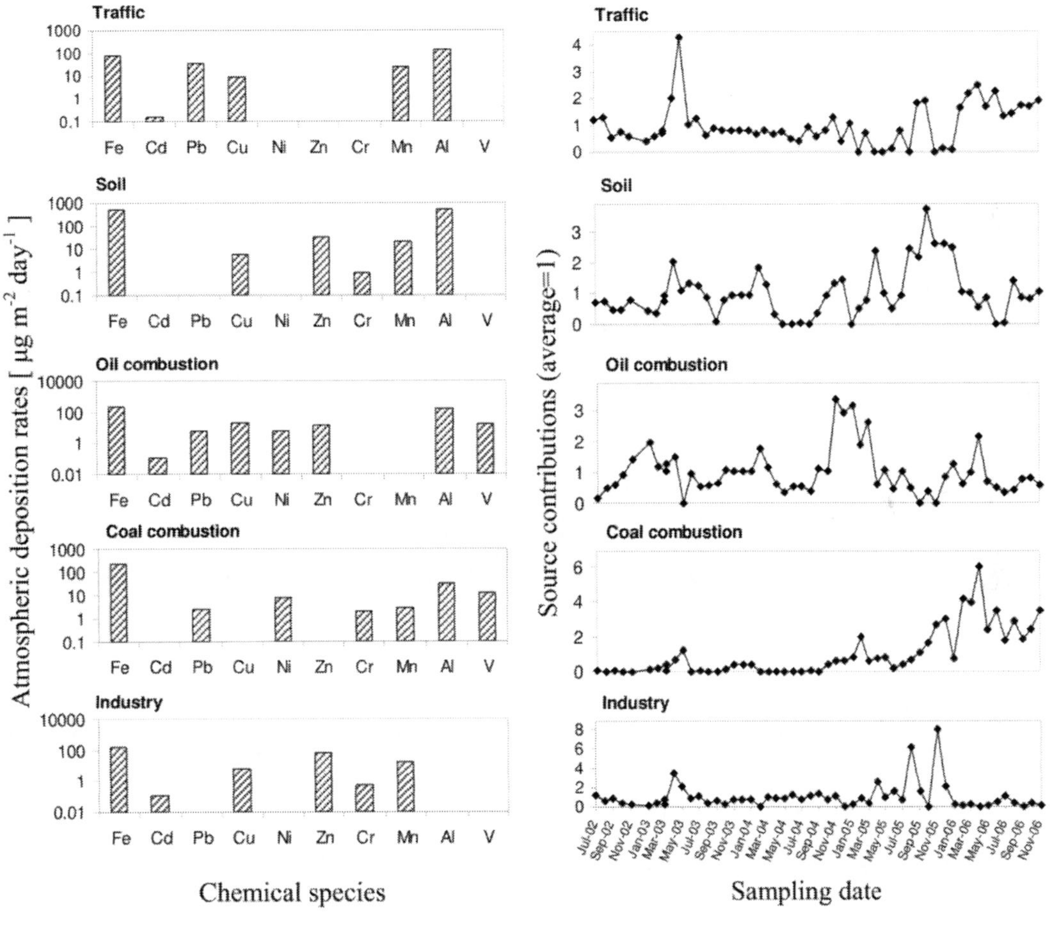

Figure 3. Source profiles and time series plot of source contributions resolved from the Botanic Garden

The soil factor (crustal) includes high loadings of the typical crustal elements such as Fe, Al, and Mn. This factor could represent local soil strongly contaminated with Zn, Cr and Pb. This factor does not show a regular season pattern, but it has a few high values at apparently random times. Soil component is fairly consistent among the RB and BB sites. This soil factor contributes with 14% and 21% of the total data set for RB and BG respectively.

The oil combustion source shows the high loadings of Ni and V, which are the fingerprint elements for fuel oil burning. This source most probably reflects urban region where residual oils are common fuels for utility and industrial sources. This factor associated to heavy oil burning has median contribution of 24%, 25% and 14% for RB, BG and FVM respectively.

The factor having high loadings of Cr and weak loadings of V, Ni and Fe can be attributed to coal combustion followed by emission from ferrous factory or other metal industry. Time series plot of this factor have the same pattern for RB and BB sites showing the significant increase in 2006 year. The time series for FVM indicate the stronger influence of coal combustion from residential heating units in surroundings compared to the contribution from industry, as the seasonal variation trend of the factor was observed.

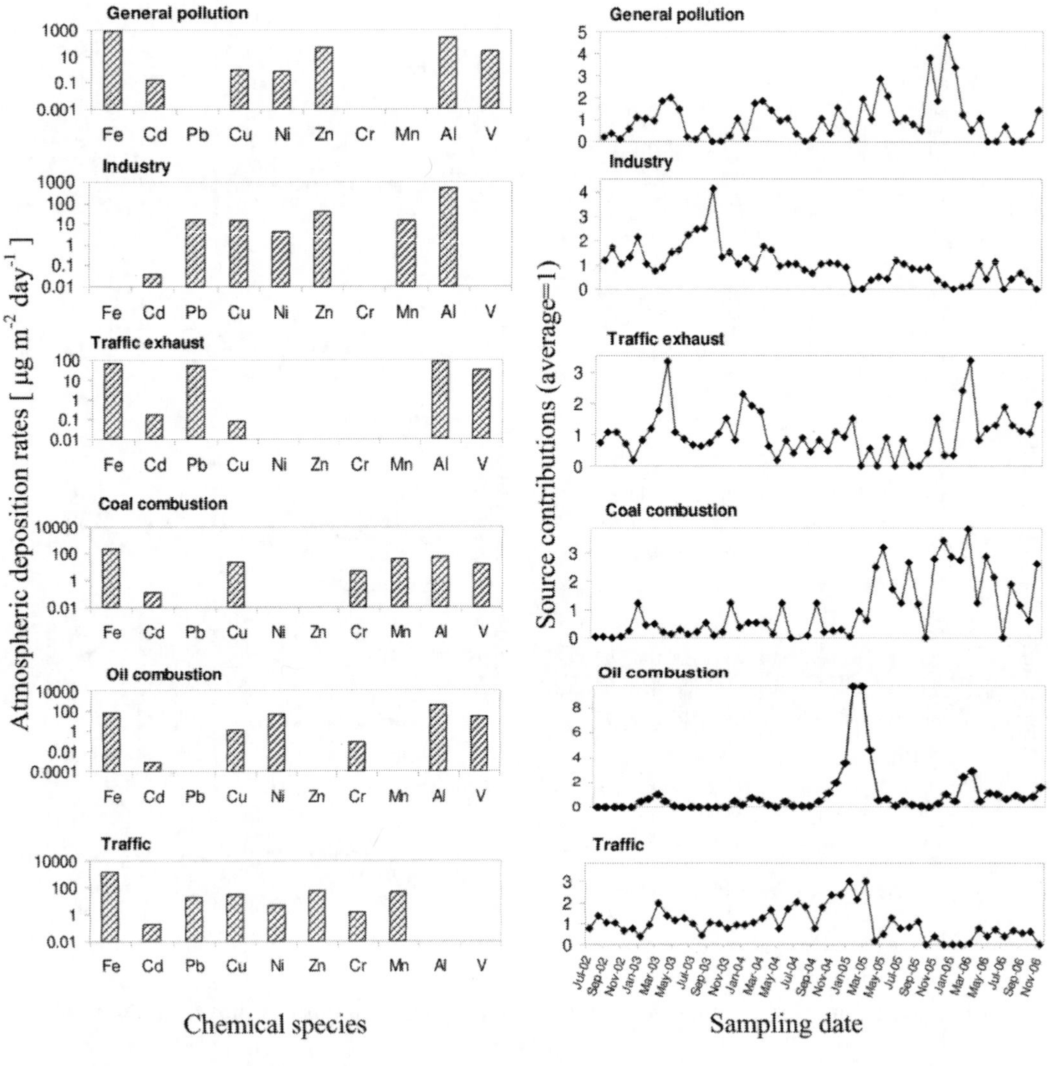

Figure 4. Source profiles and time series plot of source contributions resolved from the Faculty of Veterinary Medicine

The factor identified as industry source includes Zn, Cd, Mn, Al, Cu, and Fe, the elements associated with industrial emissions. Zink Fe, and Mn are elements characteristic of iron manufacturing; Cu and Zn may be attributed to non ferrous metallurgy; Zn comes from many pollution sources, including incinerators and smelters.

Journal of Physics: Conference Series **162** (2009) 012018 doi:10.1088/1742-6596/162/1/012018

The third factor resolved from the TD data at Faculty of Veterinary Medicine is attributed as traffic exhaust source mostly loaded with Pb, V and Cd. The factor most loaded with V, Zn, and in smaller amounts with Fe, Cd and Al was resolved at Faculty of Veterinary Medicine site and was difficult to identify. As given combination of elements might have the origin from different sources, this factor apparently is a mixed pollution signal and we named it as a general pollution factor. Median factor contributions at sampling sites, discussed previously in text, are presented in Fig.5.

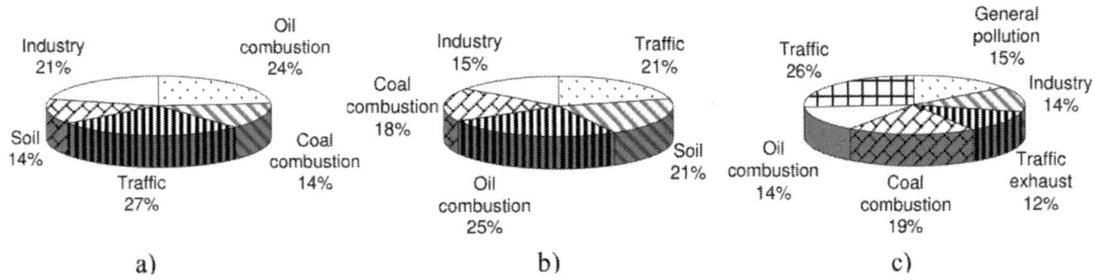

Figure 5. Median PMF factor contributions at: a) RB, b) BG, c) FVM

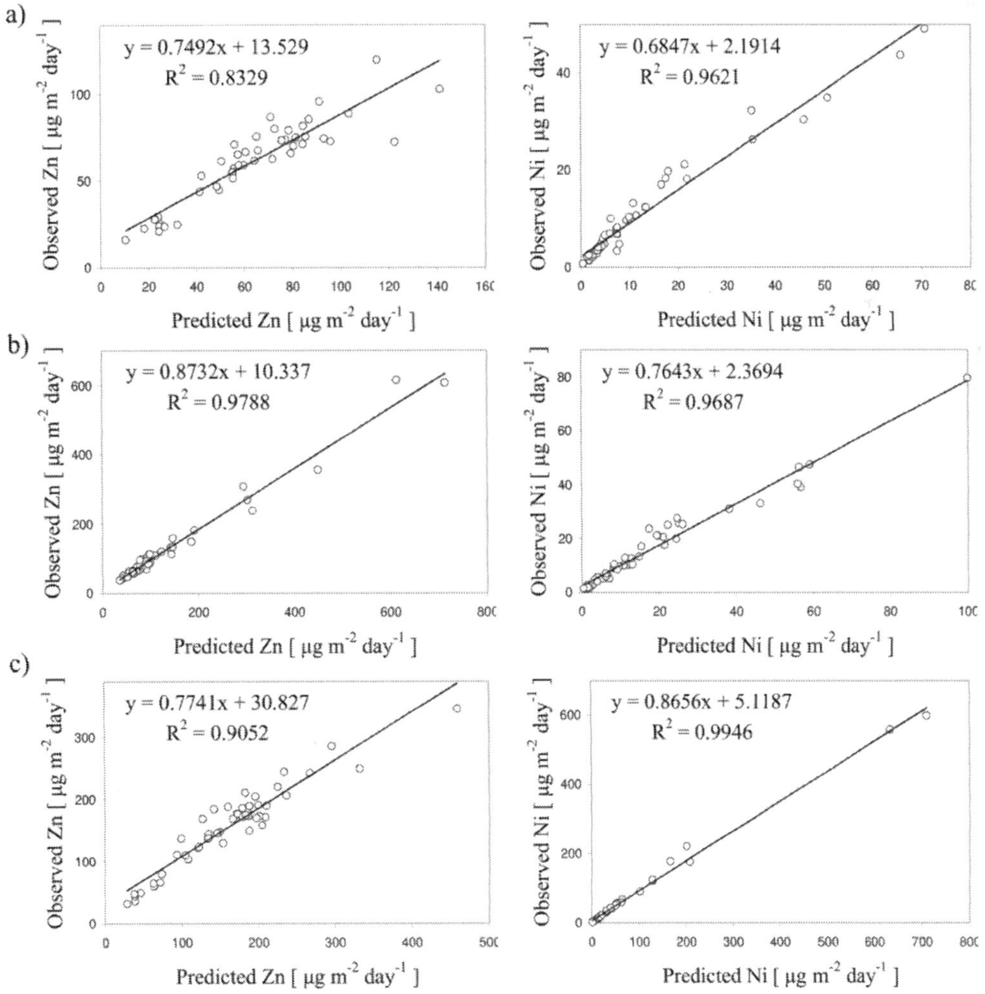

Figure 6. Observed versus predicted atmospheric depositions of Zn and Ni a) RB, b) BG, c) FVM

A comparison of the predicted bulk atmospheric depositions of Zn and Ni from all sources with atmospheric depositions for Zn and Ni measured at RB, BG, and FVM is shown in Fig. 6. The squared correlation coefficients higher than 0.90 (except for Zn at RB) indicates that the resolved sources effectively account for most of the variation in the atmospheric deposition.

4. Conclusion

Elemental bulk atmospheric deposition data obtained at three sampling sites in the Belgrade urban area in the period from June 2002 to December 2006 were subjected to multivariate receptor model Positive Matrix Factorization (PMF). PMF was used as a means of source identification and apportionment and suggested possible anthropogenic sources which explane different metal deposition patterns. The PMF method uses the estimates of the error in the data to provide optimal point-by-point weighting and permits efficient treatment of missing and below detection limit values. It also imposes the non-negativity constraint on the factors.

Five factors have been resolved at Rector`s Office building (RB) and Botanic Garden (BG) sites and six factors at Faculty of Veterinary Medicine site (FVM). Among the sources resolved at the three sites, four were common: traffic, oil combustion, coal combustion and industry. Soil factor represented by typical crustal elements as Fe, Al and Mn, followed by small amounts of Zn, Cr and Pb, was present only at RB and BG sites while at FVM pollution origin from mixed sources was identified as a general pollution factor with contribution of 15%. Traffic factor at RB and BG have almost same chemical composition, Cu, Cd, Pb, Zn and Mn, while their contribution was much smaller compared to FVM site, traffic oriented site with two resolved factors related to vehicle emission with total contribution of 38%. Oil combustion source characterized by Ni and V as typical markers, exhibit not only similar chemical compositions, but also similar seasonal variations at RB and BG. Coal combustion factor with high loading of Cr and weak loadings of V, Ni and Fe have the same pattern at RB and BG sites while the contribution from residential heating units was more pronounced as the seasonal variation trend of this factor was observed at FVM. The factor identified as industry source includes Zn, Cd, Mn, Al, Cu, and Fe. Zink, Fe, and Mn are elements characteristic of iron manufacturing; Cu and Zn may be attributed to non ferrous metallurgy; Zn comes from many pollution sources, including incinerators and smelters.

PMF modelling applied to element total atmospheric deposition data from Belgrade urban area indicated that the atmospheric deposition is mostly under the influence of anthropogenic activities.

Acknowledgment

This work was carried out within the framework of the project No 141012 funded by the Ministry of Science and Technological Development of the Republic of Serbia.

References

[1] Nriagy J.O and Pacyna J.M. 1988 *Nature* **333** 134
[2] Pacyna J.M., Martonova A., Cornille P., Maenhaut W. 1989 *Atmos. Environ.* **19** 2109
[3] Alcamo J., Bartnicki J., Olenrzynski K., Pacyna J. 1992 *Atmos. Environ.* **27A** 3355
[4] Bilos C., Colombo J.C., Skorupka C.N., Rodriguez Presa M.J. 2001 *Environ.Pollut.* **111** 149
[5] Schauer J. J., Rogge W. F., Hildemann L. M., Mazurek M. A., Cass G. R., Simoneit B. R. T. 1996 *Atmos. Environ.* **30** 837
[6] Lawlor A. J. and Tipping E. 2003 *Environ. Pollut.* **121** 153
[7] Tasić D. M., Rajšić F. S., Novaković T.V., Tomašević N.M. 2001 *Fresenius Environ. Bull.* **10** 305
[8] Azimi S., Rocher V., Muller M., Moilleron R., Thevenot S. 2005 *Sci. Total Environ.* **337** 223
[9] Morselli L., Olivieri P., Brusori B., Passarini F. 2003 *Environ. Pollut.* **124 (3)** 457-469
[10] Azimi S., Ludwig A., Thevenot D.R., Colin J-L. 2003 *Sci. Total Environ.* **308** 247-254
[11] Jickells T. 1995 *Mar. Chem.* **48** 199

[12] Guieu C., Chester R., Nimmo M., Martin J. M., Guerzoni S., Nicolas E., Mateu J.,Keyse S. 1997 *Deep Sea Research Part II: Topical Studies in Oceanography* **44**, 655
[13] Chekushin V. A., Bogatyrev I. V., de Caritat P., Niskavaara H., Reimann C. 1998 *Sci. Total Environ.* **220** 95
[14] Maneux E., Grousset F. E., Buat-M´enard P., Lavaux G., Rimmelin P., Lapaquellerie Y., 1999 *Water Air Soil Pollut.* **114** 95
[15] Golomb D., Ryan D., Eby N., Underhill J., Zemba S., 1997 *Atmos. Environ.* **31** 1349–1359.
[16] Lawlor A. J. and Tipping E. 2003 *Environ. Pollut.* **121** 153
[17] Wong C. S. C., Li X. D., Zhang G., Qi S. H., Peng X. Z., 2003 *Atmos. Environ.* **37** 767
[18] Aničić M., Tasić M., Frontasyeva M.V., Tomašević M., Rajšić S., Mijić Z., Popović A., 2008 *Environ. Pollut.* doi:10.1016/j.envpol.2008.08.003
[19] Rajšić S., Mijić Z., Tasić M., Radenković M., Joksić J., 2008 *Environ. Chem. Lett.* **6** 95
[20] Tasić M., Rajšić S., Tomašević M., Mijić Z., Aničić M., Novaković V., Marković M.D., Marković A.D., Lazić L., Radenković M., Joksić J., Assessment of Air Quality in an Urban Area of Belgrade, Serbia, In: Environmental Technologies, New Developments, Edited by E. Burcu Ozkaraova Gungor, I-Tech Education and Publishing, Vienna, Austria, www.i-techonline.com 2008, 209
[21] Song X., Polissar A.V., Hopke P.K, 2001 *Atmos. Environ.* **35**, 5277
[22] Pollisar, A.V, Hopke, P.K., Poirot, R.L., 2001 *Environ. Sci. Technol.* **35** 4604
[23] Chueinta W., Hopke P.K, Paatero P. 2000 *Atmos. Environ.* **34** 3319
[24] Elbir T., Cetin B., Cetin E., Bayram A., Odabasi M., 2007 *Environ. Monit. Assess.* **133** 149
[25] Song Y., Dai W., Shao M., Liu Y, Lu S., Kuster W., Goldan P., *Environ. Pollut.* 2008 doi:10.1016/j.envpol.2007.12.014
[26] Duan J., Tan J., Yang L., Wu S., Hao J., 2008 *Atmos. Res.* **88** 25
[27] Eberly, S. 2005 EPA PMF 1.1 user guide. Research Triangle Park, NC: USEPA National Exposure Research Laboratory.
[28] Paatero P., Tapper U. 1993 *Chem.Intell. Lab. Syst.* **18** 183
[29] Paatero P., Tapper U. 1994 *Environmetrics* **5** 111
[30] Paatero P., 1997 *Chem.Intell. Lab. Syst.* **37**, 23
[31] Wedepohl K.H.1995 *Geochim. Cosmochim. Acta* **59** 1217
[32] Taylor S.R. and McLennan S.M. 1995 *Rev. Geophys.* **33** 41
[33] Motelay-Massei A., Ollivon D., Tiphagne K., Garban B. 2005 *Water Air Soil Pollut.* **164** 119
[34] Sweet C. W., Weiss A., Vermette S. J., 1998 *Water Air Soil Pollut.* **103** 423
[35] Halstead M. J. R., Cunninghame R. G., Hunter K. A. 2000 *Atmos. Environ.* **34** 665
[36] Injuk J., van Grieken R., De Leeuw G. 1998 *Atmos. Environ.* **32** 3011
[37] Williams M. R., Millward G. E., Nimmo M., Fones G. 1998 *Mar. Pollut.Bull.* **36** 366
[38] Guieu C., Chester R., Nimmo M., Martin J. M., Guerzoni S., Nicolas E., Mateu J.,Keyse S. 1997 Deep Sea Research Part II: Topical Studies in Oceanography **44**, 655
[39] Sandroni V. and Migon C. 2002 *Chemosphere* **47** 753
[40] Environment in the City of Belgrade, 2007-2008, ed. Institute of public health of Belgrade, Secretariat for environmental protection, Belgrade The regional environmental center, Serbia

Journal of Physics: Conference Series **162** (2009) 012019 doi:10.1088/1742-6596/162/1/012019

Determination of chemical composition of individual airborne particles by SEM/EDX and micro-Raman spectrometry: a review

E A Stefaniak[1,2], A Buczynska[1], V Novakovic[1,3], R Kuduk[4] , R Van Grieken[1]

[1]Department of Chemistry, University of Antwerp, Universiteitsplein 1, B-2610 Antwerp, Belgium

[2]Department of Chemistry, Catholic University of Lublin, Al Krasnicka 102, 20-788 Lublin, Poland

[3]Institute of Physics, Pregrevica 118, 11080 Belgrade, Serbia

[4]Institute of Computer Science, Maria Curie-Sklodowska University, M. Curie-Sklodowska sqr. 1, 20-031 Lublin, Poland

E-mail: anna.buczynska@ua.ac.be

Abstract. The strategies for sampling and analysis by SEM/EDX and micro-Raman spectrometry for individual airborne particles analysis as applied at the University of Antwerp (Belgium) by the MITAC group have been reviewed. Microbeam techniques provide detailed information concerning the origin, formation, transport, reactivity, transformation reactions and environmental impact of particulate matter. Moreover, some particles of certain chemical properties have been recognized as a threat for human health and cultural heritage objects. However, the small sizes of particles result in specific problems with respect to single particle analysis. Development of equipment and software for improvement of analysis and quantification are reported.

1. Introduction

Solid and liquid phase matter suspended in the atmosphere is addressed as 'particulate matter' (PM) or 'aerosols'. PM consists of a mixture of many compounds, all in variable concentrations. The analysis of gaseous pollutants originally dominated the environmental air monitoring; nowadays, however, it is confirmed that atmospheric particles play a much larger role in environmental risk. It is now well understood that they are the major transport form of material in air, frequently contain toxic components and influence the global climate and the visibility of the atmosphere [1]. It is already a common knowledge in the field of cultural heritage and preventive conservation, that some particles (such as soot) are a threat for work arts due to soiling or blackening and constituting a medium for the adsorption and oxidation of several gases. S-rich material can cause discoloring of the paint by oxidation with H_2O to H_2SO_4; $(NH_4)_2SO_4$ can induce a bloom on varnish; $CaSO_4$ can cause blackening by adsorption of soot; Fe-rich particles play a catalytic role in the oxidation of SO_2 to H_2SO_4; etc. [2]. Without the tools suitable for single particle analysis it would be much more difficult to recognize the sources of the damage and to suggest remediation measures but also to take preventive actions. Last but not least, the impact of air quality on human health is now doubtless. Numerous studies have demonstrated associations between day-to-day variations in ambient particulate matter concentrations and day-to-day variations in health, including morbidity but also mortality [3]. Due to the highest annual mean PM concentration in Europe, Belgium has the highest estimated loss in life expectancy that can be attributed to PM, estimated to 15 months compared with 9 months as the European average [4]. The knowledge about the chemical nature of suspended particles would allow discriminating most hazardous ones and taking appropriate preventive actions.

It is a common strategy now in PM monitoring to perform bulk analyses; however, the atmospheric behavior of airborne particles is determined not only by their composition but also by their morphology. Since the atmospheric particles are chemically and morphologically heterogeneous, and the average composition and average aerodynamic diameter do not describe well their population, microbeam techniques facilitating single particle analysis must be applied. This type of analysis can be employed in studies dealing with important issues such as the source characterization and apportionment, particle transport and deposition or with various reaction mechanisms such as sulphatation and condensation [5]. However it requires a totally different analytical approach, starting from sampling methods through the whole measurement process. Due to small sizes of the particles, it is necessary first to collect and analyze a large number of particles to ensure the statistical relevance of the results. Nevertheless, much effort has been made so far to improve both the equipment and the compatible software [6-10].

This review presents mostly the research on individual airborne particles that has been done at the University of Antwerp (Belgium) by the MITAC group [11]. It includes about two decades of research aiming at single particle analysis, resulting in a number of articles, book chapters and conference presentations. The MiTAC group has a big contribution to the recognition of chemical properties of individual airborne particles worldwide; the home-made software designed by the MiTAC group enabled automatic, computer controlled measurement of particles along with subsequent numerical and statistical analysis. The effort to combine different microbeam techniques, such as electron and laser microprobe in order to obtain both elemental and molecular composition of single particles, has also proved to be successful.

2. Sampling of particulate matter for single particle analysis

Different sampling strategies of particulate matter for single particle analysis have been applied. The choice of the sampling strategy mainly depends on the desired information and/or limitations such as sampling conditions or sampling space. Generally the set up for sampling consist of a sampling unit, collecting either the total suspended particulate matter (TSP) or size segregated particles, a pumping unit and often a gas counter or flow controller. Size segregated sampling is by far the most widely applied technique in single particle analysis because it allows the separation of particles according to their cut-off diameters; however, in some cases TSP might be very useful, especially when aqueous particles (e.g. in rain) are to be analyzed.

2.1. Total Suspended Particulate matter (TSP)

Samples of single particles collected without size segregation, by means of filter units (e.g. on Nuclepore filters) may be of great importance, since due to the homogeneous distribution of particles on the filter, absolute numbers of collected particles can be calculated by extrapolation of the results of the analyzed part of the filter. This is not possible for the inhomogeneous impactor samples [12].

2.1.1. Air sampling

Total airborne individual particles were investigated in environmental issues such as analysis of the aerosols above the Lake of Baikal (Siberia) done by Van Malderen and co-workers [13] or in issues more directly connected to human health such as analysis of wood combustion particles emitted by industrial heating boilers during normal winter operation, as done by Osán et al. [14]. In both cases a

47 mm diameter, a 0.4 pore-size Nuclepore polycarbonate membrane filter (aerosol-grade) was used. The filter was placed in a plexi glass filter holder with a hat-type cover, to protect against rain or wet drops coming from waves. The cover was found to leave the aerosol size-distribution almost undisturbed [15]. The filter holder was connected to a vacuum pump, which was operated at a flow rate of approximately 50 l min^{-1}, and equipped with an automatic timer system and a flow meter. Samples were taken usually over a 2-h period (total sampling time), which resulted in a proper loading of the filter for single-particle analysis. After collection, the filters were stored in airtight Millipore Petri dishes.

2.1.2. TSP sampling from water

The seawater samples from the North Sea [16] or rain water above the North Sea [17] were firstly collected. Immediately after collection, the samples were filtered through 25-mm diameter aerosol-grade polycarbonate Nuclepore filters with 0.4-µm pore size by using a Sartorius SM 16306 25-mm glass vacuum filter holder with a 30-ml funnel and with glass frit filter support in case of rain water samples and 47-mm filters when sea water was used. To enhance the homogeneity of the particle distribution, a paper filter was placed beneath the Nuclepore filter during filtration. A suitable particle density for single particle analysis was obtained by filtering 10 ml samples from the two sampling sites close to the coast sites and 100 ml samples from the other sites. To remove crystallized salts, the filters were rinsed three times with de-ionized water. The filters were placed in petri dishes, air dried and stored in a deep-freezer until preparation for electron probe micro analysis (EPMA).

Another application where rain samples were used for single particle analysis were samples of the 'Asian Dust' that were prepared by depositing an aliquot of rainwater on a Ag foil using a micro-pipette, followed by drying the wet particles in air [18]. To asses the origin of the dust, CJ-1 (China Loess Certified Reference Material) and the local soil particles were ground to microscopic size using an agate mortar, suspended in distilled water, and deposited on the Ag foil in the same way as the Asian Dust particles. The EPMA results were then compared.

2.2. Size-segregated sampling

Various types of size-segregated impactors are commercially available; they vary starting from the number of stages and cut-off diameter of particles and ending with the dimensions of the whole sampler which are suitable for sampling in special places or conditions.

May cascade impactors (theoretical cut-offs 20, 8, 4, 2, 1 and 0.5 µm) were used inside of museum buildings, like the Sainsbury Centre for Visual Arts (Norwich, UK) with a flow rate of 20 l min^{-1} [19]. Apiezon-coated Nuclepore filters were used as collection substrates. Sampling times were up to 3 h for the impactor stages with the largest cut-off sizes. The collection substrates of the impactor stages with smaller cut-off sizes were retrieved after shorter sampling periods (7 to 30 min); since the number concentration of smaller particles in the air is larger than that of bigger particles, it is necessary to collect smaller particles in shorter sampling duration, in order to avoid creation of artificial agglomerates while overloading the sample. May cascade impactors were also used in the survey carried out to characterize the particulate matter in the Royal Museum of Fine Arts (Antwerp, Belgium) [12]. Sampling times ranged from a few minutes to a few hours, depending on the impactor stage. Apiezon-coated Nuclepore filters were also used as collection substrate.

Berner type impactors with 9 stages with cut off diameters of 0.06, 0.125, 0.25, 0.5, 1, 2, 4, 8 and 16 µm were used for different types of aerosols e.g. for indoor sampling by Brimblecombe and co-workers [19] or for sampling of a particles above the North Sea [20]. The flow rate was 29 l min^{-1}. The particles were sampled on aluminium foil. Considering the cut-off diameters of the different stages, the time available for sampling and the size limit for the detection of particles by the electron probe microanalyzer (~0.2 µm), only six stages were selected for this sampling (stages 3–8). The sampling time varied between 2 (for stage 3) and 180 (for stage 8) min, to obtain the best loading of particles in the impacted spots. The collected samples were put in plastic carriers, sealed, and stored in a desiccator. The same type of impactor was used in Sainte Chapelle, Paris [21]. However here aerosol particles were sampled on Ag foil using four stages of a Berner impactor, with aerodynamic diameter cut-offs of 4, 2, 1 and 0.5 μm for stages 4, 5, 6 and 7, respectively, at a 20 l min^{-1} sampling flow. The sampling time ranged from 10 to 180 min. Spolnik and co-workers [22] used Si wafers and Ag foils as collection substrates for Berner-impactor sampling in churches that are valuable from a cultural heritage point of view, located in Italian Alps and southern Poland. The Si wafers were used to collect the finest particles (cut-off diameters of 0.25 and 0.5 μm, corresponding to stages 3 and 4, respectively). The particles collected with cut-off diameters of 1-8 μm (stages 5-8) were deposited on Ag foils.

Another (peculiar because of its small dimensions) type of impactor for sampling of particulates is Battelle 5–stage impactor (cut-off values 4, 2, 1, 0.5 and 0.25 μm) [12], used when sampling was to be done in limited spaces, e.g. interspaces between stained glass windows and protective glazing [22] or museum show-cases. The particles were collected on special transmission electron microscopy (TEM) grids made of a very thin carbon film (15–25 nm) supported by a copper mesh.

Nuclepore filters of 0.4 pore-size and cyclone sampling of particulate matter <2.5 μm, were used for collection of wood combustion particles emitted during normal winter operation of an industrial heating boiler [14].

De Miranda and co-workers [23] collected the aerosols in the Sao Paulo Metropolitan Area using stacked filter units (SFU) which separated the aerosol as coarse (2.5–10 μm equivalent aerodynamic diameter (EAD)) and fine fraction (<2.5 μm EAD) and cascade impactors of the MOUDI-type (''micro orifice uniform deposit impactor'', [24]) which does so in 10 different stages whose nominal 50% cut-off points (D50) are: after-filter, 0.093, 0.18, 0.33, 0.56, 1.0, 1.8, 3.2, 5.6 μm. The MOUDI impactor has rotating impaction plates and provides nearly uniform deposits over circular impaction areas. Nuclepore filters with 8 and 0.4 μm pore-size were used for the SFU. The MOUDI was operated with 8 μm pore-size Nuclepore filters (Teflon-backed filters with 37mm diameter) and have a 1 μm pore-size after-filter. A portable sampler called 'Mini-Vol' was also used to monitor particles with diameters below 10 μm (PM10). Again, 0.4 μm pore-size Nuclepore filters and MOUDI impactors were used.

3. Electron microprobe measurements

Electron probe X-ray micro analysis (EPMA or EPXMA) is a microanalytical method where X-ray radiation emitted by a sample exposed to an electron beam is detected. Since a variety of signals is available that can provide various information, this technique simultaneously detects elemental (qualitative) composition and morphology of analyzed objects, even for a small interacting volume such as a single atmospheric particle [1,15,16] The X-ray photons give information on the elemental composition while the signal derived from backscattered electrons can provide the information on the roughness (topographical image) and the atomic number variation with the location in the sample (compositional image) [5]. These features are definitely applied in scanning electron microcopy combined with an energy dispersive X-ray detection (SEM/EDX). What is then the difference between EPMA and SEM/EDX? These two are nowadays practically the same; only is SEM more conceived for imaging: larger electron focusing coils, detectors at a lower angle, X-rays have a longer path through the sample, hence it becomes less quantitative. In EPMA, it is reversed – some imaging capacity is sacrificed for better quantification: the X-ray detector is positioned at a larger angle from the sample, hence there are less matrix effects. And most often, a wavelength dispersive X-ray fluorescence (WDX) detector is (or can be) used. However, during the last decade, due to technological advances, the differences between SEM/EDX and EPMA have been reduced. In this review both terms will be used interchangeably.

Studying the particulates by single particle analysis with the use of EPMA requires, however, the measurement of large population set in order to obtain statistically meaningful data [5]. For this reason the possibility of automatic measurements by analyzing the back-scattered signal from particles mounted on a filter is one of the most advantageous features of the EPMA system, inevitable for representative individual particles analysis [16]. The automation improves the accuracy of the results by allowing characterization of a large number of particles in a reasonable time. Advanced home-made software was developed in our research group, to analyze individual particles in an automated way. The automated PRC (particle recognition and characterization) system uses the following methodology. The electron beam moves across the sample while the backscatter signal is monitored. In case the digitized backscatter signal exceeds a present threshold value, a particle is assumed to be detected. The coordinates of the contour points are determined and additional information such as particle diameter, perimeter and factor shape are calculated [25]. Afterwards an energy-dispersive X-ray spectrum is collected [26,27]. The intensities of the characteristic X-ray peaks present are stored together with the size information as an object vector for the individual particle. Next, the program turns into its search mode to localize and measure the next particle. Thus the PRC program is set up in three sequential steps: localizing, sizing and finally chemically characterizing the particle. Thanks to this software around 500 particles can be measured in less then 2.5 hours under optimized working conditions [5,26,27].

An important obstacle which had to be faced in EPMA or SEM analysis was the determination of low-Z elements such as carbon, nitrogen and oxygen, in addition to the elements ($Z \geq 11$) observed by the conventional EPMA technique. These elements have to be detected if an adequate identification of low-Z elements and chemical speciation at a single particle level needs to be done. Low-Z elements are very important because they form the major mass of aerosols. This is not possible by EPMA with conventional EDX detectors, mainly because the Be-window, which is used for protection of the semiconductor detector crystal, absorbs low-energy X-rays and thus prevents detection of the low-Z elements X-rays. By the application of the modified EPMA technique, which employs either a windowless or thin-window EDX detector [2,6,28], chemical compositions, including the low-Z components, of individual particles can be at least semi-quantitatively elucidated. The determination of low-Z elements in individual environmental particles allows us to profoundly improve the applicability of the single particle analysis; many environmentally important atmospheric particles, e.g. sulfates, nitrates, ammonium and carbonaceous particles, contain low-Z elements, which have not been characterized by the conventional EPMA.

Another problem of this technique is that it needs to operate in vacuum and samples are subjected to electron bombardment; in these conditions some loss or transformation of volatile or unstable compounds such as nitrates and ammonium, could occur between the sample introduction and the analysis step. To decrease the beam-damage effect, cold sample holder at the temperature of liquid nitrogen (-193°C) was applied. This approach was found by Osán and co-workers [28] to be very efficient especially for larger particles as organic, nitrate and ammonium containing particles of diameter above 10 μm could be measured without significant damage.

3.1. Equipment

All applications reviewed here were carried out with the use of two instruments available at the Chemistry Department of the University of Antwerp. The first one is a JEOL JXA-733 (Tokyo, Japan), equipped with an OXFORD Link SATW ultrathin window energy-dispersive X-ray Si (Li) detector. The resolution of the detector is 133 eV for Mn Kα X-rays. The secondary-electron and backscattering-electron detector allow visualization and automated particle localization. The detector is equipped with a sub-micrometer thick polymer window. This apparatus is also connected to a TN-2000 X-ray analysis system, a self-modified version of a particle recognition and characterization program (Tracer Northern Middleton, WI, USA) that controls the automated analysis of the samples. The spectra are recorded by a CANBERRA S100 multichannel analyzer according to the corresponding energy [8]. All the information is stored on disk for off-line data processing. The estimated geometrical data are set as input parameters for the quantification procedure [8,29]. The vacuum system consists of oil rotary unit and oil diffusion pumps units. Detection limits for EPMA for most elements are between 100 and 1000 ppm.

Working parameters were as follows: an accelerating voltage of 25 kV [12,13,16,17], a beam current of 1 nA and 20s accumulation time for the X-ray spectra [12,13,14,16,17,21,22]. After introducing a thin-window detector, which made the determination of low-Z elements possible, the optimal experimental conditions ensuring a low background level in the spectra and a high sensitivity for light element determination, different parameters: a 10-kV accelerating voltage [8,14,20,21,22,29] and the beam current of 0.5 nA was applied [4,20,29]. In order to obtain statistically enough counts in the X-ray spectra and to minimize the beam damage effect on the sensitive particles, a typical measuring time of 20 s was used. Around 300-500 individual particles per sample were analyzed.

The cold stage of the electron microprobe allows the analysis of particulate samples at liquid nitrogen temperature and the measurements were carried out at approximately -193°C stage temperature [8,21,22,29]. Another feature of this microprobe is the application of a continuous cleaned N_2 gas flow to reduce the deposition of the oil residues from the vacuum pumps.

The second apparatus is a SEM/EDX (JEOL-JSM 6300) unit equipped with an energy-dispersive (EDX) detector (Princeton Gamma Tech, NJ, USA), a backscattered electron (BSE) detector and a secondary electron (SE) detector. The Si(Li) detector can be operated in windowless mode for identification of low-Z elements and its resolution is 170 eV for Mn K_α X-rays [6]. Aerosol samples from the Correr Museum in Venice [2] and from Sao Paulo [23] were analyzed in the automated mode. In order to analyze as many particles as possible within a reasonable time), a software program was adapted in the laboratory from the commercially available 'integrated microanalyzer for imaging and X-rays' (IMIX) program (Princeton Gamma Tech) which locates the particles on the filter, determines their size, acquires BSE and SE images and performs EDX analysis within each particle. The intensities of the characteristic peaks in the spectra are determined by the top-hat filter method. Compared to JEOL JXA-733, image processing and mapping are more efficient owing to the newer data system.

3.2. Sample pre-treatment

The aerosol particles are usually collected on substrate material such as Al [20], Ag foil [21,22,29], Si substrates [22] or Nuclepore filters [14,17,23]. Nuclepore filters, although ideal for computer-controlled EPMA because of their microscopically flat surface, contain, however, carbon and oxygen; hence it is impossible to assess the concentration of these elements. Nuclepore filters are thus more suitable when information on heavier elements is desired. The use of this filters always require a conductive coating such as carbon coating or metallic coatings [6]. Both coatings have their negative sides; carbon coating introduces large inaccuracy in the determination of carbon in the particle and metallic coatings often overlap with characteristic lines of light elements, so that the accuracy of X-ray intensities of low-Z elements is very low. Aluminum foil, although it hinders the determination of Al, can be used for measurements of low-Z elements since it doesn't require any conductive coatings. Godoi et al [42] investigated various substrates (Ag and Al foils, Si wafers, Be discs, C tape and TEM grids) with regard to surface roughness, contribution to the EDX signal and stability under the electron beam. The topography (surface roughness) is an important parameter in the visual recognition of the particles in the secondary electron image (SEI) mode. As the results showed, the C tape could not be considered as a suitable substrate. Also Al foil appeared to have the roughness that resulted in difficulties in particle recognition. This seemed not be the case when Ag foil, Si wafers, Be discs and TEM grids was used. TEM grid appears to be superior in providing almost flat background and sharp analytical signal. Fairly goods results were achieved for Ag and Be line profiles. The contribution of the substrate to the quality of the X-ray spectrum was also investigated and the TEM grid again showed the best results.

To successively analyze the individual particles collected on a given substrate, a part of a substrate containing atmospheric aerosol particles needs to be glued to a Cu–Zn samples holder and, in case of Nuclepore filters, the sample is coated by a 25 nm carbon layer [14]. Some additional pretreatment was done e.g. by Osán and co-workers [14]. The collected cyclone fly-ash and bottom-ash particles were suspended in n-hexane, pipetted onto high-purity Ag foil and dried in air. No conductive coating was necessary for the thin-window EPMA measurements of these samples [14].

Journal of Physics: Conference Series **162** (2009) 012019 doi:10.1088/1742-6596/162/1/012019

3.3. Fitting X-ray spectra

The final results obtained by EPMA, after electron beam – aerosols interaction, are stored as intensities of X-ray photons per energy channels, or, in the other words, as X-ray spectra. The analysis and deconvolution of X-ray spectra is done by the WinAxil software package, based on the internationally well-known AXIL (**A**nalysis of **X**-ray spectra by **I**terative **L**east Squares) fitting routines. A mathematical model is built using different parameters such as the background model (linear, exponential, Bremsstrahlung, smooth filter or orthogonal polynomial), the number of X-ray lines and the approximately correct values for the energy and resolution parameters of the spectrometer. The model parameters are then optimized by means of a non-linear least-squares procedure. The main goal of spectrum evaluation is to obtain unbiased, minimum variance estimates of the net intensities (under the peak area) of the fluorescence lines. Fitting an appropriate mathematical function to the spectral data can do this. The parameters of this function are chosen so that the weighted sum of squared differences between the measured spectral data y_i and the fitting function $y(i)$ are minimal. This object function is known as chi-square:

$$\chi^2 = \sum_{i=n_1}^{n_2} \frac{1}{w_i}\left[y(i) - y_i\right]^2 \tag{1}$$

The fitting function generally consists of two parts, one that describes the continuum and the other one that deals with the characteristic X-ray lines.

$$y(i) = y_{Cont}(i) + \sum_{j\ Elements} A_j \left[\sum_{k\ lines} R_{jk} P(i, E_{jk})\right] \tag{2}$$

The inner summation runs over the number of K and/or L lines of each element with R_{jk} the relative contribution of line k of element j. The outer summation runs over all elements with the linear parameter A_j being the total intensity of element j. $P(i, E_{jk})$ is a function that describes the shape of the X-ray line with energy E_{jk}. A typical implementation of $P(i, E_{jk})$ consists of a Gaussian

$$P = G(i, E_{jk}) = \frac{Gain}{\sigma_{jk}\sqrt{2\pi}} \exp\left[-\frac{1}{2}\left(\frac{E(i) - E_{jk}}{\sigma_{jk}}\right)^2\right] \tag{3}$$

in which the width and position obey the energy and resolution calibration functions of the spectrometer:

$$E(i) = Zero + Gain \times i \tag{4}$$

$$\sigma_{jk} = \left[\left(\frac{Noise}{2\sqrt{2\ln 2}}\right)^2 + \varepsilon\ Fano\ E_{jk}\right]^{1/2} \tag{5}$$

The fitting in the net peak area parameter A_j, but nonlinear in the parameters *Zero, Gain, Noise* and *Fano,* so that a nonlinear least squares procedure is required. In successive approximations, the values of the parameters are altered until the minimum in χ^2 is reached. A spectrum evaluation procedure based on these principles is very flexible since lines of one element can be grouped either as a multiplet for all lines of an element or separately for K_α and K_β lines. Escape and sum peak can be incorporated easily. Anyhow, the quality of the results depends highly on the accuracy of the model or the ability of the fitting function to describe the observed spectrum (figure 1).

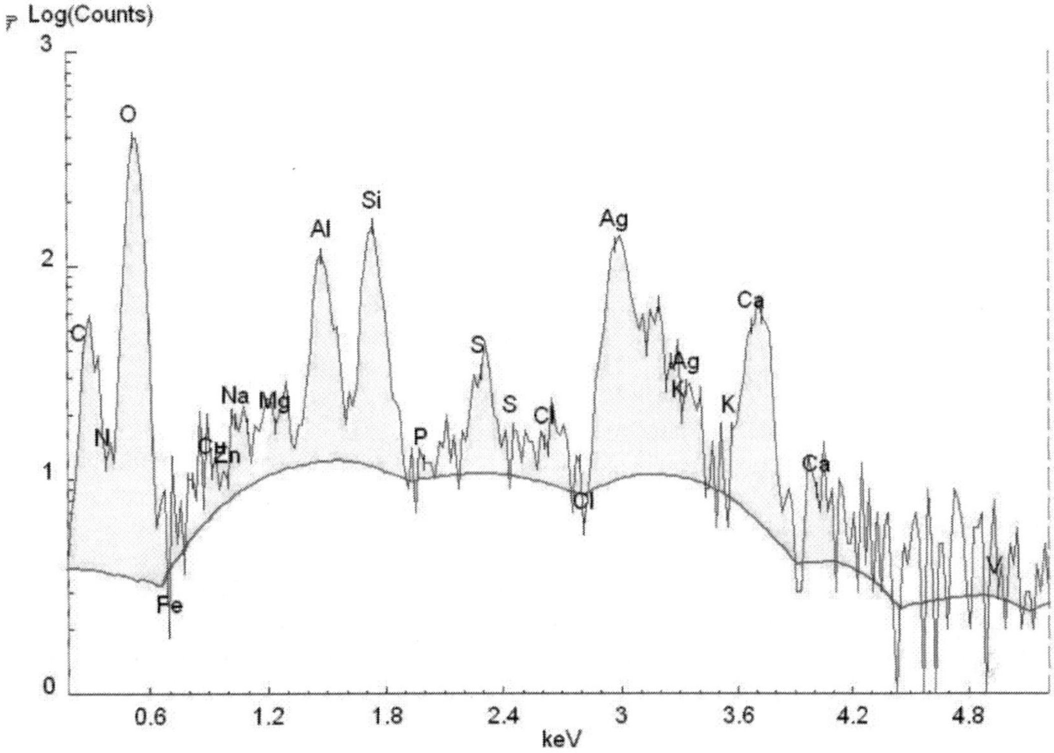

Figure 1. Typical X-ray spectrum fitted by WinAxil software

Two aspects of XRS hamper direct analytical use: matrix effects and spectral interference. By application of nonlinear least square fitting using Gauss or Voigt profiles modified with step and tail functions, overall spectra can be evaluated very accurately in a few seconds. The modification of Gaussian peak shape can be expressed by a mathematical model:

$$P(i,E_{jk}) = G(i,E_{jk}) + f_s S(i,E_{jk}) + f_t T(i,E_{jk}) \tag{6}$$

in which $G(i,E_{jk})$ is the Gaussian as in equation (3) and f_s and f_t are (linear) parameters representing the fractions of photons that gives rise to a step and tail contribution. The step and tail functions can be written as follows:

$$S(i,E_{jk}) = \frac{Gain}{2E_{jk}} erfc\left[\frac{E(i)-E_{jk}}{\sqrt{2}\sigma}\right] \tag{7}$$

$$T\left(i,E_{jk}\right)=\frac{Gain}{2\gamma\sigma\exp\left[-\frac{1}{2\gamma^2}\right]}\exp\left[\frac{E(i)-E_{jk}}{\gamma\sigma}\right]erfc\left[\frac{E(i)-E_{jk}}{\sqrt{2}\sigma}+\frac{1}{\sqrt{2}\gamma}\right] \quad (8)$$

The functions are normalized, i.e. the integral of the function over the spectrum equals unity. The parameter γ is an extra nonlinear parameter, modeling the width of the tail. To keep the number of parameters to be estimated during the least squares fit as low as possible, f_s, f_t and γ are expressed as functions of the energy. This functional relation is fitted rather than the value for each peak individually. Spectrum evaluation based on this fitting model gives more reliable results than simple Gaussians.

4. Semi-quantitative microanalysis based on Monte Carlo calculations

The Monte-Carlo method uses the same physical parameters and processes as the fundamental parameter method but at an individual microscopic level to predict the intensities of characteristic lines. Different from the fundamental parameter approach, the Monte Carlo procedure is capable of dealing correctly with coherent and incoherent scattering phenomena. The data calculated by a Monte Carlo process can be considered as a spectrum seen by a detector with infinitely good resolution [30,31].

For the estimation of the quantitative composition of the microparticles, the size and shape of the particles and the matrix effect must be considered for light elements. It was found that the excitation interactions between electrons and the atoms of the matrix and the geometric effects of fluorescence signals for light elements in individual atmospheric microparticles can be described by Monte Carlo simulation [6]. This quantification procedure is based on a modified version of the single scattering CASINO Monte Carlo program [30,31], written in ANSI-C language, which was designed for low energy beam interaction to generate X-ray and electron signals. The modified version of the CASINO program allows the simulation of electron trajectories in spherical, hemispherical and hexahedral particles located on a flat substrate. The simulation procedure determines also the characteristic and continuous X-ray flux emitted from the substrate material and the influence of the substrate material on the energy distribution of the exciting electrons. This modified version of the CASINO program is the main module of the software package Elementary [20]. It consists of three main parts: Simulation, Quantification and Interpretation. A Monte Carlo calculation program is able to simulate electron-induced X-ray interactions within a single particle during scanning electron probe analysis. The same simulation can be used iteratively to quantify elemental concentrations starting from X-ray intensities. With the additional interpretation tools, Elementary offers a versatile, integrated software package for the fast interpretation of single particle analysis data.

5. Statistical approach

As a result of automatic SEM/EDX measurements followed by a fitting procedure and/or a simulation of elemental weight concentration, we receive a large database which needs to be reduced for analytical purposes. The variables used in the database are the particle number, element and – depending on the fitting method applied –either X-ray intensity (peak area) or weight concentration. The number of analyzed particles per sample ranges from ca. 100 to ca. 400 items, thus – in order to get an overview about the types of particles and element combinations – it is necessary to reduce the database size. In the field of single particle analysis, there are several major multivariate techniques: principal factor analysis (PFA) or principal component analysis (PCA), and hierarchical (HCA) or nonhierarchical clustering (NHCA). They can be applied independently or used in a combined approach.

5.1. Hierarchical Cluster Analysis

When dealing with a large number of particles, the first question is which particles were present in the sample [13]. The first task in multivariate analysis is to identify various particle types. This was done by a hierarchical clustering method. This approach has been proved a reliable method for classifying particles into groups according to their chemical composition. The most popular software used for HCA was the program IDAS [32]. The basics of HCA are settled in presenting every particle 'i' in N-dimensional space as an object vector with coordinates according to its N-elemental composition. Here N is the number of variables and equals the number of different elements analyzed in each sample. The method starts from m objects (particles) that are to be classified. At each step the two most similar objects (particles) or already formed clusters are merged into a single cluster. The similarity between two pairs of objects was estimated from the Euclidian distance coefficient between the objects. Ward's method (error sum of squares method) was used for the calculation of the distances between newly formed clusters and the remaining objects and/or clusters [34]. Each sample composed of a number of particles (100-400) must be subjected to HCA independently. The major problem with hierarchical clustering methods is to obtain the right number of clusters. This problem can be overcome by automated "cutting" of the dendrogram based on the Akaike's criterion. More detailed information on hierarchical clustering and the automated cutting can be found elsewhere [34,35,36].

Application of HCA to determine the type of particles caught in different environments appeared in the open literature devoted both to preventive conservation and cultural heritage protection. This aspect is emphasized in case of museums, galleries and churches where priceless artworks might be exposed to hazardous air pollutants. Owing to HCA applied to computer controlled (CC) EPMA measurements of PM collected in the Correr Museum in Venice (Italy), the particle types present in the indoor atmosphere were identified [2]. Ca-rich and Ca-Si-S-Cl-rich dominated the aerosol content (added abundance >50%) in the size range >1 μm in winter and in the entire investigated size range (0.5-20 μm) in summer. Aluminosilicates constituted a third group. Minor particle types were fresh sea salt (Na-Cl-rich), aged sea salt (Na-Cl-S-rich) and marine crystallization products, as well as Si-rich, Fe-rich and gypsum particles. Indoor air quality in the Sainsbury Center for Visual Artin Norwich, England, was also the aim of investigation by Brimblecombe et al. [19]. They distinguished several clusters of different chemical characteristics in the selected size ranges by means of HCA. It became clear that the smallest indoor particles were completely dominated by S-rich and low-Z element compounds (particles for which no elements with Z<11 could be detected in the X-ray spectrum). These particles could have a purely organic composition or they might consist of low-Z inorganic species, such as NH_4NO_3. In the larger size fractions (< 2 μm), aluminosilicates (mainly soil dust but also high temperature combustion fly ash) were dominating.

Gysels et al. applied HCA to characterize PM collected inside the Royal Museum of Fine Arts in Antwerp, Belgium [12]. In this case, the indoor size range ≥4 μm was characterized by Ca-rich, Ca–Si and aluminosilicate particles. The same types were identified in the size range 2–4 μm, but sea salt gained importance as well. The particles of the size below 2 μm were completely dominated by sea salt and aged sea salt. Ca-rich, Ca–Si particles and aluminosilicates became less important, while $CaSO_4$ and particles rich in S, Cl and Ca became more abundant. Fe-rich particles were present in every investigated size range, while low-Z, S- and Na-rich particles were only identified in the smallest size fraction. Similar particle types were identified in the outdoor samples; however, in comparison with the indoor samples, the sea salt abundance was much larger in the size fraction ≥2 μm and smaller in the size range <2 μm. The abundance of Ca-rich particles was significantly lower than indoors and Ca–Si particles were practically not detected. Fe-rich particles were present in every size range in higher abundances than indoors.

Particles attacking stained glass in the historical buildings were classified by Godoi and co-workers [21], e.g. in the Sainte Chapelle chapel in Paris. In total, all sampling implied the analysis of more than 10 000 particles. Owing to the HCA procedure, it was possible to divide coarse and fine PM fractions into the following clusters: SiO_2, AlSi, organic, $CaCO_3$, $CaSO_4$, FeOx, NaCl-dust, NaCl+$NaNO_3$ and P+dust in the inside and outside air; SiO_2, AlSi, organic, $CaCO_3$, $CaSO_4$, FeOx, $NaNO_3$-NaKCl,

Journal of Physics: Conference Series **162** (2009) 012019 doi:10.1088/1742-6596/162/1/012019

organic-dust-NaCl in the interspace between the deplaced original and the new window panes. The main difference between these was the relative abundance of each type. Three types of particles that are potentially dangerous to the glass were found inside and in the interspace; however all particle types were comparable and therefore deterioration cannot be indicated. The predominant presence of $CaSO_4$ is due to the conversion of $CaCO_3$ to $CaSO_4$ via reaction with SO_2 inside of this chapel, but probably even more due to the deterioration of the plastered walls [21].

The composition of suspended PM expressed as relative abundances of various particles (clusters) gets also a special attention due to its health impact. Low-Z EPMA facilitated measurements of C-rich particles together with other elements for Z less than 11. Osán and co-workers investigated wood combustion particles, most of which fall into the inhalable fraction [14]. Their HCA classification of particles based on EPMA measurements appeared to be an efficient tool to identify the origin of fly ash and to trace the emitted particles in the ambient air. Since the majority of the particles emitted from the wood burning plant studied in that work were found to be in the respirable size range, they can cause significant damage to the human health. The health aspect of PM composition was also the aim of aerosol characterization carried out in the Sao Paulo Metropolitan Area [23]. Using hierarchical cluster analysis on the automated EPMA data, particles could be identified and grouped into Si-rich particles, metals and metal compounds, sulphates, Ca-rich particles, chlorides, organic and biogenic particles. The particles derived from the soil appeared in the coarse fraction, Ca-rich particles and chlorides appeared with intermediate diameters, and organics and sulphates were present in the fine fraction.

5.2. Non Hierarchical Cluster Analysis

After the identification of the various particle types present in each sample, the next question was to compare the aerosol composition of all the collected samples. The global interpretation over a whole sampling campaign is very difficult when using only results of a hierarchical clustering on the separate samples. Gathering all particles in all the samples from one or more sampling campaigns creates a huge database which again requires data reduction. This can be done by a classification of all examined particles in similar samples together. It is not possible by HCA but can be done using a nearest-centroid-sorting technique [13,38]. Nearest-centroid sorting is a group of nonhierarchical clustering methods, which classify objects (particles) in clusters according to their distance from the centroid of a preset training vector. The distance between every particle and the training vector (centroid) is calculated and the particles which were incorrectly classified the first time are relocated. This procedure is repeated until convergence occurs [13,38].

The sequence of double HCA followed by NHCA based on the Forgy algorithm [37] was used in numerous manuscripts devoted to single airborne particles. This way of data reduction was successfully applied by Szaloki et al. [8] to characterize aerosol particles from the North Sea. This is a fundamental work describing the way of simulating quantitative composition of a particle together with comparison of real and simulated concentrations of simple chemical compounds. Then it was applied to real aerosols from the North Sea. The presented evaluation method was applied to analyze the light element content of aerosol particles collected over the North Sea. Elemental concentrations were calculated for each individual particle. To obtain information on the possible sources of the aerosol and the possible chemical interactions between gaseous and particulate pollutants, the particles were classified into 10 groups using the NHCA of Forgy [37] for the chemical and morphological data obtained by EPMA. The initial centroids for the method were obtained by two steps of HCA carried out for each sample. As light element analysis is possible by the present method, the nitrate particle type can be distinguished. The high abundance of organic-type particles can be recognized during the sampling period. The high abundance of sea salt particles in one is a fingerprint of marine origin. Ammonium-sulfate-type particles were also observed as a possible fingerprint of sea salt gaseous sulfuric acid reaction. A large fraction of particles shows a characteristic continental origin since soil-type, fly ash, ammonium and nitrate, sulfur-containing organic and metal oxides are observed.

Wide-range investigations on the North Sea aerosols were carried out by de Hoog and co-workers [20]. The procedure of sequential clustering by HCA and NHCA described by Szaloki [8] as well as the software for estimation of elemental weight concentration was applied to particle classification. For the comparison of the sample sets collected at a different location and time, all particles for each impactor stage (size fraction) were classified into 13 groups. To simplify the graphical representation, some similar clusters were taken together into one group afterwards (e.g. several, slightly different clusters of aluminosilicates were considered as one group). The 13 groups were named according to major chemical species in the groups identified on the basis of elemental concentration data of the groups: $(NH_4)_2SO_4$, $(NH_4)_2SO_4$+NH_4NO_3, NaCl, NaCl+organic, NaCl+$NaNO_3$, $NaNO_3$, organic, organic+S, aluminosilicates, Fe_2O_3, oxides, (Ca/Mg)(CO_3/NO_3/SO_4) and biogenic.

Spolnik and co-workers investigated suspended particulate matter composition depending on the heating system in a case study of old churches [22]. For comparison of the samples collected during the operation of the investigated types of heating inside different churches, all particles were classified using NHCA. Fine and coarse particles were separately treated. The particles with diameters from 0.25 to 1 μm (stages 3-5 of the Berner impactor) were considered as fine particles, and those with diameters from 2 to 8 μm (stages 6-8) were considered as coarse particles. The average relative abundance (fine and coarse) was calculated for each set. In this manner, not only a classification of all particles in a distinct number of particle types representative for the sample can be obtained, but an allocation of an abundance value to each particle type for every sample as well. An overall analysis of the most important particle types allows determining the main sources responsible for soiling and blackening of objects displayed inside the churches.

5.3. Factor Analysis and Principal Component Analysis

The last step of the EPMA data processing is to find a correlation between the sources of the different particle types and their abundances. This can be achieved through so-called Factor Analysis (FA) or Principal Component Analysis (PCA). In FA each original variable is affected by various factors and some of them are shared by other variables in the data set (so called common factors) and the others are not shared (unique factors) [38]. The objective of FA is to take p variables (elements) and find linear combinations to produce uncorrelated new variables (factors), which are ordered in decreasing order of importance, so that the first explains the largest amount of variance. The proper number of factors should be considerably less than the number of variables, but should, on the other hand, explain a large fraction of the total variance. Only factor loadings greater than three times their standard deviation are considered as statistically significant. More detailed information can be found elsewhere [39]. In PCA there are no assumptions about the variables' structure. There are new hypothetical variables defined, independent of each other [38].

Van Malderen and co-workers [13] presented a combined way of data reduction applied to a number of PM samples collected over the total area of the Lake Baikal (Russia). They applied data combination of multivariate techniques to 52 PM samples (in total for each, 400 particles were analyzed, i.e. 20 000 spectra all together). HCA enabled to distinguish fourteen different clusters such as "Org", "Si-rich", "Ca-S", "Fe-rich", "Al-Si-rich", etc. Their abundances, however, varied between the samples, so in order to compare different samples, NHCA was performed. By gathering the particles from the samples belonging to the same location, the authors characterized the population of certain clusters in selected locations. A final step in the interpretation was to find correlations between the occurring particle groups and simultaneously link them to their various sources. Based on FA, the correlations and anti-correlations between elements were established; such as high loadings for Al, Si, K, Ca, Ti and Fe and anti-correlated organic type can comprise crust sources (different minerals).

A similar approach was applied by Worobiec et al. [40] to classify particulate matter collected in the Amazon Basin. For comparison of the different aerosol samples, the particles were grouped as coarse fraction and fine fraction and subsequently classified by NHCA based on the Forgy algorithm [41]. The initial centroids were selected by a sequence of HCA performed with the IDAS program [32]. A NHCA was performed on the fine and coarse size fractions independently. In the coarse

Journal of Physics: Conference Series **162** (2009) 012019 doi:10.1088/1742-6596/162/1/012019

particles data set, 14 different particle types were identified: aluminosilicates, P–S–Cl–K-rich, Si-rich, low-Z-rich, K–P-rich, NaCl, aluminosilicates agglomerated with fresh sea salt, K-rich, K–S-rich, CaSO4, P-rich, Na–P–S–Cl-rich, Fe-rich and Ti-rich. In the fine fraction there were following clusters identified: aluminosilicates, Fe-rich aluminosilicates, Si-rich (quartz), S-rich, low-Z-rich, P–S-rich, S-rich aluminosilicates, Fe-rich, P-rich, S–Cl–K-rich, CaSO4, Ti-rich, Cl-rich and K-rich. To investigate which samples had a similar chemical composition, the results from the NHCA were used as input for a PCA. These results proved the soil dust (Al–Si–Fe–Ti-rich species) must have been adverted from Sahara desert and also exclude the group of oceanic origin. For the fine fraction, the positive loadings of the first component represent the African dust as the source, while the negative loadings indicate the Amazon Basin (local/regional) origin. It also suggests that only two sources should be considered for the fine size fraction.

6. Micro-Raman analysis

Statistical operation of reducing the databases produced by automated SEM/EDX analysis results in a number of clusters representing chemically similar particles. The most common description given to the clusters comes from the most abundant elements in the cluster. Therefore we can find Si-rich particles or Si-Al-rich particles, S-O-N-rich particles, etc. It would be more informative and more useful to get the actual chemical composition, especially when the cluster name might raise some doubts, i.e. "C-rich". In this group, elemental carbon/organic particles as well as carbonates could be included. Furthermore, Na-S-C-O combination can be derived from a mixture of sodium sulphate and sodium carbonate but also a mixture of sulphate and soot. The technique which provides a molecular image could make the clustering more conclusive. One of the solutions is micro-Raman spectrometry (MRS). This technique is commonly used in the fingerprinting of matter due to its characteristic inelastic scattering and therefore provides valuable information on the molecular structure.

MRS is specifically suitable for single particle analysis since it is derived from a combination of confocal microscopy and Raman spectrometry. A laser spot of a size between 0.5 and 5 μm gets only a small sample volume exposed, providing information on the molecular make-up of particulate matter as small as 500 nm [42]. Figure 2 shows the visible light image of an Ag foil covered with airborne particles; this is the image recorded by a digital camera built in the MRS system, synchronized with a microscope and the motorized stage. Owing to such construction, it is possible to select individual particles and spot a laser beam on each of them, followed by recording of the Raman activity. Next to the microscope image, there is a Raman spectrum of one airborne particle presented. It is composed of the shifts belonging to $NaNO_3$ and elemental carbon, quite usual composition of PM within the size range 1-2 μm.

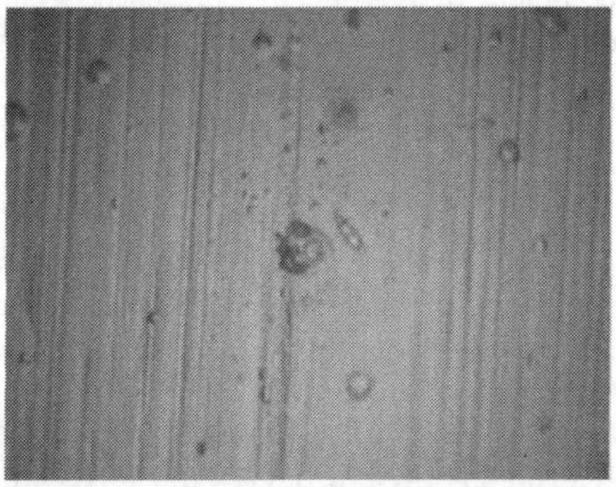

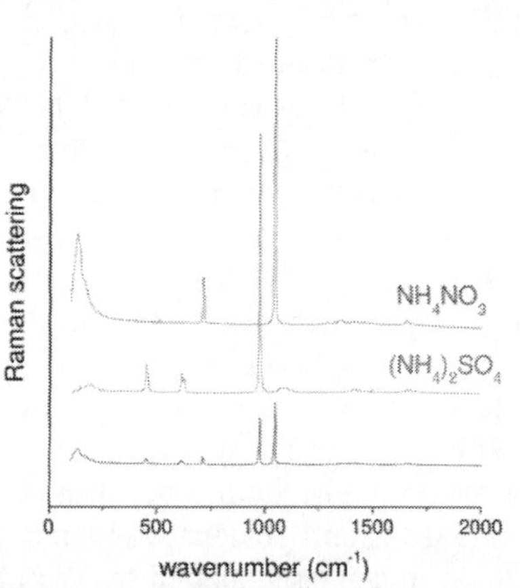

Figure 2. The visible light image of a silver foil covered with the airborne particles (magnification: 1000x) accompanied by a Raman spectrum of a particle (black line) consisted of ammonium nitrate and ammonium sulphate.

Application of micro-Raman analysis to particulate matter investigation has been the subject of some studies; however it was more adequate to geochemical research (soils, sands or sediments) instead of airborne particles [43]. However, some pioneering work on the possibilities of MRS towards analyzing individual particles was made by Potgieter and Van Grieken [44]. Nowadays MRS becomes more applicable in the field of aerosol analysis, although usually as an independent technique. Attempts to combine MRS with SEM/EDX or EPMA have also been reported [42,43]; however, the authors pointed out the limitations which have to be tackled while performing a parallel analysis with the two instruments together. Some of them are: different substrate requirements, lack of correlation between the probing depths, incompatible two images constructed in two different ways, limitations of magnification in case of optical image, carbon deposition during an electron beam scanning etc. Nevertheless, Godoi et al. described the results of parallel MRS and EPMA investigations of particulate matter collected in a museum (Rubens' house, Antwerp, Belgium). They selected several individual particles and compared their elemental composition given by low-Z EPMA, with their molecular make-up provided by MRS. C-rich particles were conclusively distinguished as soot or $CaCO_3$ or a mixture of both. The crucial issue for the two stand-alone instruments is relocation of the same particle. Once it is measured by SEM/EDX (EPMA), it has to be re-located in the optical image to be measured by MRS. Godoi et al. [42] reported TEM grids as a tool facilitating the particle relocation; TEM grids are suitable for both techniques (they are conductive and their Raman scattering is negligible) and they can create a sort of coordinate system inside which the particle position can be easily saved.

Another approach was applied by Worobiec et al. [45], who were investigating the PM composition in one of the world-famous museums - the Wawel Castle in Kraków (Poland). In this case the automated EPMA results were subjected to HCA and finally different clusters were recognized and verified by MRS. In this case two spots from the same Berner stage were analyzed independently by the two instruments, MRS and EPMA. Particulate matter sampled in Wawel Castle during the winter campaign appeared to be enormously rich with carbon. Its atmospheric origin (soot) was confirmed by MRS. Moreover; C-S-N-rich particles were recognized as a mixture of soot with ammonium sulphate, instead of organic or biogenic type.

7. Conclusion

For the last two decades, there has been extensive progress in the field of single particle analysis. It began with EPMA instrumental developments, followed by the design of computer software facilitating an automated mode of particle detection based on BSE images. Correlation of an electron beam scan with the digital image transformation, determination of the particle size from the projected area diameter and simultaneous recording of X-ray emission radiation created a new possibility of measuring a large number of individual particles in a fast automated way. The next step was the transformation of EDXRS spectra into a database correlated with abundances of elements in the particles. Since EPMA was equipped with a so-called atmospheric thin-window detector, it created a new possibility of investigating atmospheric particles composed of low-Z elements. Furthermore, development of new software for estimation of element weight concentrations (based on CASINO simulations) in each recognized particle enabled semi-quantitative analysis of single particulate matter. The first generation of this home-made software EPPROC was working in the text mode, but the next one – Elementary – was already equipped with the graphic interface. Since the atmospheric particles consist in a vast majority of low-Z elements, the chemical characterization of PM gained deeper insight. The final step of EPMA measurements is multivariate analysis leading to separation of clusters, i.e. particles of certain chemical similarity. Also in this stage the home-made programs (IDAS and Forgy) appeared irreplaceable. The whole procedure from sample preparation to recognition of the cluster types and abundances provided a lot of satisfying results about the atmospheric PM in ambient air over the whole world. The final stage of this methodological evolution is the combination of elemental composition and molecular analysis via interfacing two micro-beam techniques: EPMA and micro-Raman spectrometry. Application of both stand alone instruments and/or the hybrid version brought very promising perspectives for further single airborne particles research.

Acknowledgments
This work was supported by the OSTC (Belgian Federal Office for Scientific, Technical and Cultural Affairs) postdoctoral research program. Velibor Novakovic gratefully acknowledges a fellowship granted by the OSTC.

References

[1] Jambers W and Van Grieken R 1996 *Trends Anal. Chem.* **15** 11
[2] Camuffo D, Brimblecombe P, Van Grieken R, Busse H-J, Sturaro G, ValentinoA, Bernardi A, Blades N, Shooter D, De Bock L, Gysels K, Wieser M and Kim O 1999 *Sci. Total Environ.* **236** 135
[3] Janssen N A H, Lanki T, Hoek G, Vallius M, de Hartog J J, Van Grieken R, Pekkanen J and Brunekreef B 2005 *Occup. Environ. Med.* **62** 868
[4] Amann M, Cabala R, Cofala J, Heyes C, Klimont Z and Schöpp W 2004 Analysis report 2 http://www.iiasa.ac.at/rains/CAFE_files/CAFE-MFR3.pdf
[5] Xhoffer C, Wouters P, Artaxo A, Van Put A and Van Grieken R 1992 *Environmental Particles* Vol.1 107
[6] Ro C-U, Osán J and Van Grieken R 1999 *Anal. Chem.* **71** 1521
[7] Ro C-U, Osán J, Szalóki I, de Hoog J, Worobiec A and Van Grieken R 2003 *Anal. Chem.***75** 851
[8] Szalóki I, Osán J, Ro C-U and Van Grieken R 2000 *Spectrochim. Acta B* **55** 1017
[9] Szalóki I, Osán J, Worobiec A, de Hoog J and Van Grieken R 2001 *X-ray Spectrom.* **30** 143
[10] Van Espen P 1984 *Anal. Chim. Acta* **165** 31
[11] http://webh01.ua.ac.be/mitac1
[12] Gysels K, Deutsch F and Van Grieken R 2002 *Atmos. Environ.* **36** 4103
[13] Van Malderen H, Van Grieken R, Khodzher T H, Obolkin V and Potemkin V 1996 *Atmospheric Environment* **30** 1453
[14] Osán J, Alföldy B, Török Sz and Van Grieken R 2002 *Atmos. Environ.* **36** 2207
[15] Otten Ph, Injuk J and Van Grieken R 1994 *Sci. Total Environ.* **155** 131
[16] Jambers W, Dekov V and Van Grieken R 1999 *Mar. Chem.* **66** 17
[17] Jambers W, Dekov V and Van Grieken R 2000 *Sci. Total Environ.* **256** 133
[18] Ro C-U, Oh K-Y, Kim H K, Chun Y, Osán J, de Hoog J and Van Grieken R 2001 *Atmos. Environ.* **35** 4995
[19] Brimblecombe P, Blades N, Camuffo D, Sturaro G, Valentino A, Gysels K, Van Grieken R, Busse H-J, Kim O, Ulrych U and Wieser M 1999 *Indoor Air* **9** 146
[20] de Hoog J, Osán J, Szalóki I, Eyckmans K, Worobiec A, Ro C–U and Van Grieken R 2005 *Atmos. Environ.* **39** 3231
[21] Godoi R H M, Kontozova V and Van Grieken R 2006 *Atmos. Environ.* **40** 1255
[22] Spolnik Z, Worobiec A, Samek L, Bencs L, Belikov K and Van Grieken R 2007 J. of Cult. Herit. **8** 7
[23] de Miranda R M, de Fatima Andrade M, Worobiec A and Van Grieken R 2002 *Atmos. Environ.* **36** 345
[24] Marple V A, Kenneth L R and Steven M B 1986 *J. Aerosol Sci.* **17** 489
[25] Török Sz, Sándor Sz, Xhoffer C, Van Grieken R, Mészáros E and Molnár A 1992 *Időjárás* **96** 223
[26] Van Grieken R, Artaxo P, Bernard P, Leysen L, Otten P, Storms H, Van Put A, Wouters L and Xhoffer C 1990 *Chemia Analityczna* **35** 75
[27] Van Grieken R, Xhoffer C, Wouters L and Artaxo P 1991 *Analytical Sciences* **7** 1117
[28] Osán J, Szalóki I, Ro C-U and Van Grieken R 2000 *Mikrochim.Acta* **132** 349

Journal of Physics: Conference Series **162** (2009) 012019 doi:10.1088/1742-6596/162/1/012019

[29] Osán J, de Hoog J, Worobiec A, Ro C-U, Oh K-Y, Szalóki I and Van Grieken R 2001 *Anal. Chim. Acta* **446** 211
[30] Hovington P, Drouin D, and Gauvin R 1997 *Scanning* **19** 1
[31] Gauvin R, Hovington P and Drouin D 1995 *Scanning* **17** 202
[32] Bondarenko I, Treiger B, Van Malderen H, Van Grieken R and Van Espen P 1996 *Spectrochim. Acta B* **51** 441
[33] Worobiec A, Stefaniak E A, Potgieter-Vermaak S, Sawlowicz Z, Spolnik Z and Van Grieken R 2007 *Applied Geochemistry* **22**
[34] Bernard P, Van Grieken R and Eisma D 1986 *Environ. Sci. Technol.* **20** 467
[35] Van Malderen H, Rojas C and Van Grieken R 1992 *Environ. Sci. Technol.* **26** 750
[36] Bondarenko I, Van Malderen H, Treiger B, Van Espen P and Van Grieken R 1994 *Chemometr. Intell. Lab.* **22** 87
[37] Van Espen P, Janssens K and Nobels J 1987 *Chemometr. Intell. Lab.* **1** 109
[38] Van Malderen H, 1996, PhD thesis, *Universitaire Instelling Antwerpen*, 25-27
[39] Rojas C M, Figueroa L, Janssens K H, Van Espen P E, Adams F C and Van Grieken R E 1990 *Sci. Total Environ.* **91** 251
[40] Worobiec A, Szalóki I, Osán J, Maenhaut W, Stefaniak E A and Van Grieken R 2007 *Atmos. Environ.* **41** 9217
[41] Massart D and Kaufmann L 1983 *The Interpretation of Analytical Chemical Data by the Use of Cluster Analysis.*Wiley, New York
[42] Godoi R H M, Vermaak-Potgieter S, de Hoog J, Kaegi R and Van Grieken R 2006 *Spectrochim. Acta B* **61** 375
[43] Stefaniak E, Alsecz A, Sajó I E, Worobiec A, Máthé Z, Török S and Van Grieken R 2008 *Journal of Nuclear Materials*, **29**
[44] S.S. Potgieter-Vermaak and R. Van Grieken 2006 *Applied Spectroscopy*, **60** 39
[45] Worobiec A, Stefaniak E A, Kontozova V, Samek L, Karaszkiewicz P, Van Meel K and Van Grieken R 2007 *e-Preservation*, **3** 63

AUTHOR INDEX

Aanesland, A 96
Azharonok, V V 133
Banković, A 6
Bazylev, N B 133
Brinkmann, R P 109
Brunger, M J 52
Buckman, S J 6
Buczynska, A 212
Campbell, L 52
Caradonna, P 6
Chabert, P 96
Charlton, M 23
Czarnetzki, U 109
Djulgerova, R 73
Dlugunovich, V A 133
Donkó, Z 109, 177
Dujko, S 6
Filatova, I I 133
Fomin, N A 133
Gans, T 125, 140
Goeckner, M 148
Hartmann, P 177
Heil, B G 109
Joksić, J 200
Jones, A 6
Jovanović, J V 42
Kalman, G J 177
Kono, M 190
Kuduk, R 212
Kutasi, K 83
Loureiro, J 83
Makochekanwa, C 6
Malović, G 73
Marić, D 73
Marinković, B P 1
Marler, J P 6
Meige, A 96
Mihailov, V 73
Mijić, Z 200
Mozetic, M 157
Mussenbrock, T 109
Niemi, K 125, 140
Novakovic, V 212
O'Connell, D 125
Ogawa, D 148
Overzet, L 148
Petrović, Z Lj 6, 42, 73
Pinhão, N R 62
Pintassilgo, C D 83
Poedts, S 190
Radenković, M 200
Rajšić, S 200
Reuter, S 140
Saraf, I 148
Schaper, L 140
Schulz-Von Der Gathen, V 140
Schulze, J 109
Škoro, N 73
Slaughter, D 6
Stefaniak, E A 212
Stojić, A 200
Sullivan, J P 6
Tanaka, M Y 190
Tasić, M 200
Tsaruk, O 133
Van Grieken, R 212
Vesel, A 157
Voschula, I V 133
Vranjes, J 190
Waskoenig, J 140
White, R D 6
Zaka-Ul-Islam, M 125

Institute of Physics
Dirac House, Temple Back
Bristol BS1 6BE UK